AF778883

Hybrid MR-PET Imaging

Systems, Methods and Applications

New Developments in NMR

Editor-in-chief:
William S. Price, *Western Sydney University, Australia*

Series editors:
Bruce Balcom, *University of New Brunswick, Canada*
István Furó, *Industrial NMR Centre at KTH, Sweden*
Masatsune Kainosho, *Tokyo Metropolitan University, Japan*
Maili Liu, *Chinese Academy of Sciences, Wuhan, China*

Titles in the series:

1: Contemporary Computer-assisted Approaches to Molecular Structure Elucidation
2: New Applications of NMR in Drug Discovery and Development
3: Advances in Biological Solid-state NMR
4: Hyperpolarized Xenon-129 Magnetic Resonance: Concepts, Production, Techniques and Applications
5: Mobile NMR and MRI: Developments and Applications
6: Gas Phase NMR
7: Magnetic Resonance Technology: Hardware and System Component Design
8: Biophysics and Biochemistry of Cartilage by NMR and MRI
9: Diffusion NMR of Confined Systems: Fluid Transport in Porous Solids and Heterogeneous Materials
10: NMR in Glycoscience and Glycotechnology
11: Fast NMR Data Acquisition: Beyond the Fourier Transform
12: Cross-relaxation and Cross-correlation Parameters in NMR: Molecular Approaches

13: Contrast Agents for MRI: Experimental Methods
14: NMR-based Metabolomics
15: Modern Methods in Solid-state NMR: A Practitioner's Guide
16: Paramagnetism in Experimental Biomolecular NMR
17: Optimizing NMR Methods for Structure Elucidation: Characterizing Natural Products and Other Organic Compounds
18: Field-cycling NMR Relaxometry: Instrumentation, Model Theories and Applications
19: Hybrid MR-PET Imaging: Systems, Methods and Applications

How to obtain future titles on publication:
A standing order plan is available for this series. A standing order will bring delivery of each new volume immediately on publication.

For further information please contact:
Book Sales Department, Royal Society of Chemistry, Thomas Graham House, Science Park, Milton Road, Cambridge, CB4 0WF, UK
Telephone: +44 (0)1223 420066, Fax: +44 (0)1223 420247
Email: booksales@rsc.org
Visit our website at www.rsc.org/books

Hybrid MR-PET Imaging
Systems, Methods and Applications

Edited by

N. Jon Shah
Forschungszentrum Jülich GmbH, Germany
Email: n.j.shah@fz-juelich.de

ROYAL SOCIETY OF CHEMISTRY

New Developments in NMR No. 19

Print ISBN: 978-1-78801-074-0
PDF ISBN: 978-1-78801-306-2
EPUB ISBN: 978-1-78801-683-4
Print ISSN: 2044-253X
Electronic ISSN: 2044-2548

A catalogue record for this book is available from the British Library

© The Royal Society of Chemistry 2019

All rights reserved

Apart from fair dealing for the purposes of research for non-commercial purposes or for private study, criticism or review, as permitted under the Copyright, Designs and Patents Act 1988 and the Copyright and Related Rights Regulations 2003, this publication may not be reproduced, stored or transmitted, in any form or by any means, without the prior permission in writing of The Royal Society of Chemistry or the copyright owner, or in the case of reproduction in accordance with the terms of licences issued by the Copyright Licensing Agency in the UK, or in accordance with the terms of the licences issued by the appropriate Reproduction Rights Organization outside the UK. Enquiries concerning reproduction outside the terms stated here should be sent to The Royal Society of Chemistry at the address printed on this page.

Whilst this material has been produced with all due care, The Royal Society of Chemistry cannot be held responsible or liable for its accuracy and completeness, nor for any consequences arising from any errors or the use of the information contained in this publication. The publication of advertisements does not constitute any endorsement by The Royal Society of Chemistry or Authors of any products advertised. The views and opinions advanced by contributors do not necessarily reflect those of The Royal Society of Chemistry which shall not be liable for any resulting loss or damage arising as a result of reliance upon this material.

The Royal Society of Chemistry is a charity, registered in England and Wales, Number 207890, and a company incorporated in England by Royal Charter (Registered No. RC000524), registered office: Burlington House, Piccadilly, London W1J 0BA, UK, Telephone: +44 (0) 20 7437 8656.

For further information see our web site at www.rsc.org

Printed in the United Kingdom by CPI Group (UK) Ltd, Croydon, CR0 4YY, UK

Preface

In the latter half of the twentieth century, the advent of positron emission tomography (PET) and magnetic resonance imaging (MRI) revolutionised *in vivo* tomographical imaging. PET rapidly became an invaluable research tool and later migrated into clinical diagnostics. MRI followed the same path some years later. Notwithstanding their similar trajectories, the reasons for their general acceptance and uptake in their respective communities differed enormously, even from the outset. Whereas PET excelled as an *in vivo* metabolic technique, MRI quickly established itself as the method of choice for structural imaging, providing excellent soft-tissue contrast with exceedingly good spatial resolution, which today approaches <100 microns in ultra-high field human scanners.

Even since its initial conception, MRI has shown remarkable versatility. This largely stems from the fact that as MRI is based on NMR, it can be seen as NMR spectroscopy, with the use of gradient magnetic fields to encode space. Given this fact, it is not difficult to appreciate that many NMR techniques—which have a long and rich history—can be turned into MRI methods, with each one potentially delivering a new contrast mechanism. Moreover, the often-quoted maxim that "one man's artefact is another man's effect", has also been put to good use by the MRI community. Functional MRI is based largely on the susceptibility 'artefact' created by paramagnetic haemoglobin, for example. Furthermore, MR methods, such as diffusion weighted/tensor imaging, and chemical shift imaging, now provide connectivity and metabolic information beyond anatomy. Unlike other imaging modalities such as computerised tomography (CT), MRI does not involve ionising radiation, making it suitable for repeated scanning or, in particular, for use in the paediatric population.

New Developments in NMR No. 19
Hybrid MR-PET Imaging: Systems, Methods and Applications
Edited by N. Jon Shah
© The Royal Society of Chemistry 2019
Published by the Royal Society of Chemistry, www.rsc.org

Not to be left by the wayside, developments in PET have concentrated on the use of more and more tracers, each one specifically targeting a particular disease. This, coupled with the fact that PET has an exquisite sensitivity many orders of magnitude better than that of MRI, has led to new areas of research as well as diagnostic questions.

Naturally, the underlying physics of the two methods ensured that their development remained separate for many years, as the strong magnetic fields needed for MRI remain the antithesis of the environmental requirements for the operation of photomultiplier tubes in PET. Nonetheless, a number of groups recognised the complementarity of these two imaging modalities and began to lay the foundations for what has now become a fully-fledged field of research and development.

When one considers the complementary advantages of MRI and PET, and the success of the hybridisation of PET and CT into one combined PET/CT instrument, the benefit of combining MRI with PET in one device is obvious. Previously, PET and MRI had already been used in combination, whereby images from the two modalities had been registered and fused by using appropriate software tools, and such a combination had been particularly successful in brain imaging, where only a rigid-body registration is necessary. However, the combination of MRI and PET in one instrument with one patient bed is most advantageous in whole-body imaging. Utilising the specific advantages of MRI and PET, an integrated PET/MR-scanner or MR-PET scanner, as it is commonly called, allows a multi-parametric investigation in a "one-stop-shop" manner.

There is a lack of textbooks addressing the needs of readers wishing to gain an oversight of hybrid MR-PET. Reaching for two separate texts is not particularly helpful in this case, since the specific issues associated with bringing the two modalities together are usually not discussed in sufficient detail. This book is concerned with the underlying principles of MRI and PET and, in particular, hybrid MR-PET. In practical terms, at the level of scanner construction and operation, hybrid MR-PET is a relatively new area of research and is opening the door to novel diagnostic approaches. Whilst initial approaches were based on the work of academic research groups, the field has now moved into the commercial sector with academia pursuing further advances such as hybrid MR-PET at 7 T for human applications.

The book is divided into five parts—from A to E—and each part is further sub-divided into sections. Part A, Section I addresses the basics of MRI complete with selective applications and the up-to-date topic of ultra-high field MRI. Part A, Section II addresses the basics of PET and includes PET quantification and kinetic modelling. Part B gives a technical overview of hybrid MR-PET and addresses the issues involved once these two modalities have been brought together. Part C deals with applications in human of MR-PET dividing them into brain and whole-body applications. Part D outlines preclinical applications including a brief discussion of the hardware of preclinical hybrid scanners. Finally, PET would be nothing without tracers and this topic is therefore presented in Part E, outlining tracer production.

The book is aimed at several audiences simultaneously, as befits a book on simultaneous hybrid imaging. A novice graduate student entering the field will benefit from the whole book, whereas experts in MRI will find Part A, Section II on PET to be instructive before going on to read Parts C, D, and E on hybrid MR-PET, applications and tracer production, respectively. Conversely, the PET expert will benefit from Part A, Section I on the basics of MRI and the other parts as described above.

I wish to thank the many authors who contributed to the various chapters and sections whose names are included with each chapter.

Finally, I would like to extend my gratitude to Claire Rick for invaluable assistance throughout the editorial process and to Martina Bunn for administrative help.

N. J. Shah
Germany

Contents

Part A. Basics

Section I: Magnetic Resonance

New Developments in NMR No. 19
Hybrid MR-PET Imaging: Systems, Methods and Applications
Edited by N. Jon Shah

© The Royal Society of Chemistry 2019

Published by the Royal Society of Chemistry, www.rsc.org

Part B. Hybrid MR-PET Imaging: technical overview

Section I: Hardware

Section II: Special aspects of Data Corrections in MR-PET

Section III: Special considerations in MR-PET

Part C. Human MR-PET Applications

Part D. Preclinical Applications

Part A

Basics

Section I

Magnetic Resonance

CHAPTER 1

Introduction to Magnetic Resonance Imaging

W. A. WORTHOFF*[a], S. D. YUN[a] AND N. J. SHAH[a,b,c]

[a]Forschungszentrum Jülich, Institute of Neuroscience and Medicine – 4, Wilhelm-Johnen-Straße 1, 52428 Jülich, Germany; [b]JARA-BRAIN – Translational Medicine, Aachen, Germany; [c]Department of Neurology, RWTH Aachen University, Pauwelsstr. 30, 52074 Aachen, Germany
*E-mail: w.worthoff@fz-juelich.de

1.1 Introduction

Nuclear magnetic resonance (NMR) has expanded rapidly from its initial discovery in 1946 into one of the most important analytical techniques in modern science, with applications in both physics and chemistry. It is NMR that forms the basis of magnetic resonance imaging (MRI), which has revolutionised diagnostic medicine since the late seventies. Indeed, even today the NMR techniques invented decades ago are finding application in MRI as new imaging sequences.

The introduction of linear magnetic field gradients to impose spatial information on the measured signal, in combination with the ubiquitous Fourier transform, led to an enormous research effort in the field of medical diagnostic imaging. In the early 1980s, there was already a growing feeling within the research community that NMR imaging would be used in other applications,

New Developments in NMR No. 19
Hybrid MR-PET Imaging: Systems, Methods and Applications
Edited by N. Jon Shah
© The Royal Society of Chemistry 2019
Published by the Royal Society of Chemistry, www.rsc.org

and not solely for the study of the morphology of the body. Their prediction has indeed come to pass with the advent of functional MRI (fMRI), non-proton MRI to investigate metabolism and so on. In fact, it has been anecdotally reported that even the early NMR pioneers self-experimented. Edward Purcell placed his head in an NMR spectrometer to investigate differences in NMR signal shape dependent upon when he either concentrated hard on a specific task or freed his mind. Although no differences were measured at that time, it is apparent that these pioneers were already convinced about the possibility of applying NMR to study biological systems. A few decades later, MRI came of age.

In this chapter, even though some reference to quantum mechanics is briefly touched upon, the basics of MRI are described using a classical model, which suffices to explain standard proton-based imaging. Following on from the description of spin dynamics and the acquisition of the signal, image reconstruction methods are described, including up-to-date techniques for the acceleration of image acquisition employing parallel receive and multi-band excitation. The final section of this chapter concludes with a brief description of how sequences can be classified and, conceptually, understood.

1.2 Physics of the Dynamics of Spin

1.2.1 Spin

Nuclei, composed of protons and neutrons, possess a basic property referred to as spin angular momentum, $\vec{I}$, which originates from the intrinsic spin of the nucleons. Nucleons are fermions and have a spin of ½ in units of $\hbar$, where $\hbar$ is the reduced Planck's constant $h/2\pi$. Thus, a given nucleus has a net spin, which depends on its mass number Z (number of nucleons), atomic number A (number of protons) and neutron number N. In particular, a nucleus has a half-integer spin when the mass number Z is odd and an integer spin when the mass number Z is even and A is odd. In the case that Z and A are even, a nucleus has a zero nuclear spin. From here on, it is assumed that the nucleus under discussion has non-zero nuclear spin. In most cases, MRI is concerned with a spin ½ nucleus, namely the proton of hydrogen nuclei. Nevertheless, other MR-active nuclei, such as sodium, possess a non-integer spin and are also of interest to imaging. These will be discussed in later chapters.

Spin value determines the magnetic dipole moment $\vec{\mu} = \gamma\vec{I}$, with gyromagnetic ratio[1] given by $\gamma = ge/2m_p$. The gyromagnetic ratio depends on the nuclear g factor, the elementary charge, e, and the mass of a nucleon, m_p.[1] For reasons imposed by the uncertainty principle, the magnetic moment cannot be well defined in all spatial directions simultaneously and thus it is common to choose one axis along which the component of magnetic dipole moment is to be considered. Ordinarily, in NMR and

MRI, one is interested in the component of the magnetic dipole moment along the axis of an externally applied magnetic field, which is normally labelled z. Nevertheless, this is based solely on convention, and in principle, the choice of this axis is arbitrary.[2] The dipole moment is a quantised variable, with quantification defined through the magnetic number m_l (eqn (1.1))

$$\mu = \gamma \hbar m_l, \quad \text{with } m_l = \{-I, -I+1, \ldots, I\} \tag{1.1}$$

The macroscopic manifestation of spin is referred to as magnetisation, $\vec{M}$. The magnetisation is the vector sum of the individual magnetic dipole moments of the nuclear spin ensemble in the volume of interest. In the absence of a magnetic field, the nuclear spins are oriented randomly in all directions of space, with the distribution determined by the thermal conditions, yielding a zero-net magnetisation (see Figure 1.1).

1.2.2 Spins in a Magnetic Field

When an external magnetic field is applied to a spin ensemble, the distribution of nuclear spins becomes skewed towards the magnetic north pole, thus yielding a net magnetisation in the direction of the external field.[3] The potential energy stored in a magnetic dipole is given by the scalar product of the magnetic dipole moment and the magnetic field vector, and thus depends on the angle between the dipole and the magnetic field θ. Assuming that the

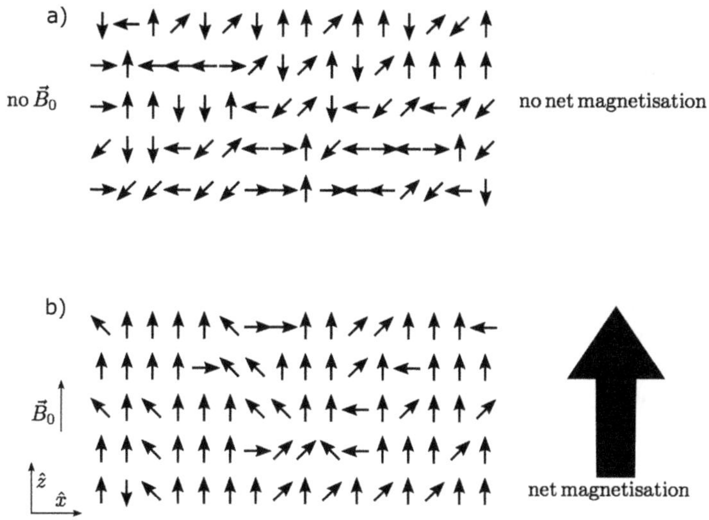

Figure 1.1 In the absence of a magnetic field, the nuclear spin is oriented randomly (a). If an external field is applied, the spins align to create a net magnetisation (b).

temperature of the system is high enough to apply the Boltzmann distribution, the angular distribution, P, is given by eqn (1.2)

$$P = \frac{1}{Z} e^{-\frac{E(\theta)}{k_B T}} \tag{1.2}$$

here k_B is the Boltzmann constant and Z is the canonical partition function, which is given by eqn (1.3)

$$Z = \sum e^{-\frac{E(\theta)}{kT}} = 2\pi R \int_0^{\pi} e^{-\frac{E(\theta)}{k_B T}} \sin(\theta) d\theta \tag{1.3}$$

The expectation value of the z-component of the dipole moment (eqn (1.4))

$$\langle \mu_z \rangle = \int P(\theta) \mu \cos(\theta) dV \tag{1.4}$$

can be approximated to (eqn (1.5))

$$\langle \mu_z \rangle = -\frac{\mu^2 B}{2k_B T} \tag{1.5}$$

This is a significant result for magnetic resonance methods, as not only does it show that there is a net magnetisation in the presence of the magnetic field, but it also implies that the available magnetisation depends linearly on the magnetic field, thus explaining the on-going endeavour to increase magnetic fields for MR imaging applications. It should be noted at this point that, for MR applications, higher equilibrium magnetisation translates to better image quality or reduced acquisition times. This should become clear to the reader during the course of this book.

Each spin state of the nucleus has an associated energy level. In the absence of a magnetic field, these levels are degenerate, but this degeneracy is lifted once an external field is applied.[2] An energy level splitting, referred to as the Zeeman effect, occurs where the energy shift is proportional to the magnetic field strength. In the classical model, which can be applied here for a better understanding, this energy splitting manifests itself in a precessional motion of the magnetisation about the magnetic field vector at a characteristic frequency, ω_0. This is referred to as the Larmor frequency and depends on the strength of the magnetic field (eqn (1.6))

$$\omega_0 = \gamma B_0 \tag{1.6}$$

This equation is of fundamental importance since it relates the frequency of precession to the total magnetic field. This can be shown by deliberately changing the static field. When the static field is changed, an alteration in the frequency can be observed. The associated difference between the energy states is $\Delta E = \hbar \gamma B_0$ (see Figure 1.2).

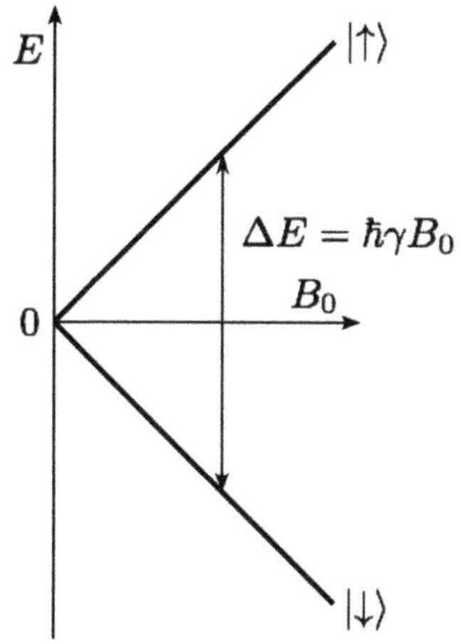

Figure 1.2 The Zeeman effect causes a splitting of the energy levels of a nucleus depending on its spin, shown here for a two-level spin-½ system.

1.2.3 Spin Dynamics

1.2.3.1 Spin Relaxation

In equilibrium, the net magnetisation of a system in the presence of an external field is oriented along the *z*-axis only, *i.e.* only longitudinal magnetisation exists. The transverse magnetisation of the system averages to zero. If the system is perturbed from equilibrium, this may no longer be the case. Transverse magnetisation might be present, and the longitudinal magnetisation might be changed or inverted. Over time, if the system is left to evolve under the influence of the static, external field, the magnetisation will return to its equilibrium state. The reason for this is that energy is transferred between individual spins and their environment.[3] In the case of protons, the dynamics follow monoexponential decays and their characteristic time constants. The longitudinal relaxation (or spin–lattice relaxation) time T_1 represents the convergence of the z magnetisation to its equilibrium magnetisation M_0 and is caused by the interaction of spins with their environment. Transverse magnetisation decreases with the time constant T_2 and is an interaction between the individual spins, thus representing an intrinsic irreversible loss of coherence. This so-called spin–spin relaxation is commonly superimposed with other effects, such as field inhomogeneity. The effective time constant T_2^* is then given by the following reciprocal relationship (eqn (1.7))

$$\frac{1}{T_2^*} = \frac{1}{T_2} + \frac{1}{T_2'} \tag{1.7}$$

with T_2' as the relaxation time due to field inhomogeneity. Although often undesirable, changes in T_2^* relaxation can be used in imaging as a marker of brain activity (see Chapter 3, Section 3.1.2) or to map effects from paramagnetic entities, such as iron (see Chapter 4, Section 4.5.1). In general, $2T_1$ is always longer than T_2, which must be longer than T_2^*.[3] However, under most circumstances, $T_1 > T_2 > T_2^*$ in biological tissue.

1.2.3.2 Bloch Equations: Static Case

The spin dynamics of ½-spin systems can easily be described by a set of three coupled first order differential equations, derived from classical physics. Nevertheless, one should keep in mind that this is an approximation, and the underlying principles are of quantum mechanical origin. Classically, at thermal equilibrium, $\vec{M}$ and $\vec{B}$ are aligned. If $\vec{M}$ is caused to point in a different direction from $\vec{B}$, then it is evident that $\vec{M}$ will precess about $\vec{B}$. Following this line of argument, the torque resulting on the net magnetisation in the magnetic field, $\vec{B}$, is $\vec{M}\times\vec{B}$. Noting that torque is the rate of change of angular momentum and, in the most basic form, ignoring the effects of relaxation and assuming a constant net external magnetic field $\vec{B}$, this set of equations takes the form of a simple vector rotation equation (eqn (1.8))

$$\frac{d\vec{M}}{dt} = -\gamma\vec{B}\times\vec{M} \tag{1.8}$$

with magnetisation $\vec{M} = \begin{pmatrix} M_x & M_y & M_z \end{pmatrix}^T$. The solution of this equation can conveniently be expressed for the case $\vec{B}_0 = \begin{pmatrix} 0 & 0 & B_0 \end{pmatrix}^T$ as a rotation about the z-axis with rotation matrix $R_z(\theta) = \begin{bmatrix} 0 & \sin\theta & 0 \\ -\cos\theta & 0 & 0 \\ 0 & 0 & 1 \end{bmatrix}$ such that (eqn (1.9))

$$\vec{M}(t) = -R(\omega_0 t)\vec{M}_{t=0} \tag{1.9}$$

thus, the magnetisation will precess about the z-axis at the Larmor frequency $\omega_0 = \gamma B_0$. In terms of the individual components of $\vec{M}$, the solution takes the form of eqn (1.10)

$$\begin{aligned} M_x(t) &= M_{x,t=0}\cos\omega_0 t + M_{y,t=0}\sin\omega_0 t \\ M_y(t) &= -M_{x,t=0}\sin\omega_0 t + M_{y,t=0}\cos\omega_0 t \\ M_z(t) &= M_{z,t=0} \end{aligned} \tag{1.10}$$

1.2.3.3 Bloch Equations

Following the tipping of the net magnetisation away from equilibrium (excitation), the transverse component decays to zero, whilst the longitudinal component returns to equilibrium. This behaviour is characterised by introducing additional phenomenological terms to represent the on-going relaxation processes. The coupled first order differential equations, commonly referred to as Bloch equations, can be obtained[4] where the two additional terms describe the spin–lattice and spin–spin relaxation processes (eqn (1.11))

$$\frac{d\vec{M}}{dt} = -\gamma \vec{B} \times \vec{M} + \begin{bmatrix} 0 \\ 0 \\ \frac{M_{\text{eq}}}{T_1} \end{bmatrix} - \begin{bmatrix} -\frac{1}{T_2} & & \\ & -\frac{1}{T_2} & \\ & & -\frac{1}{T_1} \end{bmatrix} \vec{M} \tag{1.11}$$

where M_{eq} is the equilibrium magnetisation. In terms of the individual components of $\vec{M}$, the equation can conveniently be expressed in one line: $\frac{d\vec{M}}{dt} = -\gamma \vec{B} \times \vec{M} - \frac{(M_x \hat{x} + M_y \hat{y})}{T_2} - \frac{(M_z - M_{\text{eq}})\hat{z}}{T_1}$, with unit vectors $\hat{x}$, $\hat{y}$, $\hat{z}$ pointing in the three spatial directions in the laboratory frame.

Precession about the net magnetic field is not limited to the static external magnetic field alone, but can, in fact, include effects from the temporally and spatially varying magnetic fields created by the gradient system, as well as the radiofrequency (RF) field and effects due to (usually undesired) field inhomogeneity. Understanding these fields is necessary in order to fully appreciate the imaging process, which can be conveniently described by the Bloch equations. If necessary, further terms can be added to the Bloch equation to address effects such as diffusion (see Chapter 3, Section 3.3).

1.2.3.4 The Rotating Frame of Reference

Solving Bloch equations can be an elaborate process in the laboratory reference frame, in which the observer is positioned in a stationary location and the spin system evolves and precesses accordingly. It is often convenient and useful to utilise the concept of the rotating frame of reference to solve the Bloch equations and understand spin dynamics. The rotating frame of reference is chosen such that it revolves about the *z*-axis at a given frequency, ω_r.

For now, it is assumed that the magnetisation is subject to a magnetic field, which is made up of two parts: the static field, B_0; and the effect of an oscillating linear field, $\vec{B}_1 = 2B_2 \cos(\omega_{\text{rf}} t)\hat{i}$, perpendicular to the B_0 field. The B_1 field can be expressed as a sum of left- and right-handed rotating magnetic fields. If observed relative to the precessing magnetisation, one component is rotating at a reduced relative angular frequency in the same direction as the magnetisation and the other component is rotating in the opposite direction, causing dynamic effects on a timescale not relevant to the measurement process. Omitting contributions from the latter and, for now, also the effects of relaxation, the Bloch equations reduce to a form in which they describe simple precession about the net magnetic field (eqn (1.12))

$$\frac{d\vec{M}}{dt} = -\gamma\left[B_0\hat{z} + B_1\left(\cos(\omega_{\text{rf}}t)\hat{x} - \sin(\omega_{\text{rf}}t)\hat{y}\right)\right]\times\vec{M} \tag{1.12}$$

In order to solve this set of differential equations, it is convenient to transform to the rotating frame of reference, which rotates at the same angular frequency as the RF wave. Its unit vectors are determined by eqn (1.13)

$$\begin{aligned}\hat{x}_r &= \cos(\omega_{\text{rf}}t)\hat{x} - \sin(\omega_{\text{rf}}t)\hat{y}\\ \hat{y}_r &= \sin(\omega_{\text{rf}}t)\hat{x} + \cos(\omega_{\text{rf}}t)\hat{y}\\ \hat{z}_r &= \hat{z}\end{aligned} \tag{1.13}$$

such that (eqn (1.14))

$$\frac{d\vec{M}_r}{dt} = -\gamma\vec{B}_{\text{eff}}\times\vec{M}_r \tag{1.14}$$

Here, the effective magnetic field is introduced as eqn (1.15)

$$\vec{B}_{\text{eff}} = B_1\hat{x}_r + \left(B_0 - \frac{\omega_{\text{rf}}}{\gamma}\right)\hat{z} \tag{1.15}$$

which represents a field that determines the motion of the magnetisation in the rotating frame of reference. It should be noted, that if the resonance condition, $\omega_{\text{RF}} = \omega_0$, is met, the effective magnetic field is determined by the transverse B_1 component only, *i.e.* in this case the magnetisation only evolves around the applied B_1 field. This is easy to comprehend if one considers an observer rotating about the same axis, watching an object precessing with the same angular frequency: the motion of the object would seem frozen to the observer (see Figure 1.3).

1.2.3.5 Bloch Equations in the Axial Representation

Sometimes it is beneficial to note that it is mathematically convenient to rewrite the Bloch equations in an axial representation,[5] such that the transverse magnetisation is expressed as a complex quantity $M_{xy} = M_x + iM_y$, rather than using elements of the three-dimensional magnetisation vector $\vec{M}$. By adopting a similar notation for the transverse magnetic field $B_{xy} = B_x + iB_y$, the Bloch equations (see eqn (1.10)) can be expressed by the following set of coupled differential equations (eqn (1.16))

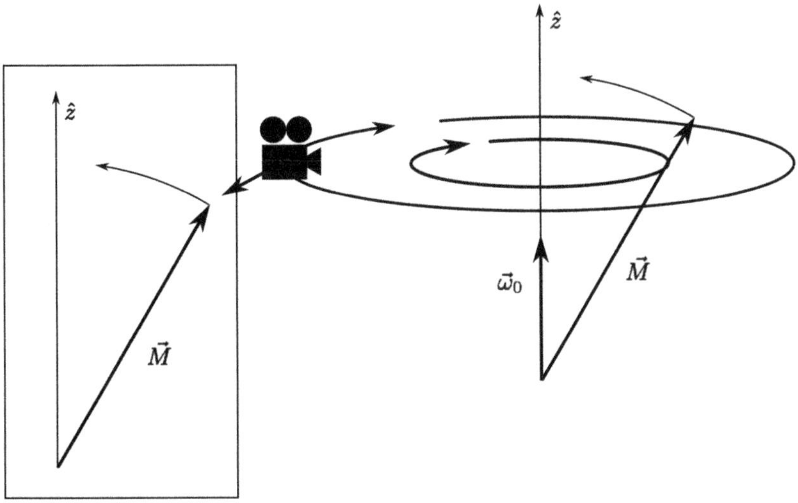

Figure 1.3 In the rotating frame of reference (indicated by the camera) the precession is frozen if the resonance condition is met.

$$\begin{aligned}\frac{dM_{xy}}{dt} &= -\gamma\left(M_{xy}B_z - M_zB_{xy}\right) - \frac{M_{xy}}{T_2}\\ \frac{dM_z}{dt} &= \frac{i}{2}\gamma\left(M_{xy}B_z^{\dagger} - M_z^{\dagger}B_{xy}\right) + \frac{M_{\text{eq}} - M_z}{T_1}\end{aligned} \tag{1.16}$$

In the axial representation, transformation to the rotating frame of reference is a simple multiplication of transverse magnetisation with the exponential $M_{xy,r} = e^{-i\omega_0 t}M_{xy}$. The longitudinal magnetisation remains equal in both the laboratory frame and in the rotating frame.

1.2.3.6 Bloch Equations: Resonant RF Pulses

In order to understand the effects of RF pulses on the magnetisation, it is convenient to use the rotating frame of reference as introduced above. Assuming a resonant RF pulse generates a magnetic field $B_{xy} = B_1e^{i\varphi - i\omega_0 t}$, it is observed as a constant field in the rotating frame of reference $B_{xy,r} = B_1e^{i\varphi}$. Considering the pulse duration τ to be sufficiently short, such that relaxation effects do not contribute to any significance, the dynamics are described by eqn (1.17)

$$\frac{d}{dt}M_{xy,r} = iM_z\gamma B_{xy,r} = iM_z\gamma B_1e^{i\varphi} \tag{1.17}$$

and eqn (1.18)

$$\frac{d}{dt}M_z = i\gamma\left(M_{xy,r}B_{xy,r}^{\dagger} - M_{xy,r}^{\dagger}B_{xy,r}\right) \tag{1.18}$$

where $M^{\dagger}_{xy,r}$ and $B^{\dagger}_{xy,r}$ denote the complex conjugates of $M_{xy,r}$ and $B_{xy,r}$, respectively.

Solving these equations for the longitudinal magnetisation yields a solution of the form (eqn (1.19))

$$M_z = C_+ e^{+i\gamma B_1 t} + C_- e^{-i\gamma B_1 t} \tag{1.19}$$

with constants of integration $C_\pm$ to be determined by the initial values. The transverse magnetisation can then be found by integration (eqn (1.20))

$$M_{xy,r} = \int iM_z \gamma B_1 e^{i\varphi} dt = e^{i\varphi}\left[C_+ e^{+i\gamma B_1 t} - C_- e^{-i\gamma B_1 t}\right] \tag{1.20}$$

Assuming that the system started from equilibrium, $M_z(t = 0) = M_{\text{eq}}$ and $M_{xy,r}(t = 0) = 0$ yields for the constants of integration $C_+ = C_- = \frac{1}{2}M_{\text{eq}}$. Thus, (eqn (1.21))

$$M_z = \frac{e^{+i\gamma B_1 t} + e^{-i\gamma B_1 t}}{2} M_{\text{eq}} = M_{\text{eq}} \cos(\gamma B_1 t) \tag{1.21}$$

and (eqn (1.22))

$$M_{xy,r} = M_{\text{eq}} e^{i(\varphi + \frac{\pi}{2})} \sin(\gamma B_1 t) \tag{1.22}$$

In the rotating frame, the magnetisation experiences a rotation about the RF field (B_1), which means that magnetisation is transferred from the z-axis to the xy-plane and *vice versa* periodically for the duration of the RF pulse.[6] The axis of rotation lies in the xy-plane, and the RF phase φ determines the orientation of the rotation axis on that plane, *i.e.* for the short RF pulse, the effective field acts as if the magnetisation experiences a constant external field in the direction of $\vec{B}_1$, about which it precesses for the duration of the pulse (see Figure 1.4 for an example of a y pulse). The angle swept out by the magnetisation during the duration of the pulse is called the flip-angle, which can be determined by $\alpha = \gamma B_1 \tau$. It should be noted that although the exponential term, including the RF phase φ, determines the direction of the motion in the rotating frame, it does not influence the flip-angle, as its magnitude is 1. Two special cases are of particular interest in MRI: the $\pi/2$ or 90° pulse and the π or 180° pulse. The former rotates the magnetisation by 90°, *i.e.* all longitudinal magnetisation is transferred to the transverse plane. The latter inverts the magnetisation and has the effect of inverting relative phases thus acting as an inversion and refocusing pulse.

1.2.3.7 Bloch Equations: Off-resonant RF Pulses

In the previous section, a perfectly resonant excitation pulse was considered; however, this condition is often not met. This is either deliberate, *e.g.* to exclude portions of the sample when excitation is only required in a slice of interest (see Section 1.3.1.1), or because resonance is simply not

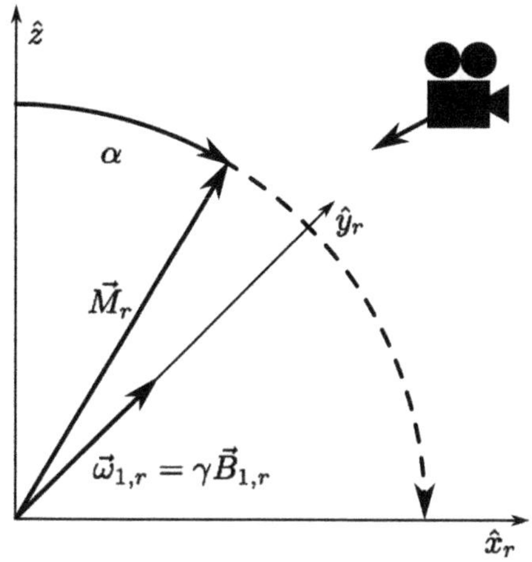

Figure 1.4 In the rotating frame of reference a resonant RF pulse causes a rotation about an axis (y in this case) into the xy plane. The flip angle α is the angle swept during the time t.

achieved technically. If the rotating frame of reference is assumed to rotate at the same frequency as the RF field $\vec{B}_1$, the Bloch equations can be rewritten as eqn (1.23)

$$\frac{d}{dt}M_{xy,r} = -i\gamma\left(B_0 - B_1\right)M_{xy,r} + iM_z\gamma B_1 \tag{1.23}$$

and eqn (1.24)

$$\frac{d}{dt}M_z = i\gamma\left(M_{xy,r} - M_{xy,r}^{\dagger}\right)B_1 \tag{1.24}$$

These equations do not, however, have an analytical general solution, but can be solved for special cases. The off-resonance term in $\frac{d}{dt}M_{xy,r}$ reflects the fact that the precession frequencies of the RF pulse and magnetisation are not equal. This manifests itself as a frequency shift and a persistent slow precession in the rotating frame. The consequence of this is that excitation occurs at a reduced efficiency, thus the achieved flip-angles are generally lower and, furthermore, a non-zero phase is introduced to the transverse magnetisation. If the off-resonance is significant, no excitation takes place at all, which may be detrimental but can also be desirable in some imaging applications.

1.2.3.8 Bloch Equations: Magnetic Field Gradients

Spatial magnetic field gradients are of utmost importance to magnetic resonance imaging. They are used to select imaging slices, to manipulate the phase of the processing spins and are essential for spatial encoding. The magnetic field is generated to vary linearly with position $\vec{r} = (x, y, z)^T$ depending on the magnetic field gradient $\vec{G}=(G_x,G_y,G_z)^T$ such that $B(\vec{r})\hat{z}=\left(\vec{G}(t)\bullet\vec{r}+B_0\right)\hat{z}$. This causes the spin dynamics to become spatially dependent. In the rotating frame of reference with frequency $\omega_r=\gamma\left(\vec{G}\bullet\vec{r}+\vec{B}_0\right)$, the transverse magnetisation evolves into eqn (1.25)

$$\frac{d}{dt}M_{xy,r}=i\gamma\vec{G}(t)\bullet\vec{r}M_{xy,r}-\frac{M_{xy,r}}{T_2} \tag{1.25}$$

The solution of this equation takes the form (eqn (1.26))[7]

$$M_{xy,r}(t)=M_{xy,r}(t=0)e^{-\frac{1}{T_2}t}e^{i\int_0^t\gamma\vec{G}(t)\bullet\vec{r}dt} \tag{1.26}$$

Decomposing $\vec{G}\bullet\vec{r}$ into its components yields eqn (1.27)

$$M_{xy,r}(t)=M_{xy,r}(t=0)e^{-\frac{1}{T_2}t}e^{i\gamma\int_0^t G_x x dt+i\gamma\int_0^t G_y y dt+i\gamma\int_0^t G_z z dt} \tag{1.27}$$

which can be rewritten in terms of spatial frequencies $k_n=\frac{\gamma}{2\pi}\int_0^t G_n dt$, with $n=x,y,z$ (eqn (1.28))

$$M_{xy,r}(t)=M_{xy,r}(t=0)e^{-\frac{1}{T_2}t}e^{i2\pi\left(k_x x+k_y y+k_z z\right)} \tag{1.28}$$

Thus, the application of a gradient creates a phase difference according to the value of $\vec{k}$. With this property, image information will be recovered. As far as the outcome of the application of the gradient is concerned, the shape of the gradient is of minor importance; it is rather the time integral (called gradient moment), which constitutes the net effect of the gradient.

It should be noted that in the context of MRI, gradients exist in three orthogonal spatial dimensions, where they describe a change of the magnetic field component in the z direction, *i.e.* an "x gradient" $\vec{G}_x=\left(\partial B/\partial x,\,0,\,0\right)^T$ generates a change in B_0 dependent on the position on the x axis. Similarly, a "y gradient" generates a change dependent on the position on the y axis and so on (see Figure 1.5).

1.2.4 Signal Formation

1.2.4.1 *Free Induction Decay*

MR devices are set up in such a way that the presence of rotating transverse magnetisation can be picked up by a radiofrequency coil. This coil couples inductively to the precessing magnetisation. Thus, if the system is in equilibrium, no transverse magnetisation is present and thus no signal is received. Assuming this equilibrium is disturbed, for example by the application of an RF pulse, and transverse magnetisation is created, the receiver coil picks up a signal. This signal possesses a carrier frequency, which is equal to the

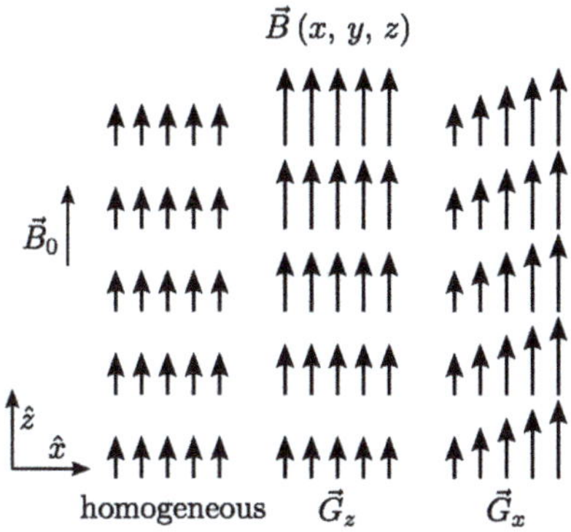

Figure 1.5 Magnetic field gradients act on the z-component of B and cause the magnetic field to vary linearly with position along the gradient direction.

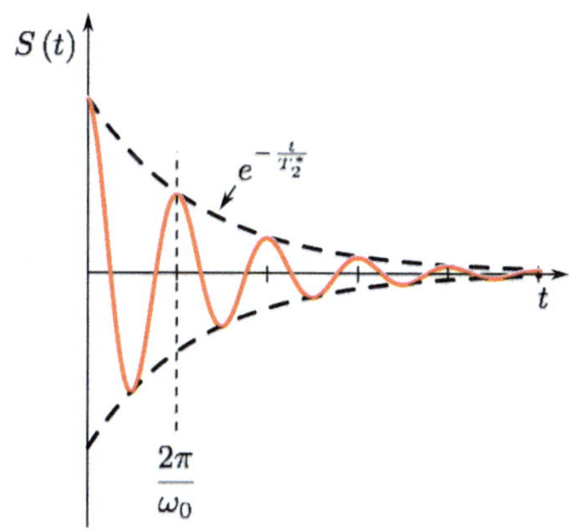

Figure 1.6 The FID (red) consists of a carrier wave of frequency ω_0 and an envelope determined by T_2^* (black).

frequency at which the magnetisation precesses, *i.e.* the Larmor frequency. The envelope of the signal is given by exponential T_2^* decay as the transverse magnetisation undergoes relaxation (see Figure 1.6). This signal, S, is commonly referred to as the free induction decay (FID). Starting from equilibrium, the highest FID signal occurs after a 90° pulse, which transfers all longitudinal magnetisation into the transverse plane.

1.2.4.2 Echoes

Imaging is rarely carried out on the FID signal itself (see Section 4.3), this is due to the fact that the time required for spatial encoding will usually cause most of the FID signal to be lost before an image can be acquired. Instead, the FID is refocused to create echoes, which are then used for imaging. The time after which an echo occurs is often abbreviated to TE, for echo time.

This time is usually measured from the centre of the RF pulse to the time at which the peak echo amplitude occurs.

1.2.4.3 Spin Echo

Spin echoes occur when existing, but dephased, transverse magnetisation is recovered by applying a refocusing RF pulse.[3] After coherent transverse magnetisation is created, the magnetisation is left to evolve for the time ½ TE, before the refocusing pulse is applied. During this time the magnetisation undergoes both intrinsic non-reversible relaxation, as well as dephasing, due to field inhomogeneity. The pulse causes the magnetisation to be mirrored along the axis of the pulse and, following the subsequent ½ TE, the reversible effects have been undone and a signal maximum occurs. Without further influences, an echo is seen which has an envelope similar to that of two FIDs facing each other, with the peak amplitude at the time TE given by the signal loss due to true T_2 (see Figure 1.7).

1.2.4.4 Gradient Echo

If the effect of a gradient is used to artificially promote dephasing along a spatial direction, an echo might be recovered if another gradient of opposite polarity, but with the same gradient moment as the first one, is applied.[8] This would cause the spins, which were initially precessing at an increased rate due to the gradient, to be slowed down, whereas those originally precessing more slowly would be sped up accordingly. Eventually, once the effect of the first gradient has been undone, an echo would occur (see Figure 1.8). This type of echo does not undo the effects of B_0 field inhomogeneity.

1.2.4.5 Stimulated Echo

Stimulated echoes occur after the application of a minimum of three consecutive RF pulses.[9] The first pulse creates coherent transverse magnetisation, which then evolves for a time, t_1. At this point, the second pulse transforms the transverse magnetisation to longitudinal magnetisation. After a time, t_2 from the second pulse, a third pulse recalls this magnetisation from the longitudinal direction into the transverse plane. Here the magnetisation refocuses, and an echo is formed. Such echo is referred to as a stimulated echo and is found at a time $2t_1 + t_2$ after the first RF pulse.

1.2.4.6 The Signal Equation and Demodulation

The imaging apparatus is set up in such a way that the receiver coil detects magnetic flux changes in the transverse direction only. The signal can be expressed in terms of the transverse magnetisation that creates the signal in

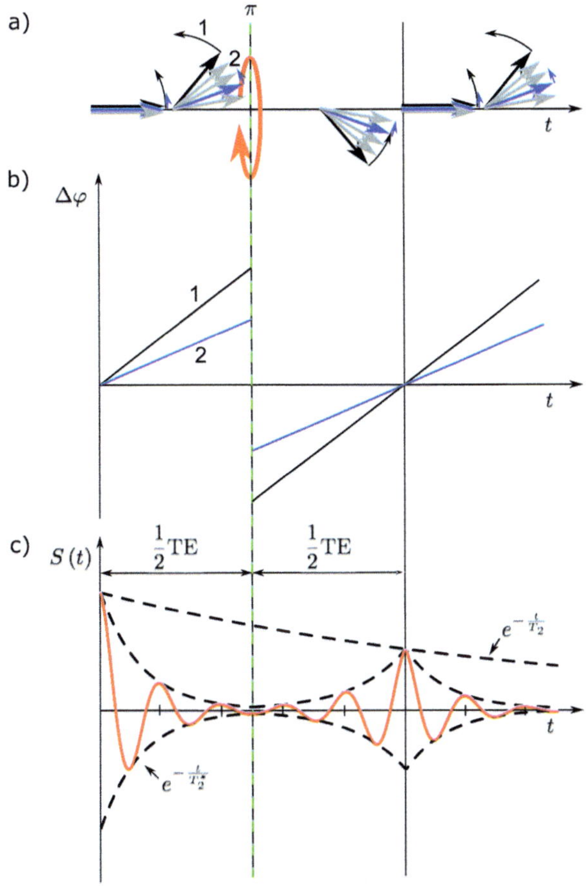

Figure 1.7 A π pulse at time ½ TE is used to generate an echo at time TE. The phase change from two independent signal sources 1 and 2 are shown in (b). The echo occurs when all signal sources are once again in phase (a). Since the phase orientations are effectively inverted at the time of the pulse π, T_2^* effects are undone. The echo amplitude is proportional to $e^{-\frac{\mathrm{TE}}{T_2}}$ since the intrinsic dephasing cannot be recovered (c).

the receiver coil. The total signal is therefore proportional to the induced voltage from all contributing spins in the receptive volume of the coil,[7] assuming the coil has a uniform sensitivity in this region (eqn (1.29))

$$S_t(\vec{r},t) \propto \iiint_V M_{xy}(\vec{r},t)\,dV \tag{1.29}$$

By substituting the appropriate form of the Bloch equations, one can predict the signal expected at any time during an MR experiment (eqn (1.30))

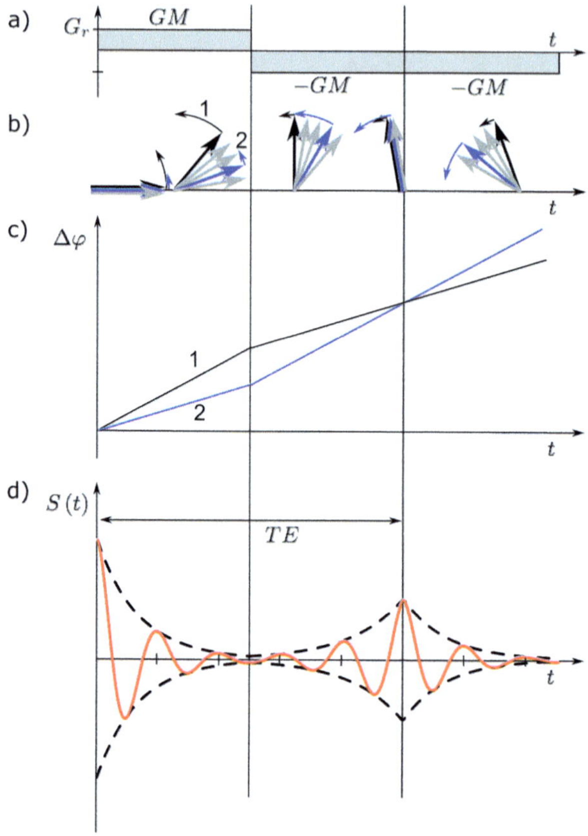

Figure 1.8 A gradient with moment GM deliberately dephases the spin system, a second gradient of equal GM but opposite polarity undoes this dephasing (a), such that an echo occurs (d). (c) Accumulated phase differences during the action of the gradients for two signal sources (1 and 2) at different locations along the gradient. (b) Influence of the gradients on the magnetisation vectors.

$$S_t(\vec{r},t) \propto \iiint\limits_V e^{i\omega_0 t} M_{xy,r} dV = e^{i\omega_0 t} \iiint\limits_V M_{xy,r} dV \tag{1.30}$$

Since the external field is assumed to be homogeneous in the region-of-interest (ROI), the Larmor frequency is not dependent on position and can be taken out of the integral, and since the detected signal is demodulated anyway, this term can be omitted entirely (eqn (1.31))

$$S(\vec{r},t) \propto \iiint\limits_V M_{xy,r} dV \tag{1.31}$$

i.e. demodulating the signal is as if one is to measure directly from within the rotating frame of reference.

1.3 Imaging

1.3.1 Spatial Encoding

MRI images are commonly acquired in all three spatial dimensions. Therefore, encoding of image information also needs to be performed in three directions. MRI uses its gradient system to achieve this. Gradients are used during pulses, during the readout and on their own to manipulate the phase of the spins and imprint information relating to the spatial location. For the purposes of illustration, simpler 2D MRI sequences can be broken down into three consecutive modules: (1) slice selection, (2) phase encoding and (3) frequency encoding. More complex methods use these modules but not necessarily in the order above and they can sometimes occur simultaneously.

1.3.1.1 Slice Selection

If a linear gradient field is applied, the Larmor frequency of the nuclei becomes spatially dependent and varies linearly with the position along the gradient. If a narrow-band RF pulse is applied whilst one encoding gradient is present, the resonance condition is only met in a subset of voxels, that is, a slice perpendicular to the applied gradient. If one considers the rotating frame of reference and a local off-resonance due to a gradient $\vec{G}_r$, the frequency offset is given by $\omega_r - \omega_0 = \gamma \vec{G}_r \bullet \vec{r}$ (see Figure 1.9). An RF pulse of the form $B_{xy,r} = B_1 e^{-i(\omega_{rf} - \omega_r)t + i\varphi}$, with the carrier frequency ω_{RF}, acts on the transverse magnetisation such that (eqn (1.32))

$$\frac{d}{dt} M_{xy,r} = i\gamma \vec{G}_r \bullet \vec{r} M_{xy,r} + iM_z \gamma B_1(t) e^{-i\omega_{rf} + i\varphi} \tag{1.32}$$

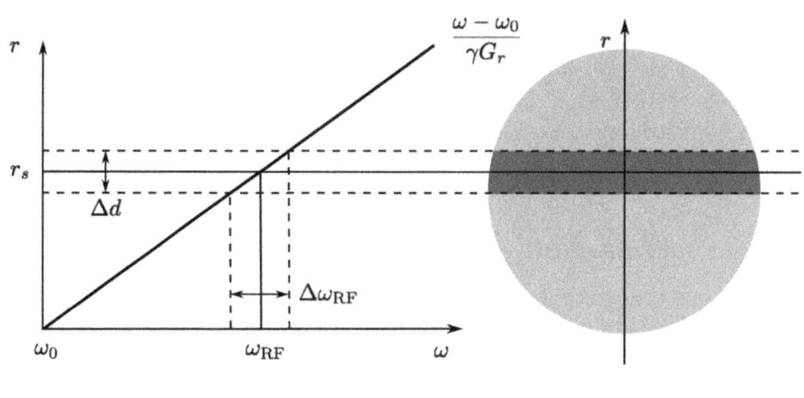

Figure 1.9 Applying an RF pulse of frequency ω_{RF}, bandwidth $\Delta\omega_{RF}$ and a gradient G_r simultaneously selects a slice perpendicular to the gradient direction centred at r_s with thickness Δd.

If the flip angle is sufficiently small and the longitudinal magnetisation is not significantly impacted by the pulse, then the longitudinal and transverse magnetisations do not couple, and the solution becomes eqn (1.33)

$$M_{xy},r(t,r) = ie^{i\varphi}M_{\mathrm{eq}}e^{-i\gamma\vec{G}_r \bullet \vec{r}t}\int_0^t e^{i\gamma\vec{G}_r \bullet \vec{r}t}\gamma B_1(t)dt \tag{1.33}$$

and for a pulse of duration τ (eqn (1.34))

$$M_{xy},r(t,r) = ie^{i\varphi}\gamma M_{\mathrm{eq}}e^{-i\gamma\vec{G}_r \bullet \vec{r}\frac{\tau}{2}}\int_{-\frac{\tau}{2}}^{\frac{\tau}{2}} e^{i\gamma\vec{G}_r \bullet \vec{r}t}B_1\left(t+\frac{\tau}{2}\right)dt \tag{1.34}$$

The integral is an FT of the envelop function of the RF pulse, which relates the time evolution of the RF pulse to the slice profile it excites in the spin system, which is under the influence of the gradient. From this, it becomes apparent that the slice has a width, which is dependent on the bandwidth of the RF pulse and the slope of the gradient.[6] In order to excite a rectangular slice, a sinc-shaped envelope function is required, since the Fourier transform of a sinc profile is rectangular. Such an envelope would require an infinitely long pulse duration, which cannot exist in reality. In other words, the pulse is truncated. This causes the slice profile to extend outside the desired slice. To prevent image artefacts, slices are not usually sampled consecutively or directly adjacent to each other. The centre of the slice is given by $r_s = \frac{\omega_{\mathrm{rf}} - \omega_0}{\gamma G_r}$, the thickness $\Delta d = \frac{\Delta\omega_{\mathrm{rf}}}{\gamma G_r}$ depends on the bandwidth of the RF pulse $\Delta\omega_{\mathrm{RF}}$, and the strength of the applied gradient $G_r = \left|\vec{G}_r\right|$ (see Figure 1.9).

In the case of arbitrarily shaped pulses, the bandwidth is usually defined as the full-width-at-half-maximum (FWHM). Further attention should be drawn to the phase term before the FT, as it causes the transverse magnetisation to be dephased after the RF pulse has been applied. In order to recover the maximum signal from the selected slice, this dephasing needs to be undone. This can be accomplished by manually inducing a phase shift in the opposite direction by applying a gradient with half the gradient moment of the slice selection gradient.[6,10]

1.3.1.2 Phase Encoding

Switching on a magnetic field gradient for a fixed amount of time, δt, causes the Larmor frequency to vary linearly with position along the applied gradient. Since the phase angle $\Delta\varphi$ is the time integral of the Larmor frequency, $\Delta\varphi$ also varies linearly with position[7,10]: $\frac{d}{dt}M_{xy,r} = i\gamma\vec{G} \bullet \vec{r}M_{xy,r}$ with a solution $M_{xy,r}(t+\delta t)=M_{xy,r}(t)e^{i\gamma\vec{G}\bullet\vec{r}\delta t}$. The phase angle after time δt is given by $\Delta\varphi=\gamma\vec{G}\bullet\vec{r}\delta t$. Phase encoding is best understood in terms of sampling the object for periodicity: for a completely homogeneous object, the signal from the whole object after application of the phase encode gradient will be very small. This is because the induced phase shifts disperse the magnetisation, leading to a small net transverse magnetisation. Now, assuming the object has a certain periodicity along the gradient direction, a gradient strength exists at which the periodicity of the object is the same as that of the phase term induced by the gradient. This causes all spins in the periodic object to acquire a phase, which is a multiple of 2π, and since the transverse magnetisation is aligned, a large signal is to be expected (see Figure 1.10). If one is to test for periodicity by assuming a different gradient strength each time and repeating the steps often enough; it is possible to detect these spatial frequencies. After each phase encoding step, it is then possible to reconstruct the measured signal and to create the corresponding MR image *via* FT. Phase encoding can be carried out in 1-, 2- and 3-dimensions, the latter is only rarely carried out, as phase encoding is demanding on measurement time.

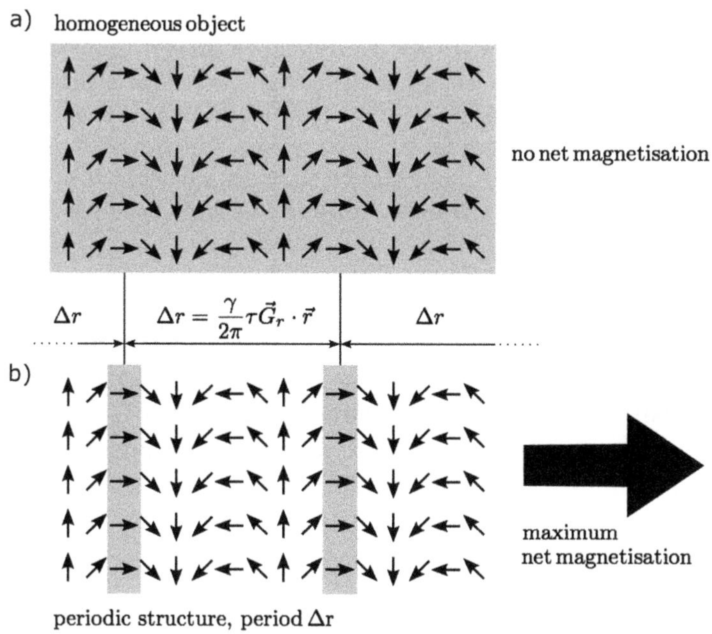

Figure 1.10 (a) Sampling a uniform object leads to no net magnetisation after a phase encoding gradient $\vec{G}_r$ is applied for a duration τ. (b) If the object has a periodicity Δr that is equal to the spatial extent of one period of the phase shift induced by the gradient, the net magnetisation is maximised.

1.3.1.3 *Frequency Encoding*

The final dimension can be determined by switching a gradient on while acquiring spectral data. This will cause the Larmor frequency of the nuclei to differ along the gradient, thus emitting signals with different carrier frequencies. Fourier transformation allows the spectral information to be recovered and converted into spatial coordinates.[8]

1.3.1.4 *2D* vs. *3D Imaging*

Two major modes of encoding exist in MRI[8]: 2D and 3D. In 2D, imaging is on a per slice basis. That is, in a slice excited *via* slice selection, one dimension is encoded *via* phase encoding and the other dimension is obtained from frequency encoding. 3D methods, however, consist of phase encoding in 2D and frequency encoding in the other. Both acquisition schemes have advantages and disadvantages. 2D methods are usually faster and are not prone to aliasing artefacts, nevertheless, slices cannot be put arbitrarily close together as the slice profiles might overlap and crosstalk between slices might introduce artefacts. 3D methods need a longer scan time due to the requirement of more encoding steps, but since no slice profile needs to be accounted for, voxels can be located directly adjacent to each other in all spatial dimensions. In some circumstances, it is beneficial to prevent excitation outside the region-of-interest. One way to achieve this is to excite a very thick slice, covering the region-of-interest, referred to as a slab. The difference between this and slice selection is that in the slab method, individual phase encoding steps are carried out in the same direction to finally yield thin slices. This method is an alternative to an oversampling in the phase encodes direction (see Section 1.3.4.6).

1.3.1.5 *The Signal Equation of Spatial Encoding*

The signal equation depends on the magnetisation of all contributing voxels at position $\vec{r} = (x, y, z)^T$. Each voxel has a specific Larmor frequency $\omega(\vec{r},t)$ and a phase offset φ, which are dependent on position (eqn (1.35) and (1.36))

$$M(\vec{r},t) = M_i(\vec{r},t)e^{i(\omega(\vec{r},t)t+\varphi(\vec{r},t))}\Lambda(\vec{r}) \tag{1.35}$$

$$\omega(\vec{r},t) = \omega_0 + \vec{G}\bullet\vec{r} \tag{1.36}$$

where $M_i(\vec{r},t)$ is the magnetisation at $\vec{r}$ before spatial encoding commences. $\Lambda(\vec{r})$ is the slice selection function, which describes the selected slice profile. In the case of an ideal, rectangular slice in a 2D sequence (or for slab selection in 3D) with a width Δz along the z direction, this function is $\Lambda(\vec{r}) = \mathrm{rect}\left(\frac{2z}{\Delta z}\right)$.

In the case of a 3D sequence, it is simply $\Lambda(\vec{r})=1$. Considering the 2D case, in which the direction of frequency encoding is x, such that $\omega(x)=\omega_0+\gamma\int_0^t G_x x dt$, phase encoding is applied for a duration τ along y, $\left(\varphi(y)=\gamma\int_0^\tau G_y y dt\right)$, and z is the direction of slice selection (eqn (1.37))

$$\begin{aligned} S(x,y,z,t) &\propto \Delta z \iint_{x,y} M_i(x,y,z_0,t) e^{i(\omega(x)t+\varphi(y))} dxdy \\ &= \Delta z \iint_{x,y} M_i(x,y,z_0,t) e^{i\omega_0 t + 2\pi i(k_x x + k_y y)} dxdy \end{aligned} \tag{1.37}$$

The time integral of the gradients $k_n = \frac{\gamma}{2\pi}\int_0^t G_n dt$, n = {x,y,z} are the spatial frequencies. In this case $k_x = \frac{\gamma}{2\pi} G_y \tau$ and $k_y = \frac{\gamma}{2\pi} G_y \tau$. They form the basis of the acquired MRI data and are commonly referred to as *k*-space. The choice of *k*-space trajectory, *i.e.* the choice of how and when gradients are applied, is the key element in MR imaging. In order to reconstruct an image from the raw MR data acquired, the correct trajectory needs to be chosen in such a way that it contains sufficient information relating to the final image.

It should be noted that, particularly when phase manipulation *via* gradients is executed, the signal equation attains the form of a FT in 2D or 3D, where the components are mapped out by the choice of gradient pulses (eqn (1.38))

$$S(x,y,t) \propto \iint_{x,y} M_{i,r}(x,y,z_0,t) e^{2\pi i(k_x x + k_y y)} dxdy \propto FT\{M_{i,r}(x,y,z_0,t)\} \tag{1.38}$$

The signal gives information on the transverse magnetisation in the region-of-interest. The aim is to acquire sufficient knowledge about the underlying magnetisation from which this signal originates to be able to reconstruct an image containing the NMR parameters[7] such as the density of the sample nucleus and relaxation times (T_1, T_2).

It should be noted that, for the FT, it is necessary to fill the matrix step-by-step, which creates the signal S, *i.e.* multiple iterations with different phase encoding gradients are necessary to populate S. In the case of 3D spatial encoding, the signal equation becomes a 3D FT. Acquisition commences stepwise for each combination of k_y, k_z (eqn (1.39))

$$\begin{aligned} S(x,y,t) &\propto \iiint_{x,y,z} M_{i,r}(x,y,z_0,t) e^{2\pi i(k_x x + k_y y + k_z z)} dxdydz \\ &= FT\{M_{i,r}(x,y,z,t)\} \end{aligned} \tag{1.39}$$

The signal matrix $S = S(k_x, k_y, k_z, t)$ is commonly referred to as *k*-space data.

1.3.1.6 Spatial Resolution and Field-of-View

Since the relationship between image space and k-space is that of an FT, simple inverse relationships between the field-of-view (FOV) and k-space line spacing Δk, as well as voxel size δ and the maximum k-space value $k_{\max}$, can be used (eqn (1.40))

$$\begin{aligned} \mathrm{FOV}_{x,y,z} &= \frac{1}{\Delta k_{x,y,z}} = \left.\frac{N_k}{k_{\max}}\right|_{x,y,z} \\ \delta_{x,y,z} &= \left.\frac{1}{k_{\max}}\right|_{x,y,z} \end{aligned} \tag{1.40}$$

Here, N_k is the matrix size. It should be noted that in order to increase resolution, *i.e.* reduce voxel size, it is necessary to achieve greater k-space values. Which, for a set encoding time, requires an increase in gradient strength, thus the maximum gradient strength is a limiting factor for the detectable special resolution of the image.[8] Another limit to the achievable resolution is imposed by the receiver bandwidth, which determines the maximum detectable frequency.

1.3.2 Image Reconstruction and Acceleration

1.3.2.1 k-Space

In MRI, spatial encoding for an imaging object is accomplished by means of gradient fields, which create location-dependent spatial frequencies in x, y or z directions. The spatial frequencies are physically denoted by 'k' and hence, the space where the acquired MR data initially exist is called 'k-space'. The k-space represents how much each spatial frequency component contributes to the image and therefore, the k-space data do not directly show anatomical structures of the imaging object (see Figure 1.11(a)). The anatomical image can be obtained by transforming the data space from the frequency domain (k_x, k_y and k_z) to the spatial domain (x, y and z). This can be achieved by FT (see Section 1.3.1.5). The corresponding spatial domain image of the example k-space data (Figure 1.11(a)) is depicted in Figure 1.11(b).

For most anatomical images, the signals in the spatial domain change relatively gradually in each spatial direction. Therefore, a large fraction of their frequency components are distributed around the low-frequency region. This can also be observed in the example k-space data above; signals in central k-space are much brighter than those in peripheral k-space. Here, the central k-space contains most of the signal-to-noise ratio (SNR) and overall contrast of the object, whilst peripheral k-space plays an important role in determining the detailed resolution of the image. This feature can be simply

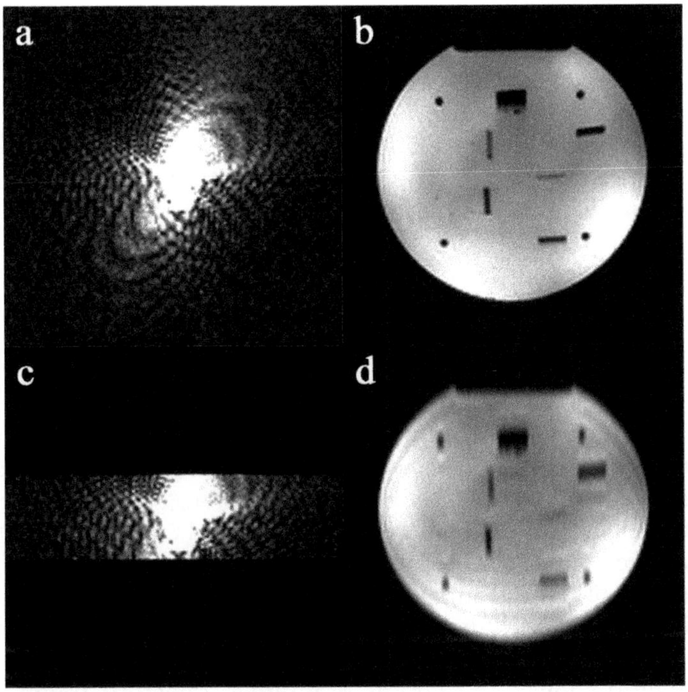

Figure 1.11 k-space and a reconstructed MRI scan. (a) Fully-sampled k-space and (b) corresponding reconstructed MRI image obtained by FT. (c) Down-sampled k-space, which presents only central quarter of the above original k-space and (d) corresponding reconstructed MRI scan.

verified by nulling the peripheral k-space data while maintaining the central k-space data. Figure 1.11(c) depicts the case where only a quarter of the original k-space remains in the k_y direction. Although the corresponding reconstructed image generally looks blurry (see Figure 1.11(d)), it can be seen that the image still preserves most of the SNR and the overall shape of the image. This fact can be effectively used for the signal compression or more efficient sampling schemes.[11–13]

1.3.2.2 Gibbs Ringing Artefact

As described in the previous section, an MR image is reconstructed by a linear combination of a set of linearly independent spatial frequency components. If high-frequency components are not sufficiently sampled, the detailed structures of the MR image cannot be well represented with the given Fourier basis set. This is more pronounced in the regions where signal discontinuities exist (*e.g.* edges)[14] as those regions require a greater number of frequency components in representing the shape than the regions where signals change smoothly. The problem results in oscillating structures and is known as the so-called "Gibbs ringing artefact". As an example, oscillating structures can be observed near the edge region of Figure 1.11(d), which is, however, an exaggerated case where many of the high-frequency components are missing.

A straightforward way to solve the problem is to simply increase the *k*-space sampling range. However, since this strategy requires an increased scanning time, there are several alternative approaches to minimise the artefacts. One simple solution is low-pass filtering. Although this is quite effective in alleviating the oscillating structures, it also causes blurring of the sharp structures. There are also more sophisticated schemes to remove the problem while maintaining the shape of the sharp structures.[14,15]

1.3.2.3 Reconstruction for Non-Cartesian k*-Space Trajectories*

A typical and widely used coordinate system in sampling *k*-space is the Cartesian grid. Its main advantage is that an imaging sequence for this grid can be easily implemented and the corresponding image can be simply reconstructed by applying the FT. However, since standard Cartesian imaging allocates an identical number of sampling points per *k*-space area regardless of *k*-space density of natural images, its sampling strategy is not efficient in terms of acquisition time and SNR. One solution to overcome this issue is to employ non-Cartesian sampling using radial or spiral trajectories.[16]

Figure 1.12(a) depicts an example of a radial trajectory. It consists of a number of radial lines for a set of rotation angles ($0° < \theta < 180°$), each of which crosses the central *k*-space point. According to the projection theorem, the *k*-space data on a radial line with rotation angle θ are equal to the FT of the projection data of the image space at the same rotation angle θ.[17] Therefore, the reconstructed image can be obtained by first performing an FT to the radial *k*-space data and then applying the back-projection algorithm. In order to account for dense sampling around the central *k*-space region, additional compensation filtering is required. This reconstruction

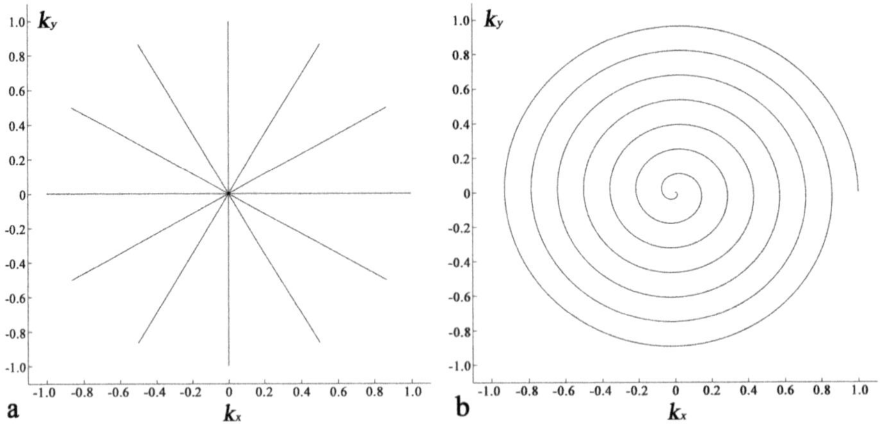

Figure 1.12 Non-Cartesian *k*-space trajectories. (a) Radial and (b) spiral trajectories.

procedure is the so-called filtered-back projection.[18] However, it is nevertheless possible to reconstruct the radial data with more sophisticated approaches such as regridding or using an inverse model. Although the approaches may require increased reconstruction time, they do yield enhanced image qualities.[18–20]

Figure 1.12(b) shows another non-Cartesian *k*-space trajectory for a conventional single-shot spiral out sequence.[2] The trajectory starts from the origin of the *k*-space and spirals outward to the peripheral *k*-space. In this way, *k*-space can be more rapidly and efficiently sampled than Cartesian sampling and it has been proven to be robust against motion and distortions.[12] Depending on the purposes of the study, several variants of the spiral trajectory have been proposed. For instance, a 'spiral-in/out' trajectory was presented to improve BOLD contrasts in the regions where signal-drop typically exists due to the susceptibility differences.[21] Spiral data can also be reconstructed by using regridding or establishing an inverse model. One major issue in spiral imaging is that image blurring, caused by the off-resonance effects, often degrades the image quality. In order to correct image blurring several methods have been proposed using a field map acquisition[22,23] or an automatic approach without any field map calibration.[24,25]

1.4 Magnetic Resonance Pulse Sequences

Ever since the first days of NMR, leaving aside the continuous wave methods, pulse sequences have been the *de facto* measurement technique even if the term was invented somewhat later. And, in essence since imaging is 'only' NMR spectroscopy with gradients, it is no surprise that pulse sequences have a rich and varied history and still proliferate today.

In order to understand the concepts employed by modern pulse sequence designers, it is important to understand a few basic elements and to understand that the purpose of a pulse sequence in imaging is to generate image contrast between two neighbouring tissue types, for example, a tumour surrounded by healthy tissue or flowing blood in vessels. One of the fundamental considerations that determines the details of a pulse sequence is the manner and speed with which *k*-space is traversed and this will determine contrast, resolution and field-of-view, for example. It is important to note that each *k*-space trajectory brings with it a different set of advantages and disadvantages. The plethora of sequences that exist in MRI arise from the fact the modules alluded to below, and the order in which they are played out, has no strict modular meaning. It is this fact that fundamentally underpins the creativity of the sequence and leads to the manifold applications and richness of MRI.

Imaging pulse sequences can be broken down very conveniently by dimensionality:

- 1D (a spectrum of profile)
- 2D (a slice or many single slices)

- 3D (selection of a '3D slab' which is then 'sliced' in the reconstruction)
- 4D (3 spatial dimensions and, *e.g.*, a frequency, resulting in spectra for each voxel in the 3D image)
- and higher dimensionalities are also possible in principle but rarely used.

Another convenient classification results from asking how is the signal, that is to be spatially-encoded, is generated and here again a number of different methods can be used:

- free induction decay (FID); usually used in conjunction with non-Cartesian *k*-space trajectories[26]
- gradient echo; gradient echo sequences lack refocusing RF pulses and generate T_2^* contrast
- spin echo; make use of a refocusing pulse to generate T_2 contrast
- stimulated echo; used in spectroscopic sequences such as STEAM.[27]

The manner in which *k*-space is traversed is also a classification parameter: Cartesian or non-Cartesian (spirals, cones, helices). As noted above, each trajectory has pros and cons.

A pulse sequence can be conceptually modularised in the following way: contrast-generation; slice selection; phase encoding; frequency encoding; and reconstruction.

1.4.1 Contrast Generation

As an example, at the start of a pulse sequence, the application of a 180° inversion pulse (an inversion recovery module) can be applied to tip all spins to point along the −z direction from where they will begin to recover through T_1 relaxation, resulting in T_1 contrast. The application of a 90° pulse followed by crusher gradients and a time delay (saturation recovery module) will also produce T_1 contrast, albeit with a reduced dynamic range. Different tissues relax with different rates and this results in image contrast. Generally, though, it is not necessary to have a separate contrast generation module. Contrast can be generated through just the interplay of flip-angle, TE, and TR. A good example of a T_1-weighted sequence is the 3D, MP-RAGE method.[28] Other conceptual examples of contrast generation include diffusion-weighted imaging,[29] perfusion-weighted imaging,[30] flow-encoded imaging,[31] *etc.*

1.4.2 Slice Selection

In its simplest form, a 2D slice can be selected by applying an RF pulse in the presence of a gradient (see Section 1.3.1.1). For a given RF pulse, a stronger gradient selects a thinner slice and a weaker gradient selects a thicker slice, also known as a slab; in 3D imaging, this slab is then 'sliced' computationally by virtue of the fact that the third direction is phase encoded. Repeating the 2D selection module for different spatial locations results in the so-called 2D multi-slice method. For many years, this simple method—an RF pulse in the presence of a gradient—remained practically the only method for slice selection and, indeed, it is still the only method used in the vast majority of pulse sequences.

At ultra-high field, the use of parallel transmit is now standard (parallel transmit is analogous to parallel receive, see Section 1.3.4.2). The use of parallel transmit brings with it the possibility of using more complex RF pulses (of discrete phase and amplitude modulation) in the presence of multiple gradient lobes to select complex volumes and not just rectilinear slices.[32]

Further, in the pursuit of reduced image acquisition time, acceleration in the slice direction using so-called multi-band pulses has also become common. Multi-band excite multiple, superimposing, 2D slices simultaneously, which are then separated in image reconstruction.

Having selected a slice, the imaging task now reduces to an encoding of the two, orthogonal in-slice directions. For one of these directions, phase encoding is used (see Section 1.3.1.2). This typically involves the application of a blipped gradient of short duration along the chosen phase-encode axis. Phase encoding imparts a phase offset to the spins dependent upon their spatial location. The FOV in the phase direction is determined by the step size in k-space in this direction, and spatial resolution is determined by the extent of k-space (see Section 1.3).

1.4.3 Frequency Encoding

The frequency encoding direction is synonymously called the readout direction and encoding in this direction involves reading the signal in the presence of a gradient, the readout gradient. The FOV in the readout direction is determined by the step size in k-space in this direction and spatial resolution is determined by the extent of k-space (see Section 1.3.1.3).

It is noted here that, for non-Cartesian trajectories, phase and frequency encoding take place simultaneously as the k-space trajectory is swept out.[33]

1.4.4 Reconstruction

Currently, the most common reconstruction method is the application of an inverse FT, of appropriate dimensionality, to k-space data acquired on a Cartesian grid. Non-Cartesian k-space data can be resampled onto a Cartesian grid and then Fourier transformed.

In the search for imaging speed, a number of other methods have been introduced that incompletely sample Cartesian k-space, reconstruct the missing data by various means before applying an FT; primary exponents in this class are GRAPPA (*k*-space-based) and SENSE (image-based). Even more recently, methods that deliberately sample *k*-space incoherently have been used in conjunction with compressed sampling (CS) to achieve higher imaging speeds. Here, clearly, it is the reconstruction scheme to be employed that dictates the pulse sequence to be used.

The two main introductory classes of sequence are: (1) gradient echo sequences, lacking refocusing pulses, in which low-flip angle pulses are repeated at short intervals, each followed with read gradients applied while data are acquired; and (2) spin echo sequences, which typically use higher-flip angle excitation pulses, with refocusing pulses to recover the spin magnetisation dephased by local magnetic field gradients.

1.4.5 Gradient Echo Sequences

Gradient echo sequences lack refocusing RF pulses and typically use low-flip angle pulses that are repeated with short TR intervals. Usually one readout per TR takes place during which data are acquired. There are three main contenders for high SNR and good grey matter/white matter contrast: (a) MP-RAGE sequence,[34] (b) a modified version using a second inversion pulse, called MP2-RAGE[35] and (c) the Modified Driven Equilibrium Fourier Transform (MDEFT) sequence, optimised by Deichmann.[36] Acquisition parameters for the MP-RAGE sequence have recently been optimised by Stöcker and Shah[37] and Bock *et al.*,[38] giving a substantial improvement in CNR per unit time.

1.4.6 Spin Echo Sequences

In its simplest form, the spectroscopic spin echo (SE) sequence can be described as follows

$$90° - \tau/2 - 180° - \tau/2 - \text{readout} - \text{relaxation delay}$$

This sequence can easily be turned into an imaging variant by the judicious application of gradients. The SE sequence has the advantage that it refocuses dephasing due to magnetic field inhomogeneities and is T_2-weighted. However, because of the use of 90° pulses, the relaxation delay is long, making the overall acquisition time long too.

An imaging variant of the CPMG sequence,[39,40] which uses multiple 180° pulses, was invented by Hennig in 1986 and called RARE.[41] The RARE sequence can be represented as

$$90° - \tau/2 - 180° - \tau/2 - \text{readout} - [\tau/2 - 180° - \tau/2 - \text{readout}]_n$$

The string of refocusing pulses following a single excitation pulse form the essence of the RARE (also known as turbo-spin-echo (TSE) or fast spin echo) greatly increased the efficiency of the spin echo sequence. Unfortunately, the increased number of 180° pulses also greatly increases SAR, making the standard TSE sequences difficult to use at field strength above 3 T. The disadvantage can be overcome by using the GRASE sequence.[42]

1.4.7 Echo Planar Imaging Sequences

The echo planar imaging (EPI) sequence, invented by Mansfield in 1977[43] can be conceptually thought of as a gradient echo sequence in which, following a single 90° pulse, the polarity of the readout gradient is continually alternated, and a train of signals is thus read. At the time point of readout gradient reversal, a blipped gradient is applied to enable traversal of *k*-space in one shot following the application of only one excitation pulse. Since gradient alternation can be done in a millisecond or so, the EPI sequence is very rapid.

The introduction of a 180° between excitation and the readout train will result in a spin-echo EPI sequence giving T_2 contrast. Further, traversal of *k*-space can be segmented to produce multi-shot EPI[44] or combined with a keyhole to merge the advantages of single-shot/multi-shot EPI.[45] EPI is the mainstay of fMRI and DWI/DTI methodology.

1.5 Acceleration in MRI Acquisition

In order to achieve high spatial resolution in MRI, it is necessary to increase the *k*-space sampling range while maintaining a sufficiently adequate sampling rate. High-resolution MRI requires a greater number of sampling points than low-resolution MRI and hence, a longer acquisition time is required. In order to address this and achieve a reduction in acquisition time, numerous acceleration methods have been presented.

Depending on the applied sequence scheme, that is, whether the sequence is based on a Cartesian or non-Cartesian trajectory, or whether the sequence is 2D or 3D imaging, some of the acceleration methods may not be possible. Here, for simplicity, each acceleration technique will be described under the condition of 2D multi-slice Cartesian imaging. For this case, the acceleration methods can be further divided into two categories: in-plane acceleration and through-plane acceleration. The in-plane acceleration technique reduces the scanning time required for encoding each slice and the through-plane acceleration technique shortens the scanning time by simultaneously exciting multiple imaging slices. These two techniques can even be combined to achieve both benefits. In the following sub-sections, several widely used methods in the community will be introduced along with their fundamental features.

1.5.1 Partial Fourier

For a conventional 2D imaging sequence such as GRE or SE, the required acquisition time for encoding each slice is mainly determined by the number of phase encoding lines. The partial Fourier technique shortens the acquisition time by skipping sampling for a part of the phase encoding lines.

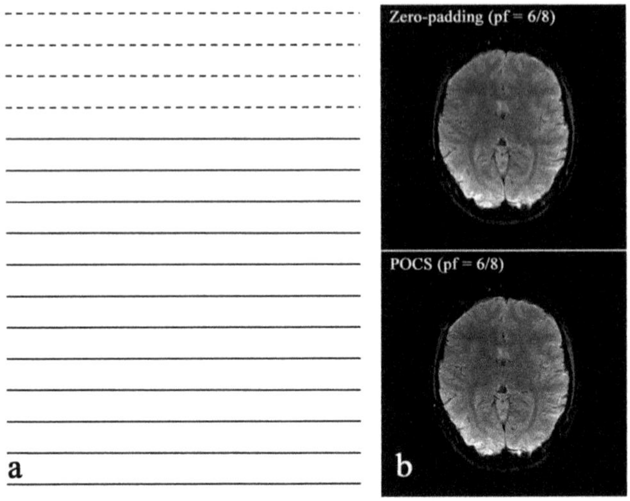

Figure 1.13 The partial Fourier technique. (a) A schematic diagram of phase encoding lines for a partial Fourier imaging sequence (pF = 6/8). The dotted and solid lines indicate the phase encoding lines to be skipped and sampled, respectively. (b) Reconstructed images obtained by zero-padding (top) and POCS methods (bottom).

Figure 1.13(a) depicts an example where a quarter of the top *k*-space region is excluded for sampling. This strategy is established based on the assumption that the final natural image only contains real values (*i.e.* no imaginary signals) and hence, its corresponding *k*-space has a conjugate symmetry (Hermitian). In theory, the missing *k*-space part can be reconstructed by simply taking the complex conjugate of the acquired part. However, in practical situations, spatial or time domain effects caused by flow, T_2 decay, B_0 inhomogeneity or eddy currents may induce imaginary components in the MR image. As a consequence, the constraint of conjugate symmetry cannot be perfectly applied for reconstruction.[46] Instead, a more sophisticated reconstruction scheme needs to be employed. Figure 1.13(b) shows the reconstructed image obtained by filling the missing *k*-space region with the projection onto convex sets (POCS) method.[47] It is observed that detailed structures of the brain are more clearly delineated in the POCS reconstructed image than in the zero-padding reconstructed image. As shown in the example figure, the partial Fourier technique is effective in reducing the scanning time whilst maintaining good image quality. However, due to reduced sampling, partial Fourier leads to an SNR reduction by a square-root of the acceleration factor. In this example, the SNR of the partial Fourier image (SNR^{PF}) can be expressed as eqn (1.41)

$$\text{SNR}^{\text{PF}} = \text{SNR}^{\text{Full}}\sqrt{0.75} \tag{1.41}$$

where SNR^{Full} indicates the SNR when full *k*-space sampling is used.

1.5.2 Parallel Imaging

To avoid aliasing artefacts when sampling *k*-space, it is important to keep the sampling interval small enough so that it satisfies the Nyquist criterion. For the given imaging FOV, the Nyquist sampling interval is given by $\Delta k^{\mathrm{Ny}} = 1/\mathrm{FOV}$. In conventional Cartesian imaging, this interval is maintained for the acquisition of each phase encoding line. Here, the *k*-space coverage (k^{cov}) is determined by an integer multiple of the interval (*i.e.* $k^{\mathrm{cov}} = \Delta k^{\mathrm{Ny}} \cdot N$) and the acquisition time increases in proportion to the total number of phase encodings (N). Similar to partial Fourier imaging, parallel imaging also accelerates the acquisition by reducing the number of phase encoding steps (*e.g.* $N' \rightarrow N/2$). However, instead of cutting the *k*-space coverage, parallel imaging accelerates the acquisition time while keeping the *k*-space coverage by increasing the sampling distance (*e.g.* $\Delta k^{\mathrm{PI}} = 2\Delta k^{\mathrm{Ny}}$). This approach allows one to maintain the image resolution but leads to aliasing artefacts in the reconstructed images. The aliasing artefacts need to be eliminated by additional processing.

For the elimination of the aliasing artefacts, several methods have been presented.[48–51] The detailed procedure is different depending on each method, but all the methods require multi-channel data to retrieve the original voxels from the aliased voxels. The multi-channel data can be obtained using a multi-channel receiver coil. Each channel has its own distinct sensitivity profile, which is presented as a multiplication factor in the image in each channel. Figure 1.14 depicts an example *k*-space trajectory for a parallel imaging scheme where the acceleration factor is two ($\Delta k^{\mathrm{PI}} = 2\Delta k^{\mathrm{Ny}}$). The figure also exhibits the corresponding multi-channel images reconstructed by Fourier transform (see Figure 1.14(c)), each of

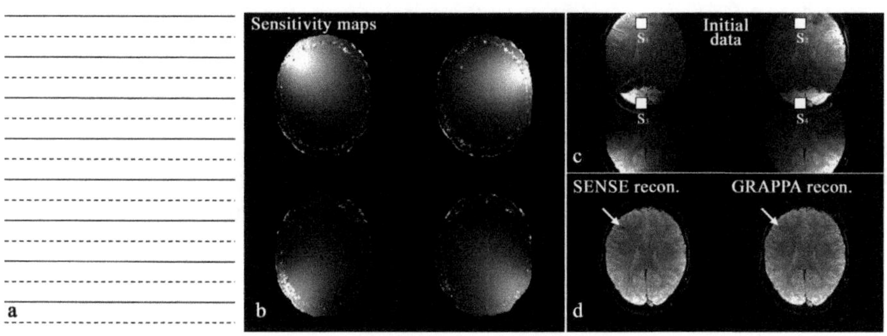

Figure 1.14 Parallel imaging technique. (a) A schematic diagram of phase encoding lines for a parallel imaging sequence ($R = 2$). The dotted and solid lines indicate the phase encoding lines to be skipped and sampled, respectively. (b) Coil sensitivity profiles for four channels, which depict distinct sensitivity weighting for each channel. (c) Reconstructed images obtained by applying Fourier transform to the reduced phase encoding set. (d) Reconstructed images obtained from the SENSE (left) and GRAPPA (right) methods.

which shows two-fold structures with its own individual sensitivity profile. From the figure, it can be seen that a voxel on the folded regions is the result of the superimposition of the original two voxels, each of which is weighted by the corresponding sensitivity profile. This can be expressed as the following formula (eqn (1.42))

$$\begin{bmatrix} C_1^a & C_1^b \\ C_2^a & C_2^b \\ C_3^a & C_3^b \\ C_4^a & C_4^b \end{bmatrix} \begin{bmatrix} V^a \\ V^b \end{bmatrix} = \begin{bmatrix} S_1 \\ S_2 \\ S_3 \\ S_4 \end{bmatrix} \tag{1.42}$$

where C_i^j, V^j and S_i denote the coil sensitivity profile of the i-th channel at the voxel location j, original (unfolded) signal intensity at the voxel location j and the obtained (folded) signal intensity of the i-th channel. The coil sensitivity profiles can be estimated from an additional reference scan obtained with full k-space sampling (see Figure 1.14(b)). Therefore, the original voxel intensity (V^a and V^b) can be computed by taking the pseudoinverse of the matrix equation. The reconstruction procedure described above is the scheme employed by SENSE[49] and its resultant image is shown in Figure 1.14(d). As shown, the method is quite effective in eliminating the aliasing artefacts, however, depending on the accuracy of the sensitivity profile estimation, the reconstructed image quality may be affected (see the region marked by the white arrow). In order to overcome this issue, another reconstruction scheme, GRAPPA, has been presented.[51] GRAPPA also requires a reference scan for calibration, but it does not need a complicated estimation procedure to obtain coil sensitivity profiles. Figure 1.14(d) shows a GRAPPA reconstructed image where the signal drop-out visible in the SENSE reconstructed image has disappeared.

Due to the reduced sampling, parallel imaging acceleration also induces an SNR reduction by a square root of its acceleration factor (R). In addition, to the spatially dependent term, the coil geometry factor (g) also constitutes a further SNR reduction which describes the ability of the multi-channel coil to separate the superimposed voxels.[49] Overall, the SNR of the parallel imaging reconstructed image ($\mathrm{SNR^{PI}}$) can be written as: $\mathrm{SNR^{PI}} = \mathrm{SNR^{Full}}/(g\sqrt{R})$.

1.5.3 Multi-shot, Readout Segmented EPI and EPIK (EPI with Keyhole)

This sub-section introduces acceleration techniques, which are particularly associated with/used in EPI techniques. Single-shot EPI is one of the fastest MRI techniques and it acquires all phase encoding lines following a single RF excitation. When the number of phase encoding lines is N, the length of the sampling window of EPI is N-times as large as that of the conventional

GRE or SE sequences (assuming equal bandwidth in the readout direction). As a result, the effects of T_2/T_2^* decay or accumulated phase errors develop significantly, which in turn causes blurring artefacts and geometric distortions in the reconstructed images. Furthermore, the increased sampling window of EPI requires a relatively long minimum TE. The acceleration techniques in EPI are, thus, of great help in not only reducing the scanning time but also in alleviating the aforementioned issues. The above acceleration techniques, such as partial Fourier or parallel imaging techniques, are also applicable to EPI. Here, segmentation-based approaches (*e.g.* multi-shot or readout segmentation) particularly designed for EPI will additionally be described.

The basic concept behind the multi-shot approach is to split the long EPI readout into several segments, each of which is acquired with an individual RF excitation.[52] As shown in Figure 1.15(b), multi-shot imaging performs

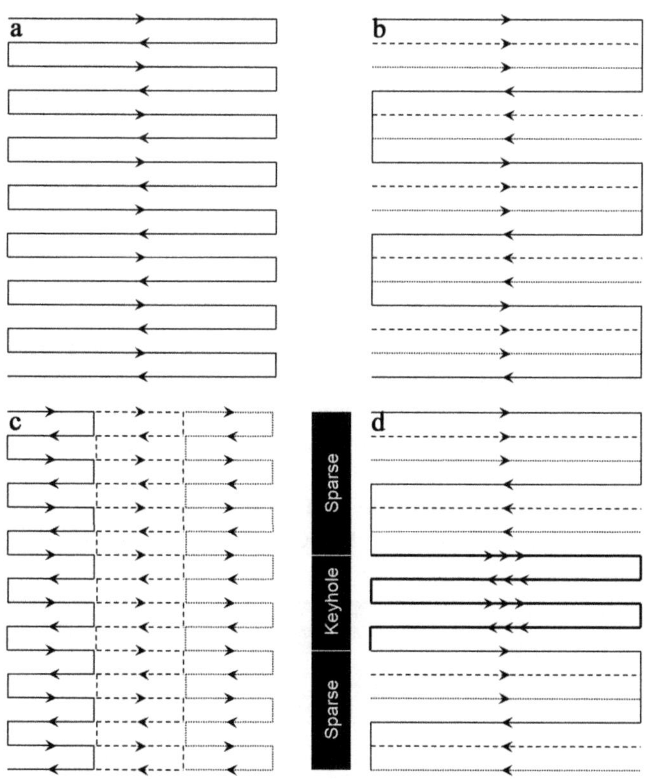

Figure 1.15 *k*-space trajectories of EPI methods. (a) Single-shot EPI, (b) multi-shot EPI, (c) readout-segmented EPI and (d) EPIK. For figures (b)–(d), the solid, dashed and fine-dashed lines indicate the sampling positions performed at every first, second and third temporal frame, respectively. In EPIK, the thick solid lines are sampled at every temporal frame (keyhole region).

segmentation along the phase encoding direction; to aid comparison, a single-shot EPI trajectory is also depicted in Figure 1.15(a). This example shows the case of three-shot EPI where only every third line is sampled at every frame and the missing lines are acquired in an interleaved way in the following two consecutive scans. Figure 1.15(c) shows another possible way to achieve *k*-space segmentation, which is termed as 'readout-segmented EPI' where segmentation is applied along the readout direction.[53]

The multi-shot EPI and readout segmentation techniques can effectively reduce the long readout duration of EPI. However, due to the increased number of excitations, both techniques require more time to completely acquire *k*-space than the single-shot technique. An alternative reconstruction scheme to overcome this issue is a sliding window reconstruction[54] where each shot is considered as a single temporal frame and the missing *k*-space data are filled by the data from neighbouring scans. Although this approach reduces scanning time, it employs the repeated use of the same *k*-space data in several temporal frames (autocorrelation).

Another major drawback of multi-shot EPI is that as shot-to-shot instabilities cause a substantial magnification of the physiological noise, its use for dynamic MRI studies such as fMRI may be limited unless a proper correction procedure is applied.[55,56] The difficulties associated with using the multi-shot scheme for fMRI can be overcome with, for example, EPIK (EPI with keyhole).[56–59] As depicted in Figure 1.15(d), the EPIK acquisition scheme resembles three-shot EPI ($\Delta ky' = 3/\mathrm{FOV}$) in terms of employing segmented acquisition, but differs from it in that the segmented acquisition is applied only to peripheral *k*-space. For the central *k*-space, full Nyquist rate sampling ($\Delta ky = 1/\mathrm{FOV}$) is instead applied, which plays the role of the 'keyhole'[60] for every single time frame. The missing *k*-space lines in the peripheral region are completed with those from two adjacent scans. Here, it is important to note that a sliding window technique[54] is used to reconstruct the images, which ensures that the keyhole and the periphery of *k*-space are continually updated, albeit at different rates.

Figure 1.16 shows an example EPIK reconstructed image. Due to the reduced readout duration, the EPIK image shows reduced geometric

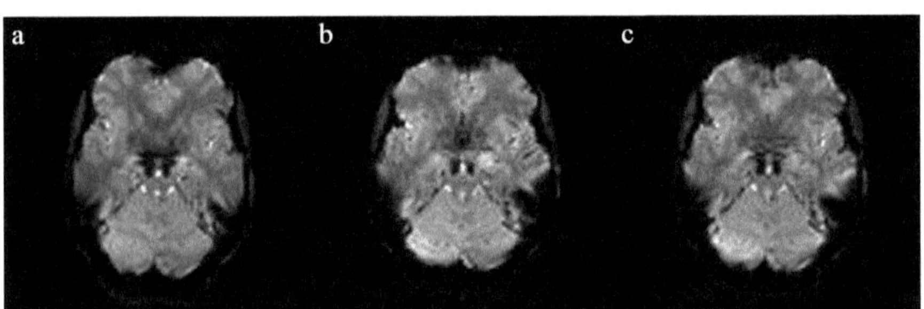

Figure 1.16 A reconstructed image from three EPI imaging methods. (a) Single-shot EPI, (b) three-shot EPI and (c) EPIK.

distortions around the frontal area when compared to the single-shot EPI image. The reduction of geometric distortions can be also observed in the multi-shot EPI image. However, in EPIK, due to full Nyquist sampling for the keyhole region at every temporal frame, the performance of EPIK in detecting dynamic functional signals is comparable to that of community-standard, single-shot EPI method[56,59] and better than that of multi-shot EPI; the performance of multi-shot EPI has been shown to be significantly degraded due to undersampling of the keyhole region.

1.5.4 Multi-band Imaging

The acceleration methods described above are in-plane (*e.g.* *x*- and *y*-direction) acceleration techniques. The multi-band technique is a different type of acceleration, which reduces the number of multiple slice loops (*e.g.* *z*-direction) in the acquisition of volume data (*i.e.* through-plane acceleration). Multi-band utilises a specially designed RF pulse to excite multiple slices simultaneously; a conventional 2D multi-slice sequence typically uses a sinc RF pulse to excite only a single slice per slice loop. Simultaneous excitation is realised by applying sinusoidal modulation to a sinc RF pulse, which yields multiple clones of the original excitation band.[61] An example RF pulse and its corresponding excitation bands are shown in Figure 1.17. Here, the sinusoidally modulated RF pulse produces three excitation bands, whereas the sinc RF pulse yields only a single excitation band (marked by white lines in each figure).

However, simultaneous excitation results in the overlapping of signals from multiple slices. Each individual slice can be reconstructed by applying a signal unfolding equation, as in the parallel imaging technique.[62] For this reason, the multi-band data also need to be acquired with a multi-channel receiver coil. Here, in order to avoid imposing similar sensitivity profiles on the simultaneously excited slices, an additional imaging gradient is applied in the slice direction so that each slice has a different distance offset in the phase encoding direction. This scheme assists a more effective signal separation from the superimposed slice signals.[62] Figure 1.17(c) displays the superimposed slices with different distance offsets and corresponding reconstructed individual slices.

1.5.5 Combination of Acceleration Techniques

High-resolution MRI is of great interest not only for clinical diagnosis but also for research activities. However, in general, high-resolution MRI requires an increased acquisition time. Furthermore, depending on particular MRI applications, the acquisition time that can be applied is often limited by the regulations, the subject or by the purpose of the study. In order to achieve an accelerated acquisition even in such circumstances, there have been many attempts to combine two or more acceleration techniques. This subsection introduces a high-resolution imaging method which achieved

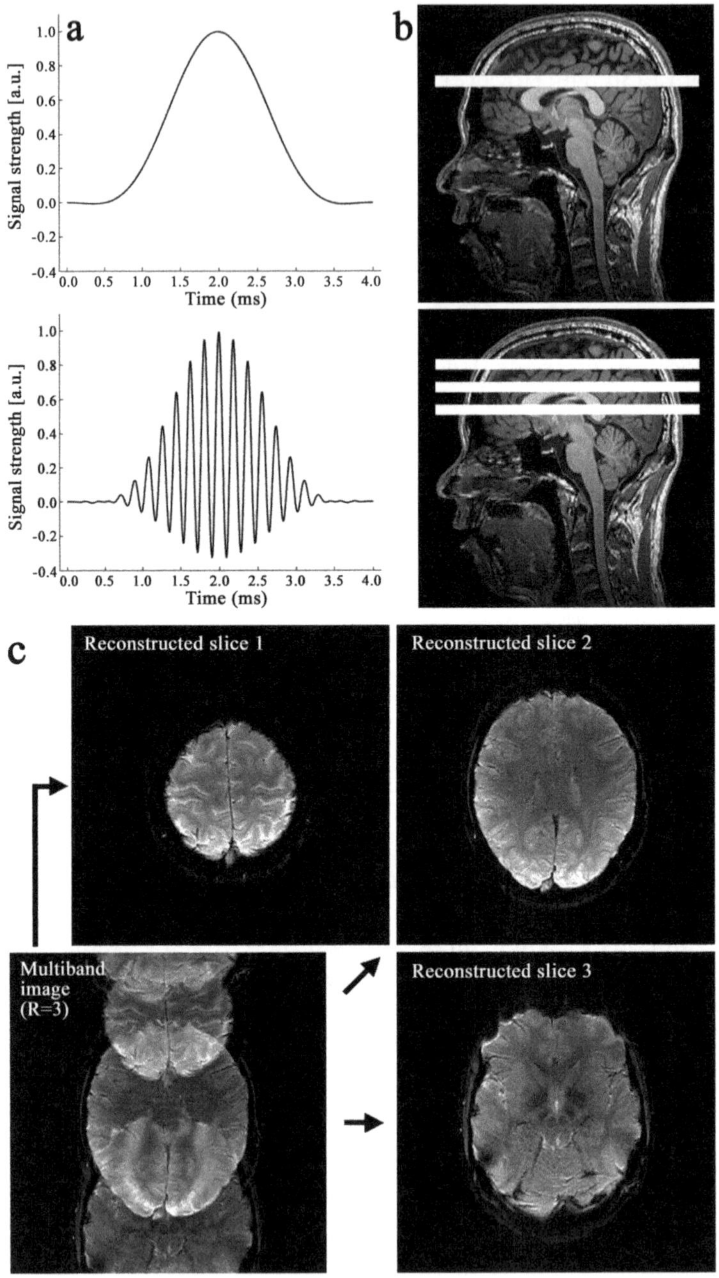

Figure 1.17 Multi-band technique. (a) A sinc (top) and multi-band (bottom) RF pulse designed for excitation of single-slice and three slices, respectively. (b) A schematic diagram of the excitation bands from the RF pulses of (a). (c) Superimposed slices (bottom left) with different distance offsets and reconstructed individual three slices.

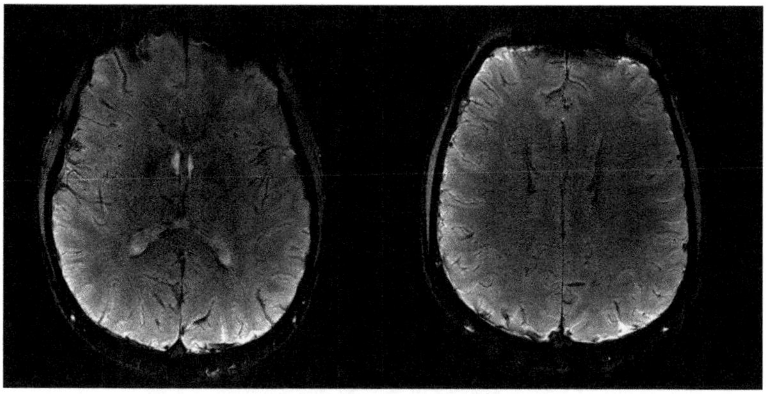

Figure 1.18 Half-millimetre resolution EPIK images.

half-millimetre in-plane resolution with a whole-brain coverage for fMRI using the EPIK technique.[63] The proposed scheme is illustrative in that it combines four in-plane acceleration techniques (*i.e.* parallel imaging, partial Fourier, EPIK and readout-segmentation techniques) and achieves a minimum required TR of around 97 ms for a single slice with an in-plane resolution of 0.5 × 0.5 mm^2. The result indicates that if further combined with a multi-band acceleration with a factor of three, the proposed method can provide whole brain coverage in a typical fMRI setting (TE/TR = 35/3000 ms). Figure 1.18 shows examples of reconstructed images where a detailed spatial representation of the anatomical structures such as gyri or sulci is observed.

The use of EPIK for high-resolution fMRI at 7 T brings a number of advantages over EPI given that the T_2^* at 7 T is reduced making of an increased number of k-space lines concomitantly more difficult to acquire with standard methods. The ultra-high fields allow one to acquire MR data with an increased SNR. This makes it possible to scan human subjects with a high spatial resolution whilst maintaining a sufficient SNR. Figure 1.19 depicts an example of sub-millimetre (0.73 × 0.73 mm^2) resolution scans obtained using EPIK at 7 T, where parallel imaging acceleration of 2 and partial Fourier factor of 5/8 were used. In Figure 1.19(a) four representative slices from a whole-brain acquisition, shown in Figure 1.19(b), are shown. The same imaging protocol was applied to a finger-tapping fMRI study using a block-based paradigm. Prior to fMRI data analysis, the standard pre-processing procedure including realignment, coregistration, normalisation and spatial smoothing was carried out for the initial fMRI time series. The first-level analysis using a generalised linear model (GLM) model was then performed to obtain functional activation as shown in Figure 1.20; a more detailed explanation of fMRI paradigm design and analysis can be found in Section 3.1. Motor activation can be clearly seen, and it elegantly follows the cortical ribbon. Details of the imaging experiments are given in the respective figure captions.

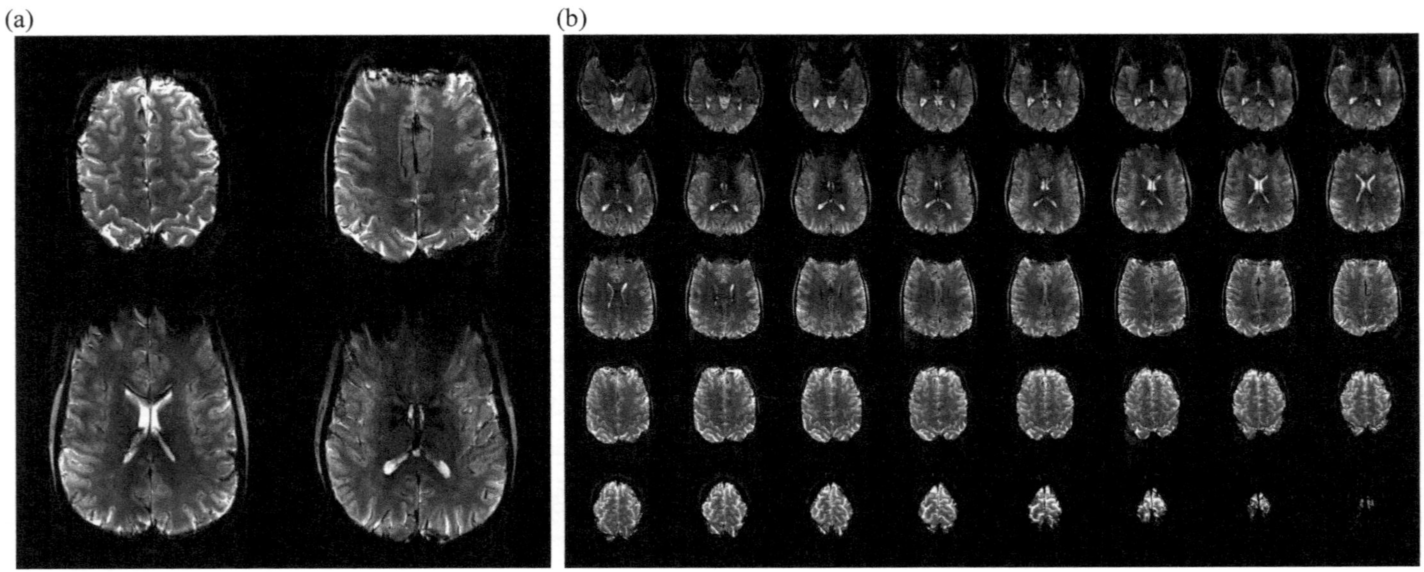

Figure 1.19 High-resolution, sub-millimetre resolution EPIK scans at 7 T. (a) Four representative slices from a whole-brain data set are shown. (b) Whole-brain acquisition obtained under a typical fMRI setting for 7 T (TR/TE = 3000/25 ms).

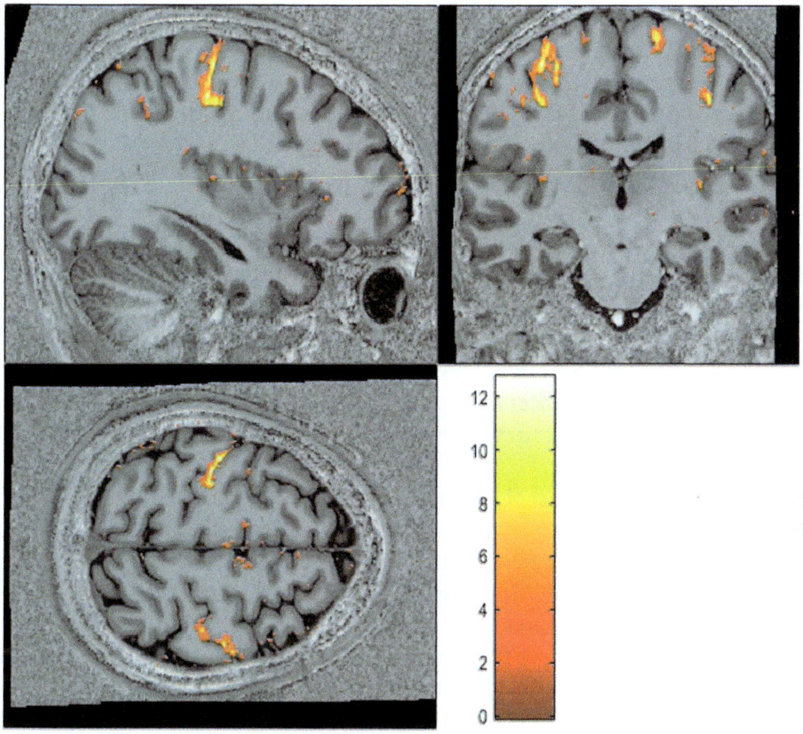

Figure 1.20 The first-level analysis using a generalised linear model (GLM) model was then performed to obtain functional activation as shown. It can be seen that motor activation elegantly follows the cortical ribbon.

1.5.6 Other Acceleration and Reconstruction Methods

The acceleration and reconstruction methods introduced in this chapter have been widely used and verified in a number of previous works. There are also other types of imaging methods such as random sampling to avoid coherent sampling patterns.[11] Depending on the given imaging sequence, more complicated reconstruction methods may need to be required, for instance, using an inverse model with a proper regularisation technique.[64] However, as describing these methods is beyond the scope of the purpose of this chapter, the reader is referred to the related works.[65,66]

References

1. K. Krane, *Introductory Nuclear Physics*, Wiley, New York, 1987.
2. B. H. Bransden and C. J. Joachain, *Quantum Mechanics*, Prentice Hall, Upper Saddle River, 2000.
3. M. H. Levitt, *Spin Dynamics: Basics of Nuclear Magnetic Resonance*, Wiley, New York, 2001.

4. F. Bloch, *Phys. Rev.*, 1946, **70**(7–8), 460.
5. E. T. Jaynes, *Phys. Rev.*, 1955, **98**(4), 1099.
6. E. M. Haacke, R. W. Brown, M. R. Thompson and R. Venkatesan, *Magnetic Resonance Imaging: Physical Principles and Sequence Design*, Wiley, New York, 1999.
7. D. G. Nishimura, *Principles of Magnetic Resonance Imaging*, Stanford University, Stanford, 1996.
8. D. W. McRobbie, E. A. Moore, M. J. Graves and M. R. Prince, *MRI from Picture to Proton*, Cambridge University Press, Cambridge, 2007.
9. D. Burstein, *Concepts Magn. Reson.*, 1996, **8**(4), 269.
10. M. A. Bernstein, K. F. King and X. J. Zhou. *Handbook of MRI Pulse Sequences*, Elsevier Science, New York, 2004.
11. M. D. Migliore, *IEEE Antennas Propag. Mag.*, 2014, **56**(2), 14.
12. D. G. Nishimura, P. Irarrazabal and C. H. Meyer, *Magn. Reson. Med.*, 1995, **33**(4), 549.
13. M. Zaitsev, K. Zilles and N. J. Shah, *Magn. Reson. Med.*, 2001, **45**(1), 109.
14. R. Archibald and A. Gelb, *IEEE Trans. Med. Imaging*, 2002, **21**(4), 305.
15. K. T. Block, M. Uecker and J. Frahm, *Int. J. Biomed. Imaging*, 2008, 184123.
16. I. Rodriguez, R. P. de Alejo, M. Cortijo and J. Ruiz-Cabello, *Concepts Magn. Reson., Part B*, 2004, **20**(1), 40.
17. Z. H. Cho, J. P. Jones and M. Singh, *Foundations of Medical Imaging*, John Wiley and Sons, New York, 1993, Chapter 3, p. 71.
18. J. I. Jackson, C. H. Meyer, D. G. Nishimura and A. Macovski, *IEEE Trans. Med. Imaging*, 1991, **10**(3), 473.
19. T. Knopp, S. Kunis and D. Potts, *Int. J. Biomed. Imaging*, 2007, **2007**, 24727.
20. L. L. Geyer, U. J. Schoepf, F. G. Meinel, J. W. Nance, G. Bastarrika, J. A. Leipsic, N. S. Paul, M. Rengo, A. Laghi and C. N. de Cecco, *Radiology*, 2015, **276**(2), 339.
21. C. S. Law and G. H. Glover, *Magn. Reson. Med.*, 2009, **62**(3), 829.
22. P. Irarrazabal, C. H. Meyer, D. G. Nishimura and A. Macovski, *Magn. Reson. Med.*, 1996, **35**(2), 278.
23. H. Moriguchi, B. M. Dale, J. S. Lewin and J. L. Duerk, *Magn. Reson. Med.*, 2003, **50**(3), 643.
24. L. C. Man, J. M. Pauly and A. Macovski, *Magn. Reson. Med.*, 1997, **37**(6), 906.
25. W. Chen and C. H. Meyer, *Magn. Reson. Med.*, 2006, **56**(2), 4572.
26. E. L. Hahn, *Phys. Rev.*, 1950, **77**, 297.
27. A. Haase, J. Frahm, D. Matthaei, W. Hanicke, H. Bomsdorf, D. Kunz and R. Tischler, *Radiology*, 1986, **160**, 787.
28. M. Brantzawadzki, G. D. Gillan and W. R. Nitz, *Radiology*, 1992, **182**, 769.
29. D. G. Taylor and M. C. Bushell, *Phys. Med. Biol.*, 1985, **30**, 345.
30. B. R. Rosen, J. W. Belliveau and D. Chien, *Magn. Reson. Q.*, 1989, **5**, 263.
31. M. A. Bernstein, A. Shimakawa and N. J. Pelc, *J. Magn. Reson. Imaging*, 1992, **2**, 583.

32. F. Padormo, A. Beqiri, J. V. Hajnal and S. J. Malik, *NMR Biomed.*, 2016, **29**, 1145.
33. N. G. Papadakis, A. A. Wilkinson, T. A. Carpenter and L. D. Hall, *Magn. Reson. Imaging*, 1997, **15**, 567.
34. J. P. Mugler, 3rd and J. R. Brookeman, *J. Magn. Reson. Imaging*, 1991, **1**, 561.
35. J. P. Marques, T. Kober, G. Krueger, W. van der Zwaag, P. F. Van de Moortele and R. Gruetter, *NeuroImage*, 2010, **49**, 1271.
36. R. Deichmann, *NeuroImage*, 2006, **33**, 1066.
37. T. Stöcker and N. J. Shah, *Magn. Reson. Med.*, 2006, **56**, 824.
38. N. A. Bock, E. Hashim, R. Janik, N. B. Konyer, M. Weiss, G. J. Stanisz, R. Turner and S. Geyer, *NeuroImage*, 2013, **65**, 1.
39. H. Y. Carr and E. M. Purcell, *Phys. Rev.*, 1954, **94**, 630.
40. S. Meiboom and D. Gill, *Rev. Sci. Instrum.*, 1958, **29**, 688.
41. J. Hennig, A. Nauerth and H. Friedburg, *Magn. Reson. Med.*, 1986, **3**, 823.
42. D. A. Feinberg and K. Oshio, *Radiology*, 1991, **181**, 597.
43. P. Mansfield, *J. Phys. C: Solid State Phys.*, 1977, **10**, L55–L58.
44. G. C. McKinnon, *Magn. Reson. Med.*, 1993, **30**, 609–616.
45. M. Zaitsev, K. Zilles and N. J. Shah, *Magn. Reson. Med.*, 2001, **45**, 109.
46. J. R. Macfall, N. J. Pelc and R. M. Vavrek, *Magn. Reson. Imaging*, 1988, **6**, 143.
47. G. McGibney, M. R. Smith, S. T. Nichols and A. Crawley, *Magn. Reson. Med.*, 1993, **30**(1), 51.
48. D. K. Sodickson and W. J. Manning, *Magn. Reson. Med.*, 1997, **38**, 591.
49. K. P. Pruessmann, M. Weiger, M. B. Scheidegger and P. Boesiger, *Magn. Reson. Med.*, 1999, **45**(5), 952.
50. W. E. Kyriakos, L. P. Panych, D. F. Kacher, C. F. Westin, S. M. Bao, R. V. Mulkern and F. A. Jolesz, *Magn. Reson. Med.*, 2000, **44**, 301.
51. M. A. Griswold, P. M. Jakob, R. M. Heidemann, M. Nittka, V. Jellus, J. Wang, B. Kiefer and A. Haase, *Magn. Reson. Med.*, 2002, **47**, 1202.
52. F. Hennel, *Concepts Magn. Reson.*, 1997, **9**(1), 43.
53. D. A. Porter and R. M. Heidemann, *Magn. Reson. Med.*, 2009, **62**(2), 468.
54. J. A. d'Arcy, D. J. Collins, I. J. Rowland, A. R. Padhani and M. O. Leach, *NMR Biomed.*, 2002, **15**(2), 174.
55. A. Lutti, D. L. Thomas, C. Hutton and N. Weiskopf, *Magn. Reson. Med.*, 2013, **69**(6), 1657.
56. S. Yun and N. J. Shah, *PLoS One*, 2017, **12**(9), e0184759.
57. M. Zaitsev, K. Zilles and N. J. Shah, *Magn. Reson. Med.*, 2001, **45**(1), 109.
58. M. Zaitsev, J. D. Arcy, D. J. Collins, M. O. Leach, K. Zilles and N. J. Shah, *Phys. Med. Biol.*, 2005, **50**, 4491.
59. S. Yun, M. Reske, K. Vahedipour, T. Warbrick and N. J. Shah, *NeuroImage*, 2013, **73**, 135.
60. J. J. van Vaals, M. E. Brummer, W. T. Dixon, H. H. Tuithof, H. Engels, R. C. Nelson, B. M. Gerety, J. L. Chezmar and J. A. den Boer, *J. Magn. Reson. Imaging*, 1993, **3**, 671.

61. D. J. Larkman, J. V. Hajnal, A. H. Herlihy, G. A. Coutts, I. R. Young and G. Ehnholm, *J. Magn. Reson. Imaging*, 2001, **13**, 313.
62. K. Setsompop, B. A. Gagoski, J. R. Polimeni, T. Witzel, V. J. Wedeen and L. L. Wald, *Magn. Reson. Med.*, 2012, **67**(5), 1210.
63. S. Yun and N. J. Shah, *Proceedings of the 25th International Society for Magnetic Resonance in Medicine*, 2017, Abstract 1485.
64. S. Yun, S. S. Oh, Y. Han and H. Park, *J. Magn. Reson. Imaging*, 2009, **29**(4), 924.
65. O. N. Jaspan, R. Fleysher and M. L. Lipton, *Br. J. Radiol.*, 2015, **88**(1056), 20150487.
66. M. Lustig, D. L. Donoho, J. M. Santos and J. M. Pauly, *IEEE Signal Process. Mag.*, 2008, **25**(2), 72.

CHAPTER 2

MRI Instrumentation

J. FELDER*[a], C.-H. CHOI[a] AND N. J. SHAH[a,b,c]

[a]Forschungszentrum Jülich, Institut of Neuroscience and Medicine - 4, Wilhelm-Johnen-Str., Jülich, 52425, Germany; [b]JARA-BRAIN – Translational Medicine, Aachen, Germany; [c]RWTH Aachen University, Department of Neurology, Pauwelsstraße 30, Aachen, 52074, Germany
*E-mail: j.felder@fz-juelich.de

2.1 Introduction

From the previous chapter, it is evident that magnetic resonance imaging (MRI) utilises the interaction of three different magnetic fields: a strong static field to align nuclear spins, three orthogonal, time-varying field gradients for spatial encoding, and a radiofrequency (RF) field for the excitation of the spin system. This chapter describes the MRI hardware and components required to generate the different magnetic fields, as well as the receiver chain that acquires MR signals from the excited spin system. Figure 2.1 shows an overview of a clinical MRI system. A detailed description of the individual components is given in the sections below.

2.2 Main Magnet

A strong and homogeneous static magnetic field (B_0) is a pre-requisite for MRI. Based on the Zeeman splitting of nuclear spin energy levels, the magnetic field generates a bulk magnetisation in the sample or subject. The

New Developments in NMR No. 19
Hybrid MR-PET Imaging: Systems, Methods and Applications
Edited by N. Jon Shah
© The Royal Society of Chemistry 2019
Published by the Royal Society of Chemistry, www.rsc.org

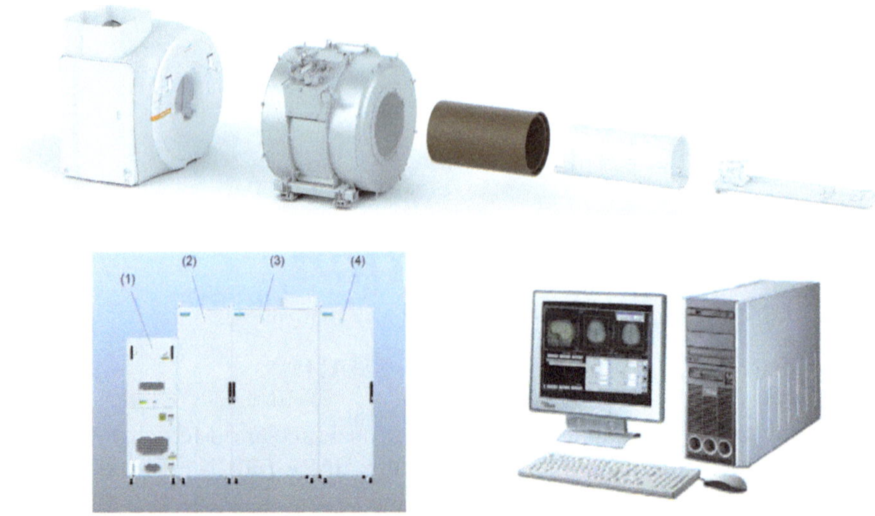

Figure 2.1 Overview of a clinical MRI system and its major components. © Siemens Healthineers.

strengths of the static fields currently used in clinical practice are typically 1.5 T or 3 T (note that 'Tesla (T)' is actually a derived unit of magnetic flux density, which is informally referred to as the magnetic field strength). This is substantially higher than the naturally occurring magnetic fields on Earth,[1] for example, a 3 T system is about 60 000 times stronger than the strength of the Earth's magnetic field. Since the signal-to-noise ratio (SNR) of MR images is roughly proportional to the main magnetic field strength,[2] researchers are constantly striving for stronger magnets. The first 7 T human system to qualify as a medical device went into operation in 2017. Research systems for human applications currently reach field strengths of up to 10.5 T[3] and an 11.7 T system is currently under construction.[4] An overview of most standard high-field magnet installations is given in ref. 5. In addition to the strength of the B_0 field, field homogeneity is also of major importance and can be expressed as the maximum deviation from the nominal value over a region in space, that is to say over a diameter of spherical volume (DSV). A typical value of the uniformity is less than five parts per million (ppm) over 45 cm DSV.

2.2.1 Superconducting Magnets

Today, superconducting magnets, as shown in Figure 2.2, are used in the majority of clinical and preclinical MRI systems. The major advantages of using superconducting magnets are associated with their capability to achieve a high field strength and homogeneity, as well as maintaining a very stable magnetic field over time (low-field drift).

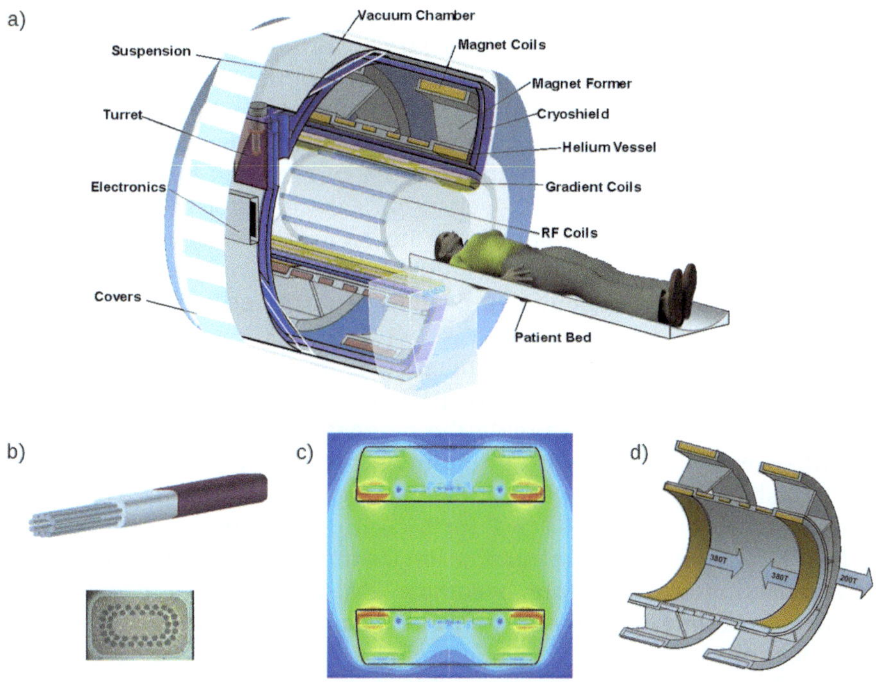

Figure 2.2 (a) Magnet assembly, (b) superconducting NbTi wire, (c) cross section with field plot and (d) axial forces of the coil elements. © Siemens Healthineers.

Due to the relative ease of fabrication and handling, MRI magnets are built using niobium–titanium (NbTi) wires embedded in a highly conductive and mechanically stable copper matrix, as shown in Figure 2.2(b). NbTi is a low-temperature superconductor, requiring immersion into liquid helium (~4.2 K at atmospheric pressure[6]). Because of the reduction of critical current density with increasing magnetic induction, NbTi superconducting wires can be used in human MRI magnets up to a field strength of approximately 10 T, depending on the design. For higher field strengths, other superconducting materials have to be employed or NbTi wires have to be operated at even lower temperatures. Currently, the feasibility of using niobium–tin (Nb_3Sn) alloys for human MRI applications is being investigated. However, due to the brittle nature of the alloy, handling is rather complex, and furthermore, the materials are much more expensive (by approximately a factor of 10 times[7]). Notwithstanding the advantages of helium-free cooling, high-temperature superconductors, requiring cooling with liquid nitrogen only, have not yet reached a level of maturity such that clinical MRI magnets can be constructed using these.

In contrast to earlier designs, today's MRI magnets do not use solenoidal field generating coils but instead use sectioned coil packages.[8]

This configuration not only reduces the length of wire required, but also allows the magnet to be tailored for larger homogeneous volumes with lower stray fields and shorter overall magnet length. The direction of the field in this type of magnet is along the magnet axis, which is commonly defined to be the *z*-axis in the context of MRI. Sectioned magnets have a usable magnetic field of approximately 3% of the total field produced,[9] with the highest field strength occurring near the outer field generating coils (see Figure 2.2(c)). It should be noted that as the sectioned coils experience large axial forces, see Figure 2.2(d), the magnet formers are not merely winding aids, but also need to be designed as sturdy support structures.

Another feature of modern MRI magnets is termed active shielding and refers to the larger diameter superconducting coil elements that decrease the magnetic fringe fields outside the magnet assembly. This is done as the MR specific IEC guideline[10] requires a permanent controlled area to be established, such that outside this area the static magnetic fringe field strength does not exceed 0.5 mT (or 5 G). In this regard, a reduction of the fringe field decreases the installation footprint and consequently costs. Active shielding has, therefore, become a *de facto* standard and even ultra-high field installations today only require a manageable amount of passive iron shielding.

Technologically, the magnet assembly compromises several vessels, as shown in Figure 2.2(a). The inner vessel contains the superconducting magnet windings, which is filled with liquid helium (~4 K). It is thermally isolated from the environment through cryoshields arranged inside a vacuum chamber. Since the supply of cryogens is a major factor in operational costs, modern magnet systems are equipped with a coldhead (or cryocooler), which recovers a part or all of the boiled-off helium gas through a re-condensation process. The coldhead is driven from compressors installed outside the Faraday shielded magnet room and requires replacement if its re-condensing margin drops below a certain threshold.

Superconducting magnets are put into operation with a procedure called ramping up. During ramping up the wires inside the magnet are slowly charged with current from a low-voltage DC power supply until the magnet reaches the desired field strength. Once ramped up, a superconducting 'switch' is closed, which short circuits the terminals of the coil windings (persistent mode of operation) and the power supply is removed. Rapid ramping can result in strong forces on the wires inside the magnet and can cause potential damage.

Ramping down a magnet is routinely carried out, for example, to replace coldheads or for other magnet related services. An abnormal termination of magnet operation is called a quench. This occurs when part of the magnet wire becomes resistive. The most common mechanisms that cause the magnet wire to turn from a superconducting to a resistive state are (locally) exceeding the critical field strength/current density or local heating, for example, through induced eddy currents. In addition, excessive mechanical force, for example, when trying to remove attracted ferromagnetic objects, may cause a dislocation of magnet wires and an ensuing quench.

During the quench, large amounts of energy are transferred as dissipated heat into the liquid helium reservoir causing a rapid boil-off. As the volume of helium expands dramatically during transition from liquid to the gas state (expansion factor ~1 : 757) a means of safely transporting the helium gas to the outside of the building has to be provided. In practice, this is achieved by using a large diameter quench pipe, which is separated from the helium reservoir through a burst disc. To prevent damage to the magnet itself, a copper matrix is used to contain the superconducting wires, which provides a low resistance path for current flow and efficient means for heat conduction. In combination with the mechanical fixations given by the magnet formers, it also provides mechanical stability against forces resulting during a quench. Means of quench protection are implemented during the magnet design, commonly using active as well as passive circuitries.[11]

Although the use of liquid cryogens is the most frequently employed cooling method today, superconducting magnets not requiring cryogen have been available for some time[12] and have matured to 1.5 T magnets for human head imaging as well as to 7 T for preclinical scanners. These magnets use coldhead–compressor systems, which can maintain a temperature low enough for the magnet coil to turn superconducting. The coldhead in these systems operates in two stages, with temperatures dropping to ~4.2 K in the first stage and to ~77 K (liquid nitrogen temperature) in the second stage. Since this design relies entirely on coldhead refrigeration, it is important to ensure a high-quality joint between the coldhead sections on one side and the magnet and shield on the other side. The term 'quench' is no longer relevant because there is no liquid cryogen inside the magnet.

2.2.2 Resistive Magnets

MRI systems using a resistive magnet are only operated by a continuous power source, allowing the system to be switched on and off. The benefits of this system are a lower investment compared to a superconducting magnet, as well as the feasibility to repair magnet damages on site. Magnets built with resistive wires, however, dissipate large amounts of heat when a significant level of current flows in the coil windings. Heat dissipation limits whole body designs to approximately 0.4 T.[8] Heating of the magnet also causes field drifts and degradation of homogeneity.

Technologically, whole-body designs have favoured the use of anodised aluminium strips of the same width as the winding section and between 0.5 and 1 mm thick. The anodised layer is a good electrical insulator, even at high temperatures and good thermal conduction is provided from the interior of the winding to the coil sidewalls, where the cooling water runs. This type of construction leads to a low total resistance, typically less than 1 Ω, but requires relatively high currents (~hundreds of amperes).[13]

2.2.3 Permanent Magnets

Permanent magnets are mostly made from the rare earth element neodymium (*i.e.* $Nd_2Fe_{14}B$) and are particularly suitable for low or ultra-low field installations or for use in geographically isolated locations with low electrical power grid stability. The major issue associated with the use of permanent magnets is the temperature-dependent field drift that leads to a deviation of the main magnetic field strength while local heating (*i.e.* in the vicinity of the gradient coil) disturbs field homogeneity. Thus, it is not sufficient to shift the Larmor frequency during an experiment to follow the overall field strength, but rather one needs to account for changes in homogeneity. The latter is a non-trivial task, requiring complex hardware or sophisticated post-processing, or both.

2.2.4 Non-conventional Magnets

Apart from horizontal cylindrical magnet designs, which are most common due to their large access aperture, several other specialised magnet layouts have been used in MRI. These include designs referred to as 'open' systems that consist of two vertically stacked magnet poles connected *via* a C-shaped flux guide and can be best compared to a horseshoe magnet. A key benefit of the open design is more available free space for the patient, allowing larger patients (*e.g.* obese population) or people suffering from claustrophobia inside conventional MRI magnets to be scanned. Open MR systems are also used in interventional procedures, such as during image-guided biopsies.[14] However, a disadvantage of the open magnet design is the low achievable field strength.[15] In recent years, and with the increase in the prevalence of obesity, MR vendors have addressed the needs of physically larger patients with the introduction of 'wide bore' systems that provide larger gantry diameters than the conventional 60 cm openings.

Non-conventional magnet designs using permanent magnets include a Halbach magnet[16] setup that has negligible stray field and can consequently be used in non-access-controlled surroundings, such as in intensive care units.[17] Another lightweight solution uses a permanent magnet rotated about the object to replace gradient encoding.[18]

For combined positron emission tomography (PET) and MRI, a fast field-cycling technique[19] can be used that allows B_0 to be changed during the measurement. In particular, it means that B_0 can be rapidly ramped down to zero at specific times, allowing the detection of the positron-emitting radioisotopes by scintillators coupled to conventional photomultiplier tube detectors. Afterwards, the magnet can be ramped back to the initial B_0 field for MR signal detection. The efficiency of the PET scanner was found to be enough for PET examinations when using the proposed technique.[20]

2.3 Gradient System

In order to obtain spatial information, MRI makes use of three orthogonal, time-varying field gradients. The strength of the field gradient depends on the position on the respective axis and the applied drive power, and the gradient field vector is oriented parallel to B_0. The field gradients are generated by three resistive coil systems, which are usually arranged between the magnet inner bore and the RF body coil. Note that in many MR-PET systems, a body coil is not implemented, and the PET system resides in the space.

This section describes two simple gradient winding geometries for illustration purposes and explains the general design goals for eddy current suppression, including linearity, maximum strength, minimum rise time, power handling capability, and shielding. The latest technical developments of the gradient coil systems are briefly introduced towards the end of the section. Another important factor in the gradient system is the gradient amplifier that drives the field-generating currents in the gradient coils. Modern switched mode gradient amplifiers are briefly described in Section 2.3.3 and their performance is discussed.

2.3.1 Gradient Coils

The coil system that creates a field gradient along the *z*-axis most simply is the Maxwell pair. The original construction, using three solenoid coils arranged on the same axis, has been modified to create magnetic field gradients in MRI by removing the element in the middle.[21] As opposed Helmholtz pair, which produces a homogeneous field between two coils, the Maxwell gradient set is driven by equal currents but the currents flow in opposite directions. As a result, the magnetic field generated by the left-hand loop halfway between the two loops is exactly cancelled by that from the right-hand loop; in the region between the two loops, a linear magnetic field gradient is generated. In practice, the solenoid coils of a Maxwell pair are connected in series, so that exactly equal but opposite currents flow in each half.

An early implementation of the *x*- and *y*-dependent (transverse) field gradients was based on the Golay set[22] and used a four-saddle coil configuration per axis. Because of the limited linearity of the original Golay set, a target field method,[23] also termed 'fingerprint' design, requiring distributed current pathways, was subsequently introduced,[24] as shown in Figure 2.3. Although much higher linearity over the space of the field gradients can be achieved in this way, the four-coil configuration per axis remains.

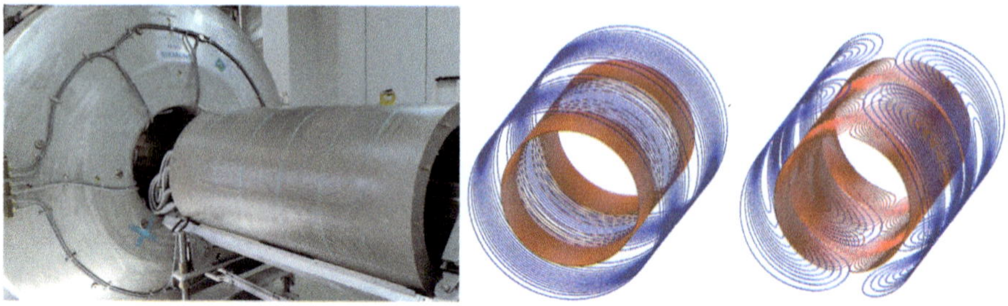

Figure 2.3 Gradient coil set picture (left) and schematic diagram. Right reproduced from ref. 38 with permission from Elsevier, Copyright 2016.

The design of a gradient coil set needs a complex optimisation process to address several functional and practical constraints. A typical gradient tube of a whole-body setting, which contains the gradient coils baked into epoxy resin, has an outer diameter of ~90 cm and an inner one of ~60 cm. The length of the magnet, which should be short for increased patient comfort, restricts the length of the gradient system. Although a shorter distance between the *z*-gradient coils increases the achievable gradient strength, this is usually limited by the peripheral nerve stimulation (PNS) threshold (compare with Section 2.3.3) and comes at the expense of a smaller field of view, less homogeneity, and increased local heating in the transverse gradient coil sets.[25]

The maximum gradient strength, maximum slew rate, and gradient power duty cycle are major considerations in practically all modern, whole-body MR systems. These are limited by power losses causing (local) heating, which requires the gradient coils and amplifiers to be cooled, usually with flowing water. The overall power loss of a single axis may be described as (eqn (2.1))

$$\frac{\overline{P}}{T_R} = \frac{1}{T_R}\int_{T_R} i(t)v_L(t)\mathrm{d}t + \frac{1}{T_R}\int_{\Delta f} I(f)^2 R(f)\mathrm{d}f + P_0 \tag{2.1}$$

where T_R is the repetition time, $i(t)$ the gradient current, $I(f)$ its spectral distribution, $v_L(t)$ the voltage drop across the active amplifier output transistor, $R(f)$ the frequency dependent coil resistance, and P_0 the standby power loss of the gradient system. The gradient power duty cycle, η, usually reaches a steady state if integration is carried out over one repetition time (eqn (2.2))

$$\eta = \frac{\int_{T_R} G(t)^2 \mathrm{d}t}{T_R G_{max}{}^2} \times 100\% \tag{2.2}$$

Current state-of-the-art machines achieve a maximum gradient strength of between 40 and 80 mT m^{-1} switched with slew rates of around 200 T m^{-1} s^{-1} and a duty cycle of 10%. Note, that the value of the duty cycle is usually given for a single operation axis and excessive heating may still occur when all three axes of a gradient system are simultaneously driven with a high duty cycle.

Shielding the gradient field was not common until the mid-1980s when fast imaging techniques were severely hampered by eddy currents induced through gradient actions on the cryostat. Active shielded gradient coils,[26,27] which are similar in concept to the active shielding in superconducting magnets, use a primary (inner) and secondary (outer) coil to generate the field gradients inside the patient bore and to cancel out magnetic flux outside the gradient insert.

Several advanced gradient coil designs have been recently introduced and include ultra-high strength gradient coil (300 mT m^{-1}) for brain connectome studies,[28] asymmetric coils for a short, open bore MRI system,[29] and split gradients for hybrid MR-PET[30,31] or MR-Linac systems.[32]

2.3.2 Gradient Amplifiers

While analogue gradient amplifiers were very common in the early days of MRI, all modern systems now use switched amplifiers as they are more efficient. The three separate switched mode audio frequency amplifiers of the gradient system are usually configured in a multi-stage H-bridge layout and are controlled by a pulse width modulated[33] signal generated in the MRI console. Generally, insulated gate bipolar transistors are used as active devices due to their capability of handling the considerable amount of power required to drive whole-body gradient coils. As is evident from fundamental physics, because of the large inductance of the gradient coils, it is obvious that high voltages are required for fast switching, while strong gradients require a high current. It is, therefore, common to utilise high power gradient amplifiers driving a few kilovolts at a few hundred amperes.

Inductive loading of amplifier outputs is prone to oscillations, making the integration of controller circuits into the amplifier necessary. A proportional-integral-derivative controller that compares desired and realised output current and digital pre-emphasis is often used to achieve faster rise times as well as to compensate for eddy currents and cross terms. Eddy currents refer to the circulating currents induced preliminarily in the magnet vessel when rapidly changing the gradient amplitude and cross terms arise from imperfect isolation between the three sets of gradient coils. Compensation of gradient imperfection is necessary to avoid severe image distortions, which are especially visible in fast imaging sequences.

The gradient amplifier system also incorporates a number of safety features. It supervises ohmic losses in the gradient coils and maximum pulse duration to prevent the gradient system from overheating. A secondary thermal supervision usually employs temperature sensors (*e.g.* PT100 resistors) in the gradients for a thermal shutdown when an over temperature condition is detected.

2.3.3 Limiting Effects of Fast Gradients

Aside from technical limitations, the rapidity with which gradient switching can occur is limited by possible cardiac and peripheral nerve stimulation (PNS) effects.[34] Cardiac stimulation is defined as the induction of an ectopic beat or cardiac arrhythmia, while PNS refers to the sensation of activation in the nervous systems generally manifested by twitching in the leg muscles or magneto-phosphenes. Both effectively limit the rate of change of the gradient fields (dB/dt) and their limits are given in ref. 10.

Since gradient switching is associated with relevant acoustic noise emission (the gradient coils and the static magnetic field form a potent loud speaker system by virtue of the Lorentz forces) noise levels may potentially limit gradient slew rates. However, since modern gradient systems are acoustically dampened and electrically shielded, this is rarely the case in practice. Despite many efforts to reduce acoustic noise emission from the gradient system by modifying sequences[35] *etc.*, patients being scanned still require hearing protection. The maximum noise emission allowed is also limited by the regulatory bodies.[10] Noise in the gradient system is usually highest when the mechanical eigenmodes of the gradient coils are excited. The eigenmode frequencies depend on the geometrical dimensions and material properties of the gradient coil system.[36–38] Recently, a classical-music-based MR acquisition was introduced using the MR fingerprinting method to improve patient comfort.[39] With this method, the patients feel that they are listening to music rather than being exposed to the gradient switching noise.

2.4 Shim System

As described above, uniform B_0 over the volume-of-interest is important in MRI. However, there are three major factors that contribute to degradation of homogeneity: (1) physical magnet design limitations and imperfections in practical construction; (2) perturbations from ferromagnetic/paramagnetic objects within the fringe field of the magnet; and (3) magnetic susceptibility variation of tissue and air.[40] In order to restore B_0 homogeneity, various methods can be employed. These can be classified as being either passive or active shimming, and most magnets use a combination of both methods.[41]

Passive shimming uses magnetically susceptible materials that influence the magnetic field distribution of the main magnet. The most prominent passive shim method uses small ferromagnetic tiles placed around the circumference of the magnet bore after the magnet is ramped up in its final installation location. The amount of iron inserted at the predetermined locations is computed from the initial field plot of the magnet. Since susceptibility is temperature dependent, passive shims should ideally be maintained at a constant temperature. While the above method provides a static shim, passive shimming of B_0 inhomogeneity, induced from inserting biological tissue into the magnet, can also be dynamically shimmed using relocatable magnetic shim materials.[42]

Active shimming usually refers to the use of resistive or superconducting sets of shim coils designed to generate magnetic fields that emulate the spatial variation of the spherical harmonic functions. While superconducting shim sets are used for static shimming, resistive shim sets are used to dynamically null out the majority of the field distortions arising from the insertion of biological tissue. Resistive shim sets are typically integrated with the gradient coil set and are actively powered by individual shim current amplifiers. They commonly provide selected shim terms up to the second, and sometimes the third, order. Higher order shim terms have been implemented for dynamic shimming in resistive shim inserts,[43] as have modified 32-channel receive-only RF array coils, where the shims are not necessarily based on spherical harmonic decomposition.[44] However, these higher order terms are usually only required for specialised applications or at ultra-high field.

2.5 RF System

The RF system generates excitation RF pulses and receives the MR signal. Transmit and receive power differ by several magnitudes—typical excitation pulses are in the kW range while the receive signal only has a few μW. The transmit RF system compromises digital signal generation with subsequent digital-to-analogue conversion (DAC), frequency up-conversion, an RF power amplifier, a transmit/receive (T/R) switch, and an RF transmit coil. The receiver employs one or more receive coils, a T/R switch, frequency down-conversion, analogue-to-digital converters (ADC), and digital signal processing. For optimum SNR performance, most receive coils employ integrated low-noise preamplifiers (LNAs). Figure 2.4 shows an overview of a typical RF system used in MRI scanners.

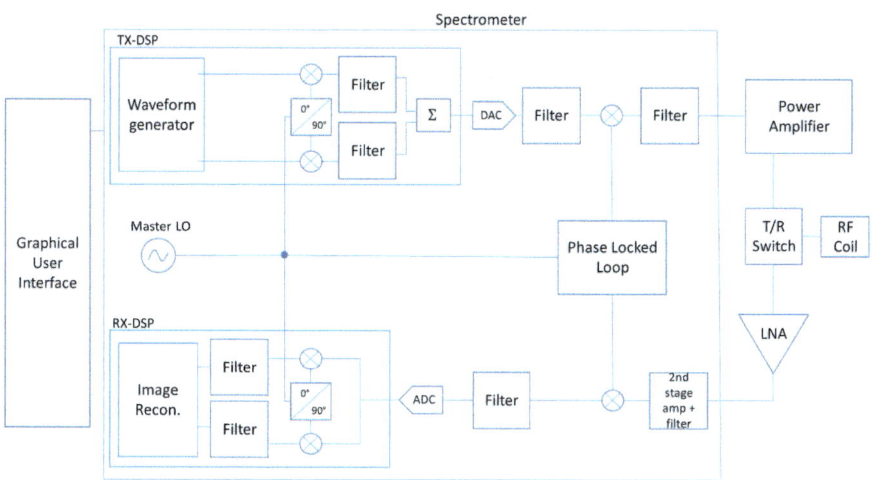

Figure 2.4 Block diagram of a simplified RF system including RF coils and chain.

2.5.1 RF Coils

MRI employs a large zoo of antennas, termed RF coils, according to the imaging task to be performed. The classification of RF coils is usually oriented around their function (transmit and receive, receive only) and RF magnetic field (B_1 field) properties (homogeneous or inhomogeneous). Generally, coils producing homogeneous B_1 field distributions are employed as transmit elements, as they will create a homogenous flip angle throughout the sample. Due to this reason, they are often also referred to as volume coils. For signal reception it is advantageous to employ smaller coils which can be positioned close to the object as they provide an inherently higher SNR,[45] have an excellent filling factor but are inhomogeneous. Finally, it should be noted that only the circularly polarised component of the RF magnetic field, perpendicular to the static magnetic field, is MR active. The counter-clockwise rotating field (B_1^+) is the transmit field, which excites a spin system, while the clockwise rotating field (B_1^-) is used for signal reception.

The selection of an RF coil is chosen based on the desired applications. Major aspects are designated imaging field-of-view (FoV) and spatial resolution, required B_1 homogeneity, parallel imaging capability, and specific absorption rate (SAR) constraints. The development of new coils has to take into account additional properties such as B_0 direction, SNR, element coupling, optimum element arrangements, penetrations depth and the target imaging area. Figure 2.5 shows a number of implementations of RF coils that have been built in-house. Items that require special attention when designing RF coils for combined MR-PET systems are addressed further in Chapter 10.

2.5.1.1 RX Arrays

To overcome the limited FoV of small coils while maintaining their beneficial SNR, the use of multiple receive elements in a 'phased array' configuration[45] has become the *de facto* standard. Moreover, coil arrays allow for

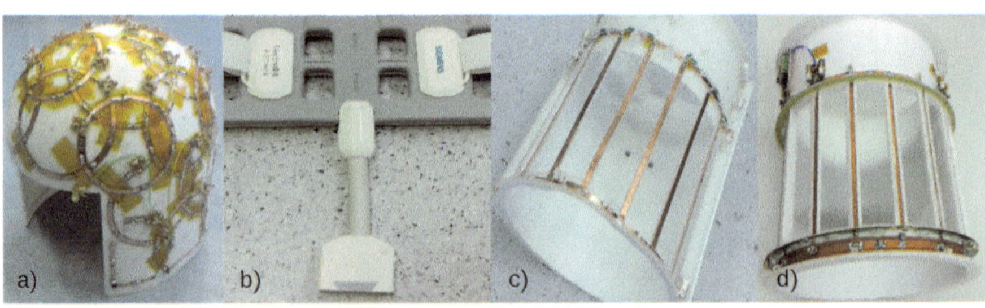

Figure 2.5 Images of different RF coils ((a) 32 Ch RX array, (b) flexible loop, (c) birdcage, (d) double-tuned coil).

accelerated image acquisition by exploiting their sensitivity variation in the imaging volume and by applying *k*-space subsampling strategies.[46]

In order for RX arrays to be applicable for parallel reception techniques, the sensitivity of each element needs to be sufficiently different from that of the other elements in the array. This can usually be achieved by the careful geometric arrangement of the elements. Another important feature in an array design is element decoupling. As SNR averaging efficiency depends on the fraction of correlated to uncorrelated noise, the decoupling of receive antennas is essential. In an optimum array, correlated noise would only originate from the sample being investigated, while noise generated in other array elements would be suppressed. Several methods exist to cancel out noise correlation from neighbouring elements, such as overlapping loop elements to minimise mutual inductance, using inductive[47] or capacitive components,[48] and/or employing preamplifier decoupling.[49] The latter method uses low input impedance preamplifiers, which by correct phase shifting, represent a high impedance at the coil terminals and thus reduce current flow on the coil. By observing that (eqn (2.3))

$$V_{\text{induced,coil\#1}} = M_{12} i_{\text{coil\#2}} \tag{2.3}$$

nulling the mutual inductance M_{12} by overlapping or by minimising current flow in neighbouring coil elements (*e.g.* through a high impedance termination of the antenna terminals) noise coupling is reduced. In Figure 2.6, a high impedance between the coil terminals is achieved by inserting a suitable

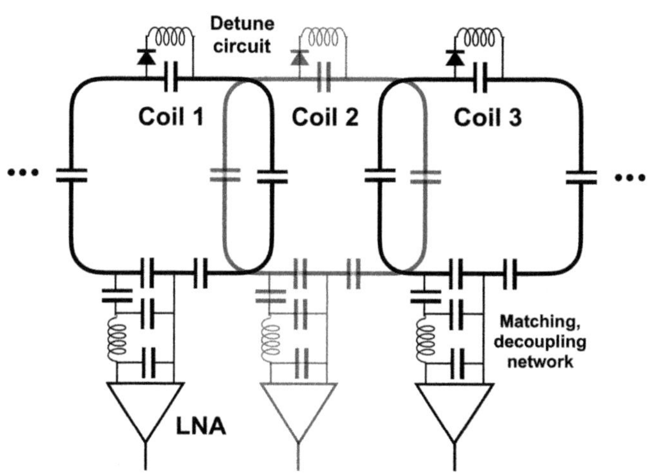

Figure 2.6 Detuning in TX mode and preamplifier decoupling in RX mode for multi receive loops.

phase shifting circuit (*e.g.* a π-filter) between the coil terminals and the low input impedance preamplifier.

Figure 2.6 also contains an 'active' detuning unit in the form of a PIN-diode, which switches between high and low impedance for RF currents using a DC bias. Detuning the RX elements is crucial, as the resonance of the receive loop would otherwise interfere with the resonance of the transmission coil during excitation. This, in turn, detunes the transmit coil and potentially makes the receive loop pick up substantial currents that can lead to elevated local SAR.

2.5.1.2 *Multi-tuned RF Coils*

The MRI of X-nuclei (non-protons) is an emerging topic that allows information to be obtained from nuclei directly involved in metabolic processes.[50] In combination with PET, a wide variety of applications can be envisaged. The full information content of a hybrid system may include the imaging of targeted radiotracer distribution using PET, the depiction of metabolic activity using X-nuclei in MRI, and the co-registering of all information with anatomy using proton MRI.[51] However, it must be emphasised that X-nuclei have an inherently low MR sensitivity, often factors of magnitude below that of protons in natural abundance, and this needs to be taken into account. Thus, X-nuclei measurements are another key driver for high B_0 field strength.

X-nucleus measurements frequently require the double- or multi-tuning of RF coils so that a single physical antenna can pick up signals at the proton frequency as well as at that of the X-nucleus resonance frequency. Multiple resonances can generally be achieved by 'switching' or 'blocking' the resonance frequency of a single-tuned structure using PIN-diodes or trap circuits.[52] A third method is to use multiple structures, for example, one for ^1H and another for X-nuclei, that produce orthogonal B_1 fields and can be said to be geometrically decoupled.[53]

2.5.2 RF Electronics

In most respects, the RF chain of an MRI system is comparable to a transmitter system that is capable of amplitude and phase modulation in combination with a phase sensitive detector system.

2.5.2.1 *Transmit Chain*

As shown in Figure 2.4, transmit pulses are generated digitally and are then converted to analogue by a DAC. Frequency up-conversion is carried out by a mixer stage that is driven from a highly stable, low phase-noise, local oscillator. The mixer output is filtered so as to obtain a single sideband signal, which is further amplified in the power amplifier. The transmit signal is emitted from the RF coil and perturbs the spin system.

As the transmit frequency of an MR system can vary in a large range when measuring nuclei with different Larmor frequencies, as well as in subtle steps (*e.g.* when exciting off-centre slices), an agile frequency control is required. Many modern systems, therefore, use direct digital synthesis methods[54] for frequency control. Since the excitation of a spin system requires a significant amount of power, the RF power amplifier (PA) of an MR system is very potent and may deliver output powers of 8 kW or more at duty cycles of up to 10%. In order to allow excitation pulses to be played out in rapid sequences, it is important that higher duty cycles can be reached when less output power is generated. While the standard proton PA uses narrowband architecture, X-nucleus experiments require additional hardware in the form of a broadband PA that spans the X-nucleus frequency range.

For safe operation, the SAR has to be continuously monitored. While a variety of implementations are possible, and many systems use multiple SAR monitoring methods for 'sanity checking', measurements based on directional couplers are quite common. Here, power travelling to the antenna and reflected due to impedance mismatches is sampled. The difference between the two values equals the power accepted from the transmit coil and is converted into an SAR burden using a coil specific factor. It is common practice to base the conversion factor on the SAR value (*e.g.* local SAR, exposed body SAR or global SAR) that is most critical.

With the increased availability of ultra-high-field systems, a technique called parallel transmission has been introduced in MRI.[55] Similar to the operation of phased array antennas, this method uses multiple independent transmit channels in combination with transmit gradient encoding. It enables a homogeneous excitation profile to be achieved by allowing both the tailoring of RF-pulses and transmitting gradient encoding during transmit, despite B_1 field interferences from the different coil elements that increase with deceasing wavelength.

2.5.2.2 *Receive Chain*

MR image reconstruction requires the amplitude and phase information of the received signal in order for the receiver chain to convert the single sideband signal to a complex representation suitable for the digital post-processing units. Given that signal acquired with the receiver coils is weak but has a high dynamic range,[56] SNR performance, as well as the maintenance of the high dynamic range all the way through AD conversion, are two major design concerns.

Generally, signal conditioning starts with an LNA, which in order to minimise losses, is placed as close to the receive coil as possible. The preamplifiers used for MRI have noise figures in the range of ~0.5 dB and a gain of 25 dB or more. They are operated in a noise-matched condition—as opposed to power matching of power amplifiers in the transmit path—and usually exhibit a low input impedance for receive coil decoupling (compare Section 2.5.1).

As also shown in Figure 2.4, the pre-amplified MR signals are down-converted to an intermediate frequency (IF) by means of a frequency mixer and subsequent low-pass filtering. In all modern receiver architectures, the IF signal is directly sampled by ADCs, which avoids issues caused by analogue quadrature baseband detection. The required ADC resolution depends on the dynamic range of the MR signal, and contemporary ADCs are capable of sampling at 10 MHz and at resolutions of 16–24 bits. Finally, quadrature demodulation of the IF signal is carried out digitally.

Recently, techniques to digitise the analogue MR signals within the receive coil set using optical or wireless solutions[57–59] have been introduced. The major driving factor for this development is the increased use of modern, high-channel-count receive arrays. These arrays give rise to a number of technological and application challenges. For example, they require a high number of receive cables, making handling bulky and cumbersome, while at the same time, each receive cable potentially requires a cable trap to avoid shield currents. In addition to overcoming the practicalities associated with modern, high channel count arrays, direct sampling on the coil is a way to standardise receiver architectures, thus potentially reducing production and development costs.

2.6 Console/Spectrometer

Modern spectrometers are based around a realtime controller that executes a pulse sequence and sends commands to the peripheral units described above. The controller also handles numerous safety checks and the hardware supervision necessary on clinical machines. Image reconstruction can either be carried out offline, following the MR acquisition, or be integrated into the real time framework. The latter allows sequence adaption during run-time. One example of this is a FoV that follows patient movements. Final images from most MRI systems are stored in the DICOM format, which allows easy data transfer to medical imaging archives (PACS). The MR system is usually operated from a console computer to enable the desired pulse sequences for the examination to be set up and the imaging results to be displayed.

2.7 MRI Suite

As described above, MR systems provide a very sensitive receive system for electromagnetic waves. The system operates in the very high-frequency band, ranging 30–300 MHz, which is also used for a variety of other radio services, including television, amateur radio, radar, radio navigation, and air traffic control. The MR signal itself may be easily swamped from these unwanted radio transmissions, which makes shielding an MRI system from electromagnetic interferences necessary. The MR room is shielded from the outside using an RF screen (Faraday cage). The RF screen attenuates RF signals at the Larmor frequency by 90 dB or more. Any desired electrical connections from the outside into the Faraday cage must be filtered or transmitted over non-conducting connections such as fibre-optic cables *via* dedicated wave-guides. Desired electrical connections do not only include MR-related signals, but also energy for room lights and auxiliary equipment inside the MR room.

References

1. World Health Organization, *Static Fields*, World Health Organization, Geneva, 2006.
2. Z. Cao, J. Park, Z.-H. Cho and C. M. Collins, *J. Magn. Reson. Imaging*, 2015, **41**, 1432.
3. J. T. Vaughan, L. DelaBarre, J. Tian, S. Sohn, D. Shrivastava, G. Adriany and K. Ugurbil, *2013 IEEE MTT-S International Microwave Workshop Series on RF and Wireless Technologies for Biomedical and Healthcare Applications (IMWS-BIO)*, 2013, p. 1.
4. T. Schild, G. Aubert, J. Belorgey, C. Berriaud, A. Bourquard, P. Bredy, O. Dubois, G. Gilgrass, F. P. Juster, H. Lannou, F. Molinié, M. Nusbaum, F. Nunio, A. Payn, L. Quettier, L. Scola, A. Sinanna, V. Stepanov and P. Vedrine, *IEEE Trans. Appl. Supercond.*, 2016, **26**, 1.
5. R. Warner, *Supercond. Sci. Technol.*, 2016, **29**, 094006.
6. J. Wilks, *The Properties of Liquid and Solid Helium*, Clarendon P., Oxford, 1967.
7. *Applied Superconductivity: Handbook on Devices and Applications*, ed. P. Seidel, Wiley-VCH, Weinheim, Germany, 2015.
8. J. Overweg, *Proceedings 14th Scientific Meeting, International Society for Magnetic Resonance in Medicine*, Seattle, USA, 2006.
9. M. Mallett, *Proceedings 23rd Scientific Meeting, International Society for Magnetic Resonance in Medicine*, Toronto, Canada, 2015.
10. International Electrotechnical Commission, *IEC 60601-2-33*, Geneva, Switzerland, 2015.
11. Q. Wang, Y. Dai, B. Zhao, S. Song, C. Wang, L. Li, J. Cheng, S. Chen, H. Wang, Z. Ni, Y. Li, C. Cui, X. Hu, H. Wang, Y. Lei, K. Chan, L. Yan, C. Wen, G. Hui, W. Yang, F. Liu, Y. Zhuo, X. Zhou, Z. Yan, J. Chen and T. Xu, *IEEE Trans. Appl. Supercond.*, 2012, **22**, 4400905.
12. J. M. Oort, E. T. Laskaris, P. S. Thompson, B. Dorri and K. G. Herd, in *Advances in Cryogenic Engineering*, ed. P. Kittel, Springer, Boston, 1998, ch. 15, pp. 139–147.
13. R. C. Smith and R. C. Lange, *Understanding Magnetic Resonance Imaging*, CRC Press, Boca Raton, Fla, 1998.
14. C. R. Weiss, S. G. Nour and J. S. Lewin, *J. Magn. Reson. Imaging*, 2008, **27**, 311.
15. D. Hailey, *Issues Emerg. Health Technol.*, 2006, 1.
16. H. Soltner and P. Blumler, *Concepts Magn. Reson., Part A*, 2010, **36**, 211.

17. C. Z. Cooley, *Portable Low-cost Magnetic Resonance Imaging*, Massachusetts Institute of Technology, Department of Electrical Engineering and Computer Science, 2014.
18. C. Z. Cooley, J. P. Stockmann, B. D. Armstrong, M. Sarracanie, M. H. Lev, M. S. Rosen and L. L. Wald, *Magn. Reson. Med.*, 2015, **73**, 872.
19. D. J. Lurie, S. Aime, S. Baroni, N. A. Booth, L. M. Broche, C. H. Choi, G. R. Davies, S. Ismail, D. O. HOgain and K. J. Pine, *C. R. Phys.*, 2010, **11**, 136.
20. H. Peng, W. B. Handler, T. J. Scholl, P. J. Simpson and B. A. Chronik, *Nucl. Instrum. Methods Phys. Res., Sect. A*, 2010, **612**, 412.
21. R. Pascone, T. Vullo, P. T. Cahill and A. Tankhiwale, *Nuclear Science Symposium & Medical Imaging Conference*, 1993, vol. 1–3, p. 1349.
22. M. J. E. Golay, *Rev. Sci. Instrum.*, 1958, **29**, 313.
23. R. Turner, *J. Phys. D: Appl. Phys.*, 1986, **19**, L147.
24. J. F. Schenk, M. A. Hussain and W. A. Edelstein, *Transverse Gradient Field Coils for Nuclear Magnetic Resonance Imaging*, US4646024A, 1987.
25. F. Schmitt, *Proceedings 21st Scientific Meeting, International Society for Magnetic Resonance in Medicine*, 2013.
26. R. M. Bowley, B. L. W. Chapman, P. Mansfield and R. Turner, *Magnetic Field Screens*, EP0365065 B1, 1986.
27. P. Roemer, W. A. Edelstein and J. Hickey, *Proceedings 5th Anual Meeting*, SMRM, Montreal, Canada, 1986.
28. K. Setsompop, R. Kimmlingen, E. Eberlein, T. Witzel, J. Cohen-Adad, J. A. McNab, B. Keil, M. D. Tisdall, P. Hoecht, P. Dietz, S. F. Cauley, V. Tountcheva, V. Matschl, V. H. Lenz, K. Heberlein, A. Potthast, H. Thein, J. Van Horn, A. Toga, F. Schmitt, D. Lehne, B. R. Rosen, V. Wedeen and L. L. Wald, *NeuroImage*, 2013, **80**, 220.
29. Y. Wang, F. Liu, Y. Li, F. Tang and S. Crozier, *J. Magn. Reson.*, 2016, **269**, 203.
30. L. Liu, H. Sanchez-Lopez, M. Poole, F. Liu and S. Crozier, *J. Magn. Reson.*, 2012, **222**, 8.
31. M. Poole, R. Bowtell, D. Green, S. Pittard, A. Lucas, R. Hawkes and A. Carpenter, *Magn. Reson. Med.*, 2009, **62**, 1106.
32. L. Liu, A. Trakic, H. Sanchez-Lopez, F. Liu and S. Crozier, *Phys. Med. Biol.*, 2014, **59**, 233.
33. H. Takano, S. Watanabe and M. Nakaoka, *Proceedings of the 25th Annual Conference of the IEEE, IECON '99*, 1999, vol. 2, p. 785.
34. P. R. Harvey and P. Mansfield, *Magn. Reson. Med.*, 1994, **32**, 236.
35. S. Alibek, M. Vogel, W. Sun, D. Winkler, C. A. Baker, M. Burke and H. Gloger, *Diagn. Interv. Radiol.*, 2014, **20**, 360.
36. L. Haiying and L. Junxiao, *Proceedings 4th Scientific Meeting, International Society for Magnetic Resonance in Medicine*, New York, USA, 1996.
37. R. Hedeen and W. Edelstein, *Proceedings 4th Scientific Meeting, International Society for Magnetic Resonance in Medicine*, New York, USA, 1996.
38. S. A. Winkler, F. Schmitt, H. Landes, J. DeBever, T. Wade, A. Alejski and B. K. Rutt, *NeuroImage*, 2018, **168**, 59.

39. D. Ma, E. Y. Pierre, Y. Jiang, M. D. Schluchter, K. Setsompop, V. Gulani and M. A. Griswold, *Magn. Reson. Med.*, 2016, **75**, 2303.
40. K. Wachowicz, *Res. Rep. Nucl. Med.*, 2014, **2014**, 1.
41. C. Juchem, *Proceedings 21st Scientific Meeting, International Society for Magnetic Resonance in Medicine*, Salt Lake City, USA, 2013.
42. K. M. Koch, P. B. Brown, D. L. Rothman and R. A. de Graaf, *J. Magn. Reson.*, 2006, **182**, 66.
43. W. F. B. Punchard, L. Kai-Ming, P. M. Starewicz and H. Hetherington, *Shim Insert for High-field MRI Magnets*, US20110260727A1, 2013.
44. J. P. Stockmann, T. Witzel, B. Keil, J. R. Polimeni, A. Mareyam, C. LaPierre, K. Setsompop and L. L. Wald, *Magn. Reson. Med.*, 2016, **75**, 441.
45. P. B. Roemer, W. A. Edelstein, C. E. Hayes, S. P. Souza and O. M. Mueller, *Magn. Reson. Med.*, 1990, **16**, 192.
46. D. J. Larkman and R. G. Nunes, *Phys. Med. Biol.*, 2007, **52**, R15.
47. N. I. Avdievich, J. W. Pan and H. P. Hetherington, *NMR Biomed.*, 2013, **26**, 1547.
48. C. Von Morze, J. Tropp, S. Banerjee, D. Xu, K. Karpodinis, L. Carvajal, C. P. Hess, P. Mukherjee, S. Majumdar and D. B. Vigneron, *Concepts Magn. Reson., Part B*, 2007, **31**, 37.
49. G. C. Wiggins, C. Triantafyllou, A. Potthast, A. Reykowski, M. Nittka and L. L. Wald, *Magn. Reson. Med.*, 2006, **56**, 216.
50. W. A. Worthoff, A. Shymanskaya, C.-H. Choi, J. Felder, A.-M. Oros-Peusquens and N. J. Shah, in *Quantitative MRI of the Brain: Principles of Physical Measurement*, ed. M. Cercignani, N. G. Dowell and P. Tofts, CRC Press, Taylor & Francis Group, Boca Raton, FL, 2nd edn, 2018, ch. 13, pp. 235–248.
51. N. J. Shah, A.-M. Oros-Peusquens, J. Arrubla, K. Zhang, T. Warbrick, J. Mauler, K. Vahedipour, S. Romanzetti, J. Felder, A. Celik, E. Rota-Kops, H. Iida, K.-J. Langen, H. Herzog and I. Neuner, *J. Magn. Reson.*, 2013, **229**, 101.
52. S. Ha, M. J. Hamamura, O. Nalcioglu and L. T. Muftuler, *Phys. Med. Biol.*, 2010, **55**, 2589.
53. P. A. Bottomley, C. J. Hardy, P. B. Roemer and O. M. Mueller, *Magn. Reson. Med.*, 1989, **12**, 348.
54. J. Vankka, *IEEE Tran Ultrason Ferr*, 1997, **44**, 526.
55. U. Katscher, P. Bornert, C. Leussler and J. S. van den Brink, *Magn. Reson. Med.*, 2003, **49**, 144.
56. R. E. Gabr, M. Schar, A. D. Edelstein, D. L. Kraitchman, P. A. Bottomley and W. A. Edelstein, *J. Magn. Reson.*, 2009, **198**, 137.
57. K. Aggarwal, K. R. Joshi, Y. Rajavi, M. Taghivand, J. M. Pauly, A. S. Poon and G. Scott, *IEEE Trans. Med. Imaging*, 2017, **36**, 574.
58. D. O. Brunner, B. Sporrer, C. Vogt, J. Reber, J. Marjanovic, L. Bettini, L. Wu, T. Burger, G. Troester, Q. Huang and K. P. Pruessmann, *Proceedings 24th Scientific Meeting, International Society for Magnetic Resonance in Medicine*, Singapore, 2016.
59. J. Yuan, J. Wei and G. X. Shen, *IEEE Trans. Med. Imaging*, 2008, **27**, 1432.

CHAPTER 3

Selective Applications of MRI for the Brain

S. D. YUN*[a], E. FARRHER[a], F. GRINBERG[a], A. M. OROS-PEUSQUENS[a] AND N. J. SHAH[a,b,c]

[a]Forschungszentrum Jülich, Institute of Neuroscience and Medicine – 4, Wilhelm-Johnen-Straße 1, 52428 Jülich, Germany; [b]JARA-BRAIN – Translational Medicine, Aachen, Germany; [c]Department of Neurology, RWTH Aachen University, Pauwelsstr. 30, 52074 Aachen, Germany
*E-mail: s.yun@fz-juelich.de

3.1 Functional MRI (fMRI)

3.1.1 Introduction

Historically, there have been numerous attempts to identify the different functional regions of the brain and to localise them on a topological representation of the brain. In 1909, the German anatomist, Brodmann suggested a classification for brain regions (now referred to as the 'Brodmann areas') by means of a cell staining method.[1] Here, the cerebral cortex of the human, or other primates, is subdivided into 52 distinct areas. Essentially, the Brodmann areas were obtained based on the cytoarchitectural organisation of neurons, but each area is known to be closely linked to particular cortical functions such as motor, somatosensory, visual or auditory functions.

New Developments in NMR No. 19
Hybrid MR-PET Imaging: Systems, Methods and Applications
Edited by N. Jon Shah
© The Royal Society of Chemistry 2019
Published by the Royal Society of Chemistry, www.rsc.org

This type of approach has the merit that anatomical or functional structures of the brain can be characterised in a more direct way. However, it has a drawback in that it is invasive and requires surgical procedures to obtain direct contact with the cells. This is a significantly challenging issue when applying the method *in-vivo*. To avoid this, less invasive, or non-invasive, neuroimaging techniques have been developed, such as electroencephalography (EEG), positron emission tomography (PET) and functional MRI (fMRI). The development of EEG dates back to the first half of the 20th century and its application has become a standard diagnostic method in clinical neurology and research. When compared with the other methods, it provides superior temporal resolution—on the ms scale—but has rather limited spatial resolution to localise and to distinguish spots of increased neural activity. This issue was partially overcome by the introduction of functional PET imaging, which offers much better spatial resolution but at the expenses of reduced temporal resolution (in the minute range) and invasiveness (injection of radioactivity and blood sampling). In contrast to EEG, which directly detects the electrophysiological activation, PET (and fMRI) are indirect methods, which aim to measure the effects of the increased blood supply demanded in activated brain areas. Such methods contrast the rise of activation in functional brain areas, which are stimulated by study-specific neuroscientific tasks, with the base level activation of the brain (resting state). In the case of PET imaging, increased cerebral blood flow can be measured following the administration of radiolabelled water,[2] $H_2{}^{15}O$, which accumulates in regions of enhanced oxygen demand caused by motor, sensory, or neuro-activation.[3–5] Prior to the emergence of fMRI, a countless number of studies using water-PET accomplished mapping of the different cortical locations of neuro-activations. For the analysis of such data sets, statistical parametric mapping (SPM) was invented and finally became a major tool for the analysis of fMRI data. However, functional brain mapping with PET has several drawbacks. First, because of the short half-life of ^{15}O (2 min), PET requires an on-site cyclotron and radio-synthesis, which is costly and not widely available. Furthermore, the number of sequential and repetitive measurements for studying versatile neuro-activations is limited due to the dangers associated with radiation exposure to the subject. Finally, water-PET is inferior to current fMRI with regard to spatial and temporal resolution.

By reason of the previously mentioned properties of PET, fMRI has broadly superseded PET-based functional imaging and has become one of the most widely used methods for measuring brain activity. This is due to that it enables the acquisition of neuronal data with a relatively high spatial and temporal resolution. The tremendous progress in modern functional neuroscience was made possible to a large extent by the success of fMRI. Moreover, along with the functional data, MRI also allows one to acquire high-resolution anatomical images, which can be used in the localisation of functional data for each subject.

The fMRI technique has been employed in the study of various brain functions. For more precise and robust mapping, there have also been technical advancements in the imaging technique to achieve far higher spatial and temporal resolution.[6–12] Recently, ultra-high field (>7 T) MRI has been available for clinical or research purposes. In order to attain higher signal-to-noise ratio (SNR) or contrast-to-noise ratio (CNR) for fMRI, inherent to ultra-high field, a number of technical innovations were required to make the use of fMRI feasible at such high fields.[13–16] In the following sections, the basic principles of fMRI and its data analysis will be described.

3.1.2 Blood-oxygenation-level-dependent (BOLD)

Figure 3.1 depicts a schematic representation of capillaries carrying red blood cells with the oxygen transport protein haemoglobin. Oxyhaemoglobin (oxygen is bound) and deoxyhaemoglobin (oxygen has been released) exist concurrently at a certain ratio during the resting state. If a putative brain area is activated by a given internal or external stimulus (*e.g.* finger movement or tactile touch), the higher regional demand for oxygen is overcompensated by the increased influx of blood, making the decreasing the relative concentration of deoxyhaemoglobin in this activated brain region and thus the relative concentration of oxyhaemoglobin increases. As a result, the regional T_2/T_2^* shortening effect caused by the paramagnetic property of deoxyhaemoglobin is reduced, leading to a higher MR signal from the activation state. Therefore, differences in the signal level between the resting and activation states are created and this signal contrast is referred to as the BOLD contrast.

The use of this effect for the detection of functional brain activity was first demonstrated by Ogawa *et al.* in 1990.[17] Since then, most fMRI has been performed using gradient-echo (GRE) BOLD, as it can offer a sufficiently good CNR and is easy to implement.[18] The GRE BOLD signals are created based on regional T_2^* changes caused by the intravoxel dephasing induced from the field distortions by the presence of the paramagnetic deoxyhaemoglobin. At relatively low fields, the GRE BOLD signals are predominantly obtained from large veins. As the field strength increases, contributions from the large vessels decrease and instead, contributions from

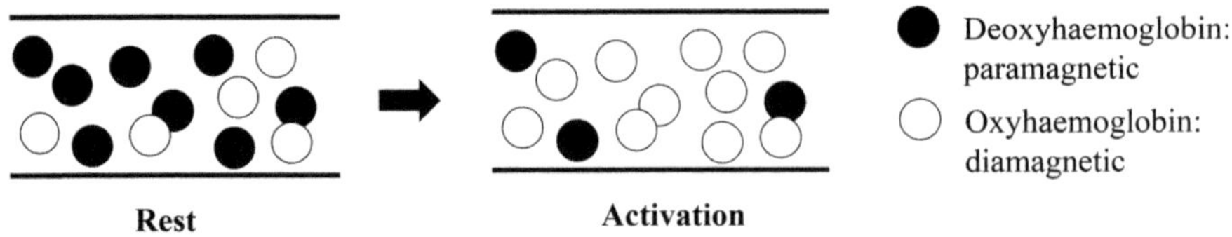

Figure 3.1 A schematic representation of the state of capillaries during rest and activation in a putative brain region showing the change in the ratio of oxyhaemoglobin to deoxyhaemoglobin. The relative concentration of oxyhaemoglobin increases during the activation.

capillaries increase. The contributions from capillaries can be effectively detected by spin-echo techniques.[19] The spin-echo BOLD signals originate from the diffusion-dependent T_2-signal loss, which is also induced from the local field distortions produced by the change of relative concentration of deoxyhemoglobin.[18] Since the spin-echo BOLD signals can identify the signals from the capillaries, they are considered to have a higher spatial specificity with which to localise the activation regions. For this reason, several high-resolution fMRI studies have been performed based on spin-echo BOLD signals.[18–20]

The BOLD signals are known to be increased as the field strength increases.[21] For this reason, many fMRI studies have been performed at ultra-high fields,[13–16] despite there being a number of technical challenges at higher magnetic fields, such as increased B_0/B_1 inhomogeneities or shorter T_2/T_2^*.

3.1.3 Haemodynamic Response Function (HRF) and fMRI Paradigm

For the analysis of detected BOLD signals in fMRI studies, there have been a number of attempts to estimate the haemodynamic response function (HRF).[22–24] The HRF is defined as the resultant response when an impulse stimulus input is given to the brain. In general, the HRF can be different depending on the subject and the brain region.[25] However, for most fMRI data analyses, a canonical HRF is utilised and its shape is defined in several fMRI analysis software packages, such as statistical parametric mapping (SPM) (Wellcome Department of Imaging Neuroscience, UCL, London, UK), FSL (available at https://fsl.fmrib.ox.ac.uk) or Brainvoyager (available at http://www.brainvoyager.com/). Figure 3.2a depicts the canonical HRF obtained from the SPM. As shown here, the response slowly rises and reaches a maximum at around 6 s. Subsequently, it falls slowly until around 15 s and still exists weakly for the next few tens of seconds. This observation infers that the haemodynamic response changes relatively slowly and there is a substantial timing delay between the start of its rise and the time when maximum signal is attained.

For task-based fMRI, there are two main types of fMRI paradigms to deliver the stimulus to subjects. One is the block-based paradigm and the other one is event-related paradigm; a mixed paradigm of the two types can also be used, which allows one to perform a better characterisation of neuronal responses.[26] An example design for each type is exhibited in Figure 3.2b. The corresponding haemodynamic responses are also depicted in Figure 3.2c, which can be obtained by convolving the paradigm and the canonical HRF. In the block-based paradigm, the stimulus is continuously given during the 'stimulus' block, which lasts, illustratively, for a few tens of seconds. In contrast, in the event-related paradigm, discrete events of short-duration are presented with a constant or randomised timing interval;[27] for simplicity, each

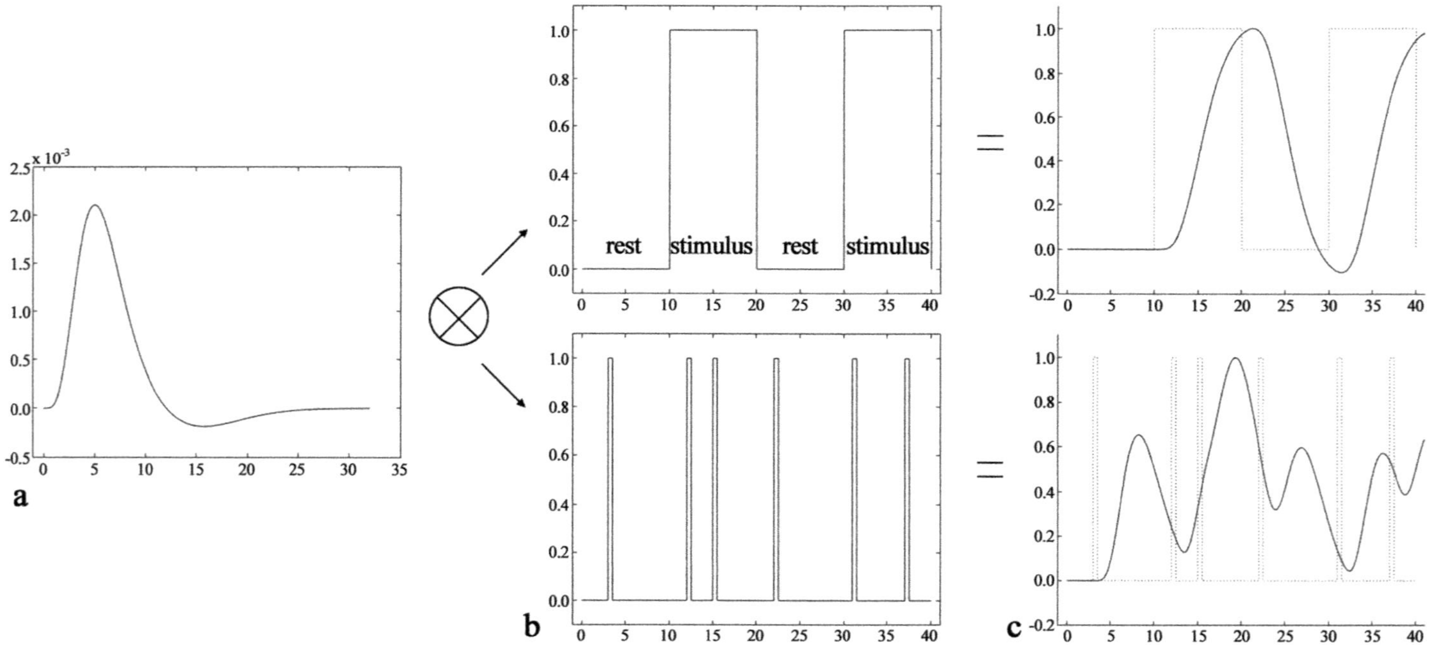

Figure 3.2 Canonical HRF and haemodynamic responses for the given type of paradigm. (a) Canonical HRF; (b) example block-based (top) and event-related paradigms (bottom); and (c) the corresponding haemodynamic responses (solid line) with the paradigm overlaid on them (dotted line). The maximum of each haemodynamic response was scaled to one for illustrative purposes.

sub-figure of Figure 3.2b depicts the presentation of only a single stimulus during the entire paradigm. The block-based paradigm is relatively simple to design, and it is straightforward to analyse the resulting data. The relatively long period of stimulation is also beneficial for the detection of neural signals with a rather high SNR, which increases statistical power.[28] However, the fact that the presentation of each block can be predicted induces habituation effects.[29] This issue can be effectively overcome with the use of the event-related paradigm. Moreover, it can detect the individual response from each stimulus trial, which is not possible with the block-based paradigm.[30] Thus, the event-related paradigm generally allows for more flexibility in designing the paradigm, however, as a consequence, it requires more complicated analysis.

As shown in the above, in task-based fMRI, an external stimulus (*e.g.* visual, auditory or tactile stimulus) is given to the subject to derive task-specific responses. Alternatively, non-task-based fMRI can be also performed without giving any stimulus inputs to the subject. This is referred to as 'resting-state' fMRI, which investigates the functional behaviour of the brain during the resting state only. The method detects spontaneous and low-frequency fluctuations (<0.1 Hz) in the BOLD signal and identifies various resting state networks (RSNs), such as the default mode network.[31,32] The default-mode network can be interrogated as a function of disease or physiological state. A detailed description of how to perform the experiment and data analysis can be found in the relevant literature.

3.1.4 Generalised Linear Model (GLM) and Data Analysis

The behaviour of actual haemodynamic responses is assumed to be similar to that of the ideal responses as depicted in Figure 3.2. Based on this assumption, the generalised linear model (GLM) establishes a regression equation that represents the real observed data with a linear combination of a set of linearly independent bases.[33] The equation below is an example GLM model for the block-based paradigm shown in Figure 3.2b. The equation (eqn (3.1)) expresses the real observed data, $\mathbf{Y}$ with three different basis vectors ($\mathbf{X}_1$, $\mathbf{X}_2$ and $\mathbf{X}_3$), each of which is multiplied by a coefficient (β_1, β_2 and β_3, respectively)

$$\mathbf{Y} = \beta_1 \cdot \mathbf{X}_1 + \beta_2 \cdot \mathbf{X}_2 + \beta_3 \cdot \mathbf{X}_3 + \varepsilon \quad (3.1)$$

The block-paradigm has two different states: resting and active. Here, the predicted responses (*i.e.* convolution of the canonical HRF and the paradigm) coming from each state become the basis vectors (*i.e.* $\mathbf{X}_1$ and $\mathbf{X}_2$) for the above equation. In addition to these two bases, the additional basis vector ($\mathbf{X}_3$) is included to take account of a DC offset; as such, the elements of $\mathbf{X}_3$ consist of only one. Based on this linear regression model, the three coefficients are estimated so that the expression best matches the obtained real observed data (**Y**), that is, the residual term (ε) is minimised. The residual is known to have a normal distribution and thus, the least-square approach can be a 'good' solution to estimate the coefficients.[34] The solution is given as follows (eqn (3.2))

$$\mathbf{B} = (\mathbf{X}^{\mathrm{T}}\mathbf{X})^{-1}\mathbf{X}^{\mathrm{T}}\mathbf{Y} \tag{3.2}$$

where **B** and **X** is given by $[\beta_1\ \beta_2\ \beta_3]^{\mathrm{T}}$ and $[\mathbf{X}_1\ \mathbf{X}_2\ \mathbf{X}_3]$, respectively. Accordingly, the fitted data and estimated noise can be expressed as follows (eqn (3.3) and (3.4))

$$\mathbf{Y}' = \mathbf{XB} \tag{3.3}$$

$$\varepsilon = \mathbf{Y} - \mathbf{Y}' \tag{3.4}$$

After estimation, hypothesis tests are performed using *t* or F-statistics. For instance, using *t*-statistics, the difference between the means of two populations (resting and stimulus states) can be statistically measured. Here, in order to see how much the signals are more active in the activation state than those in the resting state, the contrast vector is defined as $\mathbf{C} = [-1\ 1\ 0]$. For the given contrast vector, the corresponding *t*-statistics are calculated with the following equation (eqn (3.5))

$$t = \mathbf{C}^{\mathrm{T}}\mathbf{B} \Big/ \sqrt{\left(\sigma^2 \bullet \mathbf{C}^{\mathrm{T}} \left(\mathbf{X}^{\mathrm{T}}\mathbf{X}\right)^{-1} \mathbf{C}\right)} \tag{3.5}$$

where σ is the standard deviation of the residual, ε. Here, if the *t*-score is significantly high, the voxel where the observed data (**Y**) are extracted is considered as an 'activated voxel'. The entire activation regions can be obtained by repeating the above procedure for all other voxels of interest.

Before the above procedure (*e.g.* estimation and statistical tests) is performed, there are several preceding pre-processing steps that need to be applied to the initial observed data. The typical pre-processing steps include realignment, co-registration, segmentation, spatial normalisation or smoothing. The realignment is to realign the time-series data to remove subject motions that might exist during the entire fMRI scan. Subsequently, co-registration between the fMRI scans and a high-resolution anatomical scan is performed. This is so the high-resolution anatomical scan which can be used as the reference image to show the activated voxels on it. Next, the brain is segmented into several parts: grey matter, white matter or cerebrospinal fluid regions. This segmentation information is effectively used in the following step, spatial normalisation. The purpose of the spatial normalisation step is to re-map the brain voxels of each subject on the standardised coordinate so that the location of detected voxels can be recognised in the reference brain atlas. Importantly, this step facilitates a group fMRI study with different subjects or a cross comparison between individual fMRI studies. One of the widely used standardised spaces is the Montreal Neurological Institute (MNI) space. Lastly, the spatial normalisation is the step of applying a Gaussian smoothing kernel to the data to suppress noise effects and detect activated voxels more robustly. Depending on the purpose of the study, some of the pre-processing steps can be skipped or additional processing steps may be included.

3.1.5 An Example Visual fMRI Study

Figure 3.3 shows example functional maps obtained at 3 T using a visual stimulus. Here, in order to elicit circumscribed activation in the visual cortex, a visual checkerboard block-based paradigm was employed (see Figure 3.3a). The corresponding GLM model is also shown in Figure 3.3b. The results shown in the figure were obtained from SPM12. As expected, Figure 3.3c reveals that the activations are largely identified around the visual cortex. However, due to the relatively low spatial resolution (2.5×2.5 mm^2), it can

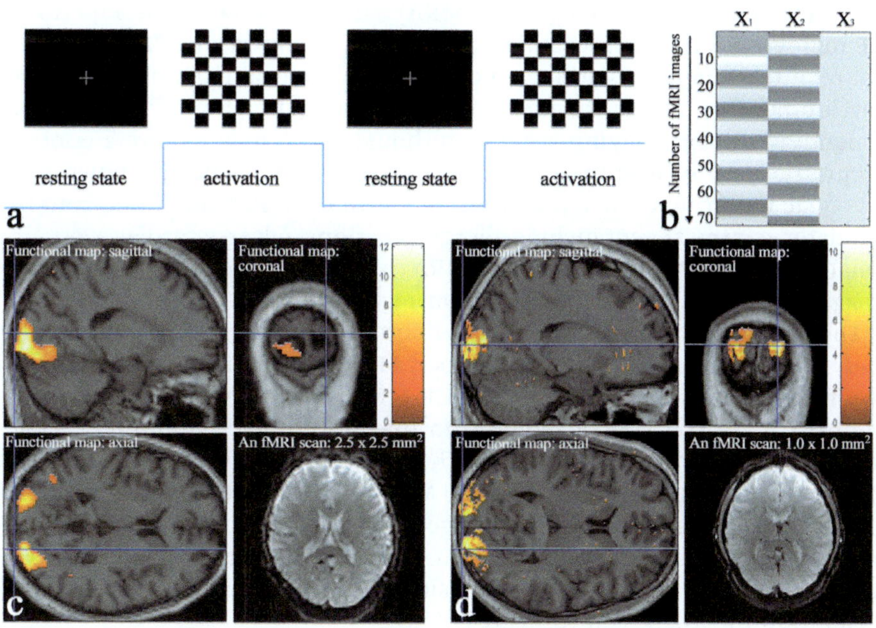

Figure 3.3 Example visual fMRI study at 3 T. (a) A visual checkerboard block-based paradigm; (b) GLM model; (c) and (d) functional maps obtained with a low-resolution technique (2.5×2.5 mm^2) and a high-resolution technique (1.0×1.0 mm^2). The activated voxels were obtained from SPM12 with an uncorrected p-value < 0.001 (t-value > 3.73). Functional maps are shown with three different views (sagittal, coronal and axial) at an MNI coordinate of (18, −104, 8). The bottom right figure of (c) and (d) is a representative fMRI scan obtained from the corresponding imaging sequence.

be observed that the functional activations are distributed quite broadly, and the exact locations of activations are not clearly seen. Recently, there have been numerous attempts to develop high-resolution fMRI techniques. One of those works has achieved a spatial resolution of 1.0×1.0 mm^2 with a whole-brain coverage.[12] An example visual functional map using this technique is exhibited in Figure. 3.3d. In contrast to low-resolution fMRI, the results of high-resolution fMRI reveal that the activation regions are more locally distributed along with the cortical areas.

3.2 Angiography and Perfusion

3.2.1 MR Angiography

Angiography is an application which particularly investigates blood vessels by visualising them more distinctly than other stationary tissues. It is of great importance in clinical diagnosis for the assessment of vascular diseases such as aneurysms, arteriovenous malformation or atherosclerotic diseases.[35] Angiography can be performed using both MRI and computerised tomography (CT). MR angiography (MRA) has been shown to be superior to CT angiography (CTA) in several respects. First, in contrast to CTA, MRA does not require ionising radiation, which carries biological risks such as cancer development or hereditary effects in the future generations.[36] In addition, depending on the imaging scheme, although MRA may require a contrast agent, the contrast agent is gadolinium-based chelate, which is known to be less toxic than the iodinated contrast agent in CTA. Possible risks from the iodinated contrast agent include allergic reactions, adverse cardiac events, or renal toxicity.[37,38] Most importantly, MRA can offer a completely non-invasive imaging method without the use of contrast agents.

The MRA methods are mainly divided into two categories: dark blood MRA and bright blood MRA. The bright blood MRA can be further classified into two types: contrast-enhanced-MRA (CE-MRA) and non-contrast-enhanced-MRA (NCE-MRA). CE-MRA requires an injection of gadolinium-based contrast agents to visualise vessels with brighter signals, whilst NCE-MRA entirely relies on the different physical MR properties between flowing blood and stationary tissues to create image contrasts. Typical NCE-MRA methods include the time-of-flight (TOF) and phase-contrast (PC) methods. The next sub-sections will describe the basic principle of each MRA method and review how the image contrasts are created.

3.2.1.1 Time-of-Flight (TOF) Method

The signal contrasts generated in this method are based on flow-related enhancements. Here, RF pulses are successively applied to a slice of interest with a relatively short TR. As a result, the longitudinal magnetisation in the stationary tissues is not fully recovered and hence, signals reach a steady state at a rather low level. This phenomenon is so-called 'magnetic

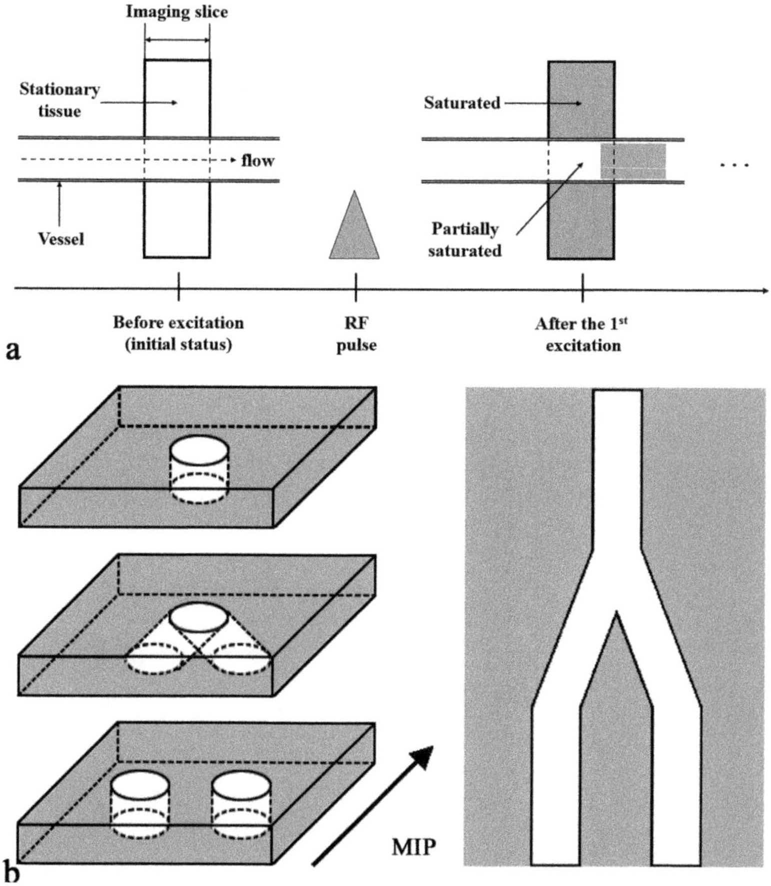

Figure 3.4 Schematic representations of (a) signal enhancement by flow and (b) MIP on the acquired TOF data.

saturation'. However, in contrast to stationary tissues, blood moves through vessels. Therefore, if fresh blood—that has not been saturated by the successive RF pulses—flows into the imaging slice, it produces higher signals than the saturated stationary tissues.

This process is depicted in Figure 3.4a. The figure presents a schematic diagram of the MR signal level that can be achieved by the saturation pulse. As depicted in the figure, the signal level of the stationary tissues drops quickly after the first excitation. However, for the vessels, the signal levels do not immediately go to a saturated value, since fresh blood comes in continuously. Therefore, this flow enhancement is maximised when the imaging slice is perpendicular to the vessel directions. This condition often restricts its use in complex geometries such as the thoracic aorta or renal arteries.[39] For a more effective depiction of the vasculature, the image is reconstructed by means of maximum intensity projection (MIP), rather than being presented with each imaging plane.[35] This procedure is depicted in Figure 3.4b.

3.2.1.2 Phase-contrast (PC) Method

This scheme utilises the fact that when a bipolar encoding gradient (see Figure 3.5) is applied, the phase evolutions experienced by different tissues, stationary tissues *versus* flowing blood, are different. Without loss of generality, when an x-gradient is applied, the phase evolution experienced by the stationary tissue can be written as follows (eqn (3.6))

$$\Theta(t) = k(t) \cdot x_0 \tag{3.6}$$

where Θ, t, k and x_0 denote phase, time, k-space value and distance of the stationary spin from the iso-centre, respectively. Here, it is important to note that the distance does not change with respect to time, since spins do not

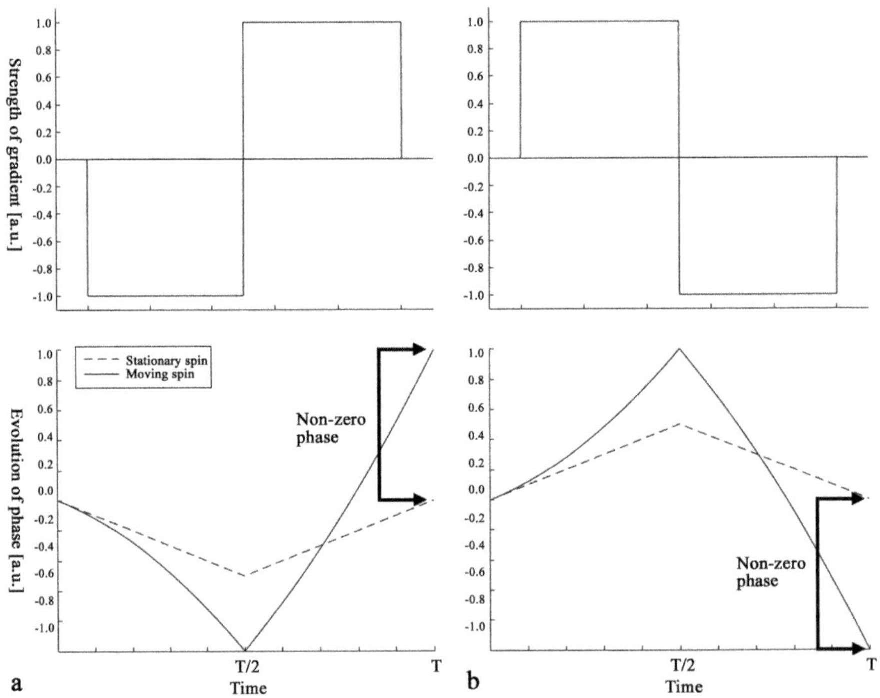

Figure 3.5 Phase evolution by a bipolar gradient where (a) and (b) represents the case of different gradient polarity where the top and bottom rows show the gradient waveform and phase evolution, respectively. In the phase evolution plots, the solid and dotted lines depict the case of moving spins and stationary spins, respectively. It can be seen that subtraction of the two data sets will lead to signal nulling of stationary spins, thereby creating contrast between stationary tissue and flowing blood.

move in the stationary tissues. If the bipolar gradient has a constant amplitude for each respective duration (see Figure 3.5), the equation can be further expressed as follows (eqn (3.7))

$$\Theta(t)=\begin{cases} -\gamma\bullet\left|G_{\text{amp}}\right|\bullet x_0\bullet t & ,0\le t\le T/2 \\ \gamma\bullet\left|G_{\text{amp}}\right|\bullet x_0\bullet(t-T) & ,T/2\le t\le T \end{cases} \tag{3.7}$$

where, γ, G_{amp} and T indicate gyromagnetic ratio, gradient amplitude and total gradient duration respectively. Therefore, for the stationary tissues, the phase changes linearly with time and the net phase gain after the end of the bipolar gradient is zero.

In contrast, for the moving spins, since their distances change as time elapses, the equation can be written as follows (eqn (3.8))

$$\Theta(t)=k(t)\bullet x(t) \tag{3.8}$$

Here, if the spins have a constant velocity, υ, the above equation can be further represented as follows (eqn (3.9))

$$\Theta(t)=\begin{cases} -\gamma\bullet\left|G_{\text{amp}}\right|\bullet(x_0+\upsilon t)\bullet t & ,0\le t\le T/2 \\ \gamma\bullet\left|G_{\text{amp}}\right|\bullet\left[(x_0+\upsilon t)\bullet t-(x_0+\upsilon T/2)\bullet T\right] & ,T/2\le t\le T \end{cases} \tag{3.9}$$

In this case, the net phase gain at the end of the gradient has a non-zero value. As indicated by the equation, the magnitude of the non-zero value is determined by the gradient amplitude, gradient duration and velocity of the moving spin. The first two parameters can be varied and are operator dependent and this can be used to regulate the degree of the sensitisation to the moving spins.

Signal contrasts can be created by acquiring two data sets, each of which has an opposite gradient polarity from the other (see Figure 3.5). For the stationary spins, since the net phase is always zero regardless of the polarity of the gradient, in ideal situations (*e.g.* no motion), the two data sets would yield exactly the same results. However, for the moving spins, as the non-zero net phase also depends on the polarity of the given gradient, a subtraction of the two data sets creates signal contrast by nulling the signal from stationary tissue and emphasising that from moving spins.

3.2.1.3 CE-MRA

Differing from the above two methods, CE-MRA employs an injection of contrast agent to create contrast for vessels. The contrast agent, based on gadolinium chelates, has an effect on the T_1 relaxation time of the blood which is decreased by the presence of the agent. As a consequence, blood is imaged with higher signal values than stationary tissues[40] because, through the rapid application of RF pulses, stationary tissue with its longer T_1 is preferentially saturated. Here, the signals of stationary tissues can even be further subtracted out using the reference image that is obtained without the use of contrast agent injection.[41]

For successful imaging of the vasculature, it is important to know when the contrast agent bolus arrives in the region-of-interest. If the scanning is performed too early or too late, the effect of the contrast agent may not be fully appreciated, which results in poor visualisation of the vasculature. Since the transit time of the contrast agent bolus from the infusion point to the region-of-interest changes depending on the medical condition of the patient (*e.g.* heart rate, cardiac output or age), an accurate estimation of the transit time is required for a more precise visualisation of the vasculature.[42]

The MR scanning is typically performed with a 3D T_1-weighted gradient-echo sequence with a relatively short TR (5–6 ms) and TE (1–2 ms).[40] The total scan time has a range from 10–30 s, indicating that reduced motion artefacts are expected when compared to NCE-MRA methods.[40,43] Moreover, CE-MRA offers higher SNR and thus, can be oriented and reformatted in any desired plane.[40] The method is further free from the concerns of the NCE-methods such as in-plane saturation or turbulence-induced signal loss.[40]

3.2.2 MR Perfusion

Similar to MRA, MR perfusion is another MR application related to blood flow. However, MR perfusion focuses on the measurement of the circulation of nutrients at the capillary level, rather than visualisation of relatively large vessels, as is the case in MRA.[44] Perfusion MRI can be also used for clinical diagnosis and the management of brain diseases (*e.g.* acute stroke, dementia or tumours) by quantifying physiological parameters.[44] The quantified parameters are different depending on the perfusion imaging method. The methods are mainly divided into two groups: (1) exogeneous methods with dynamic susceptibility contrast MRI (DSC-MRI) or dynamic contrast enhanced MRI (DCE-MRI) and (2) endogeneous methods with arterial spin labelling (ASL). The following sub-sections will review the basic features of each method and describe how the corresponding physiological parameters can be quantified.

3.2.2.1 DSC-MRI

This method relies on the decrease of T_2 or T_2^*, caused by a paramagnetic contrast agent. The corresponding signal decreases can be recorded by spin-echo (T_2-weighted) or gradient-echo (T_2^*-weighted) acquisitions, each of which has its own distinct advantages in terms of sensitivity to capillary size vessels and relatively higher signal-to-noise ratio, respectively.[44,45] In order to track the first pass of the contrast agent, DSC-MRI requires a dynamic acquisition with a relatively high temporal resolution. For this

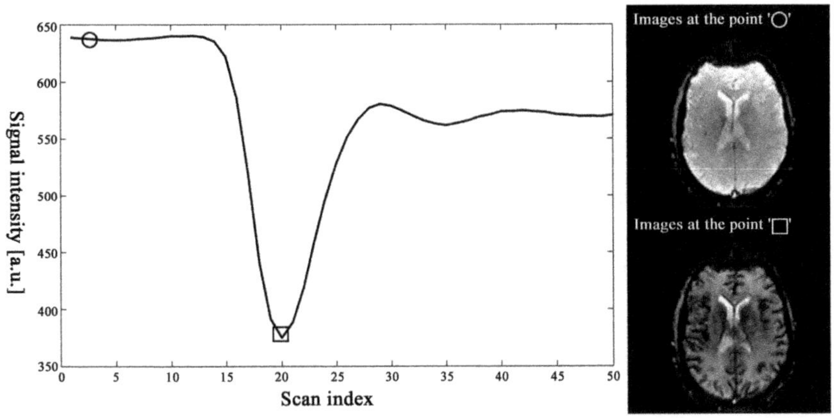

Figure 3.6 Evolution of signal intensity with respect to time in DSC-MRI. The left plot shows how signal intensities change as time elapses. The top and bottom rows on the right-side show images extracted from the time points indicated by 'o' (during the baseline phase) and '□' (during the first passage phase), respectively.

reason, EPI has been widely used in DSC-MRI. Figure 3.6 depicts an example curve obtained from gradient-echo EPI time series data, which shows typical haemodynamic signal changes. The curve mainly consists of three phases: baseline, first passage and recirculation. During the first phase, the signal stays nearly constant since the bolus of contrast has not yet arrived. In the next phase, as the bolus of contrast passes through, the signal starts to quickly drop until it reaches a minimum. Thereafter, the signal level starts to rise but it does not go up to its original baseline due to the recirculation effect. This effect is mainly generated by the second wave of gadolinium passage originating from recirculation of the gadolinium distributed to other organs (*e.g.* thyroid, kidneys and lymph nodes) when the injection was performed.[46] As shown, the recirculation effect also creates an additional small signal drop during the last phase.

Based on the obtained time–course plot, several semi-quantitative parameters can be estimated, such as cerebral blood flow (CBF), cerebral blood volume (CBV), mean transit time (MTT), time to peak (TTP), bolus arrival time (BAT), maximum peak (MP) and so on. In order to quantify CBF, deconvolution of the following equation (eqn (3.10)) is required[46,47]

$$C(t) = \mathrm{CBF} \cdot AIF(\mathrm{t}) \otimes R(t) \tag{3.10}$$

where $C(t)$ is the concentration time course (CTC) of contrast agent measured in the tissue, $AIF(t)$ is the arterial CTC and $R(t)$ is the tissue residue function which describes retention of contrast agent within the capillary bed. The formula suggests that for quantifying CBF with deconvolution, estimation of $C(t)$ and $AIF(t)$ should be performed.[45] $C(t)$ can be calculated using the obtained signal intensity plot. The formula is given as follows (eqn (3.11))

$$C(t) = -1/\mathrm{TE} \cdot \ln\left(S(t)/S_0\right) \tag{3.11}$$

where $S(t)$ and S_0 denote the signal intensity of the given time index and the initial time point, respectively. In contrast, CBV can be determined relatively easily. For an intact blood brain barrier, CBV is proportional to the area of $C(t)$ normalised by the area of $AIF(t)$.[47] Once CBF and CBV are determined, MTT can be calculated by using the central volume theorem: MTT = CBV/CBF.[48] Other parameters such as TTP, BAT or MP can be obtained without consideration of the physiological relationship above. These are also known as 'summary parameters'.[44,47]

3.2.2.2 DCE-MRI

DCE-MRI resembles DSC-MRI in terms of employing contrast agents to derive haemodynamic signals but differs from it in that DCE-MRI relies on the shortening of T_1 relaxation time (not T_2 or T_2^* shortening). The corresponding signal increases are measured by a dynamic T_1-weighted imaging sequence. Similar to DSC-MRI, time-course data need to be acquired before, during and after administration of a contrast agent.

After converting the initial time-course data into gadolinium concentrations, a pharmacokinetic model is applied to perform data fitting. One widely used model is the Tofts-Kermode Model from which several principle parameters can be derived, such as transfer constant (K^{trans}), rate constant (k_{ep}), fractional volume of the extravascular extracellular space (v_{e}) and fractional volume of the plasma space (v_{p}). Here, the first three parameters have the following relationship: $k_{\text{ep}} = K^{\text{trans}}/v_{\text{e}}$.[49,50] The transfer constant (K^{trans}) is one of the most important parameters in DCE-MRI. It represents the influx volume transfer constant of gadolinium from blood plasma to extravascular extracellular space (EES), which depends on the balance between capillary permeability and blood flow in the tissue.[50] The rate constant (k_{ep}) is the reflux rate of the gadolinium from the EES back to the blood plasma. A schematic representation of the exchange of gadolinium is depicted in Figure 3.7a.

3.2.2.3 Arterial Spin Labelling (ASL)

ASL can offer a completely non-invasive way to measure cerebral blood flow. Although this is a unique advantage over the two exogeneous agent methods described above, its SNR and CNR are relatively low when compared to exogeneous agent methods.[46,51] In addition, ASL requires a substantially prolonged scanning time, which limits its utility for imaging patients (*e.g.* acute stroke or neurodegenerative disease) who have difficulties in controlling their movements.[51] Nevertheless, its property of non-invasiveness allows one

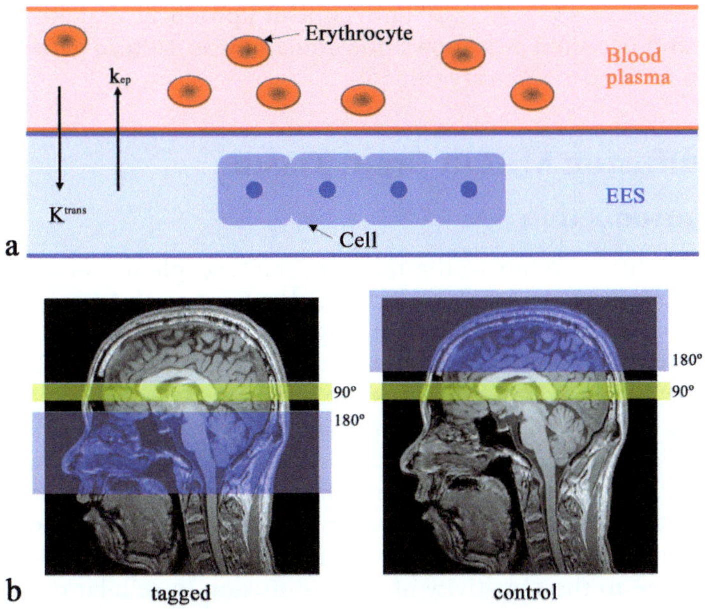

Figure 3.7 Schematic diagrams representing (a) exchange of gadolinium between the EES and blood plasma based on the Tofts-Kermode Model in DCE-MRI and (b) the imaging scheme of EPISTAR where the left and right figure show how the saturation/inversion pulses are applied in the 'tagged' and 'control' imaging, respectively.

to perform unique applications, such as functional ASL,[52] repeated perfusion measurements,[53] pharmacological applications—which study various drugs without any effect of additional compounds[54] or paediatric applications[51]—where intravenous access is difficult.

There are a number of variants for ASL imaging techniques. These are, in general, composed of two acquisitions: one with magnetic labelling of arterial water spins (tagged) and the other with no labelling (control). ASL was first demonstrated with a continuous ASL (CASL) method.[55] However, the use of relatively long, continuous RF excitation (2–4 s)[54] for the spin labelling is coupled with high tissue energy deposition, which may restrict its use for *in-vivo* subjects. In order to overcome this issue, pulsed ASL (PASL) methods have been presented.[56–58] The concept was first demonstrated with 'EPISTAR'.[56] Its acquisition starts with a 90° pulse to initially saturate the selected imaging plane (see Figure 3.7b). Next, an adiabatic 180° pulse with a relatively large thickness is applied below the slice of interest to invert inflowing spins in the blood (tagged). After the preparation module (*i.e.* saturation and inversion), a standard EPI acquisition starts to acquire data. The 'control' data are also acquired with the same procedure; however, the inversion pulse is applied above the slice of interest to minimise the magnetisation transfer effect.[54] As the inverted inflowing spins induce reduced signals, a subtraction of the two data sets (*i.e.* 'tagged' and 'control') produces regions where signal reduction occurred and shows how much it was. Here, a general kinetic model can be used to quantify CBF.[59]

3.3 Diffusion MRI in Brain Tissue

3.3.1 Introduction

Molecular diffusion refers to the random, microscopic movement of water and other small molecules due to thermal collisions and is ubiquitous in nature. The introduction of diffusion magnetic resonance imaging (dMRI)[60–62] in the early 1980s and its application to the *in-vivo* human brain[63] has opened new unprecedented horizons in neurosciences. During the typical observation times in dMRI experiments, molecules traverse distances of about 5–10 μm while being subjected to various restrictions, hindrances, and interactions caused by cell membranes, myelin sheaths, intra-cellular organelles, macromolecules, *etc.* As such, water molecules 'probe' the local microstructure on the length scale well beyond the typical voxel-size ($2 \times 2 \times 2$ mm^3). Due to the sensitivity of water diffusion to cellular organisation, dMRI has evolved to become a powerful modality for non-invasive brain visualisation[64,65] both in fundamental research and in clinical practice with unique applications in diagnostics of acute stroke,[66] tumours[67] and various neurological disorders.[68] dMRI also allows the non-invasive examination of global white matter (WM) organisation over the whole brain size (~20 cm), as well as the correlations between brain connectivity and function. Important goals of brain dMRI include understanding the biophysical background of the diffusion-weighted (DW) signal and the development of biomarkers of healthy and pathological tissue states. This subsection describes the basic principles of dMRI in brain tissue and provides an overview of some selected applications.

3.3.2 Diffusion Basics and Methods

Molecular diffusion, also known as Brownian motion, was discovered in 1827 by the Scottish botanist Robert Brown[69] and described in 1855 by German physiologist Adolf Fick in the form of two differential equations known as Fick's first and second laws of diffusion.[70] Half a century later, in 1905, Albert Einstein explained Brownian motion based on the molecular-kinetic theory of heat.[71] He described it as a stochastic random process in terms of the diffusion equation for a conditional probability that a particle initially located at a certain position in space will move to another position during the time interval t. For unrestricted diffusion in isotropic media, the solution of the diffusion equation for the ensemble average propagator is given by the Gaussian function (hence Gaussian diffusion) whose second moment, $\langle r^2 \rangle = 6Dt$, is known as the Einstein equation for diffusion and provides a fundamental relation between the mean square displacement, $<r^2>$, and time linked *via* a macroscopic quantity, the diffusion coefficient, D. Thus, the statistical diffusion process involving many elementary pathways can be quantified macroscopically by means of D. In isotropic bulk liquid, D is determined only by the molecular size, temperature and viscosity of the medium.

A typical dMRI experiment involves measuring DW images with the help of diffusion-sensitising magnetic field gradients combined with a suitable spatial encoding scheme as shown, for example, in Figure 3.8. In MRI scanners, the strength of diffusion weighting is usually measured in terms of the so-called b-factor,[63] which is an adjustable parameter (the larger b, the stronger the attenuation of the signal amplitude due to diffusion). In the simplest case of Gaussian diffusion, the DW signal represents a monoexponential function of b such that $S(b) = S_0 \exp(-bD)$, where S_0 is the signal for $b = 0$. Figure 3.8 shows typical DW images of the human brain, demonstrating contrast between the three major types of tissue: WM, grey matter (GM) and cerebrospinal fluid (CSF).

3.3.3 Water Diffusion in Brain Tissue and Diffusion Tensor Imaging

Neuronal tissue is characterised by an enormous complexity on various length scales, Figure 3.9. Therefore, it is impossible to infer a complete microscopic picture of water diffusion in such an environment. However, two main features play key roles for the diffusion mechanisms of water molecules in the brain, Figure 3.9 (bottom): (a) diffusion is restricted and/or hindered on the microscopic length scale due to the cellular organisation and (b) diffusion is globally anisotropic on the macroscopic length scale. Since motional barriers hinder the random-walk-like mechanism inherent to free bulk water, the mean square displacements and, consequently, the measured diffusion coefficient appear reduced by two to ten times (compared to free water). Therefore, it is frequently referred to as the apparent diffusion coefficient (ADC).[63,72,73] The ADC in brain tissue retains valuable integrative information about the displacements of water molecules in the tissue.

Anisotropic and coherent ordering of axons on length scales exceeding the voxel size is an essential feature of WM tissue organised in multiple fibre bundles.[74] This gives rise to anisotropic diffusion of water molecules, which appears faster along the axonal main axis than perpendicular to it. The sources of anisotropy include, among others, axonal membranes, myelin sheaths, and intra-axonal microtubules.[75] The simplest mathematical framework describing anisotropic diffusion in the brain was suggested by P. Basser *et al.*[76,77] in 1994 and gave a boost to the technique called diffusion tensor imaging (DTI).[78–80] During the following decades, DTI revolutionised non-invasive visualisation of WM in the *in-vivo* human brain and has been associated with a tremendous success.[81–86]

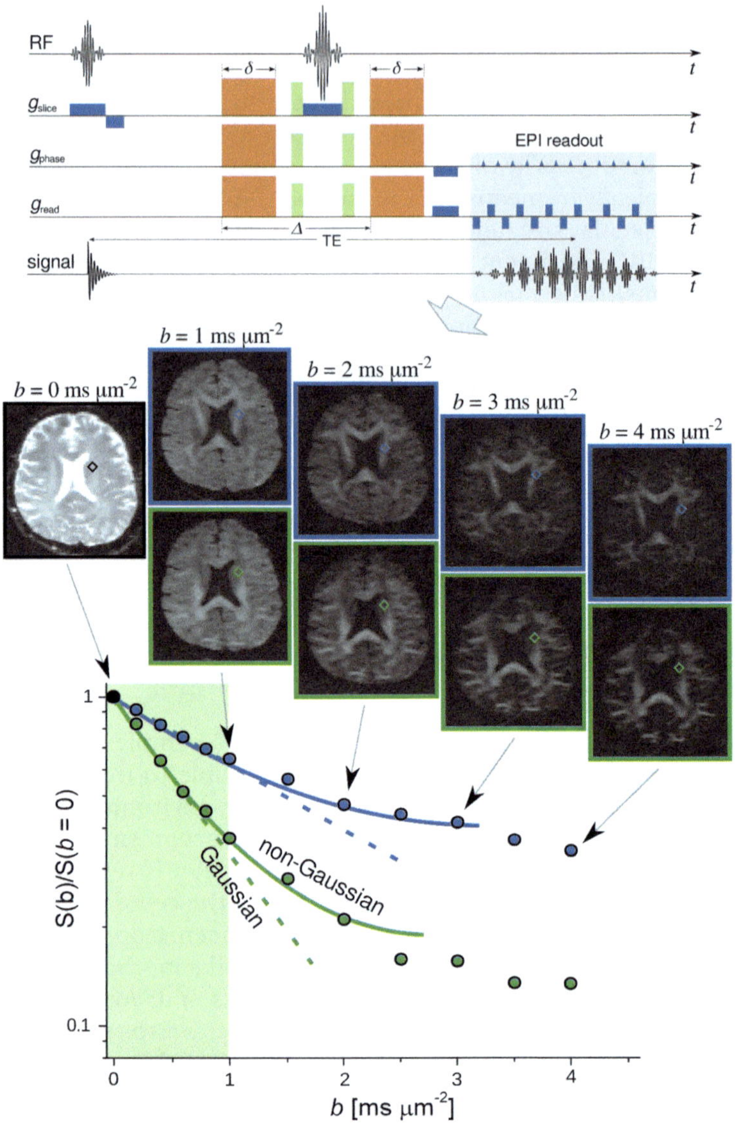

Figure 3.8 A schematic representation of the two-dimensional spin-echo EPI pulse sequence with diffusion weighting field gradient pulses (orange) (top). The imaging and crusher gradients along the three axes are shown in blue and light green, respectively. The EPI readout is shown in the grey box. After excitation, all spins ideally have the same phase. The effect of the first field gradient pulse is to dephase spins, *i.e.* spread their Larmor frequencies. Following the 180° refocusing pulse, a second field gradient pulse is applied to restore spin phases. If the spin positions along the field-gradient direction at the echo-time differ from the initial positions, a net attenuation of the spin-echo amplitude will be observed as a function of the *b*-factor. Examples of DW images of the human brain for five increasing *b*-factors and two directions of the field gradients (middle). Normalised DW signal as a function of the *b*-factor for a selected voxel in WM for the same gradient directions (bottom). Dashed lines and solid curves represent monoexponential (*i.e.* predicted by the Gaussian approach) and non-monoexponential (non-Gaussian) fits, respectively. The valid *b*-factor range for DTI is shown in light green.

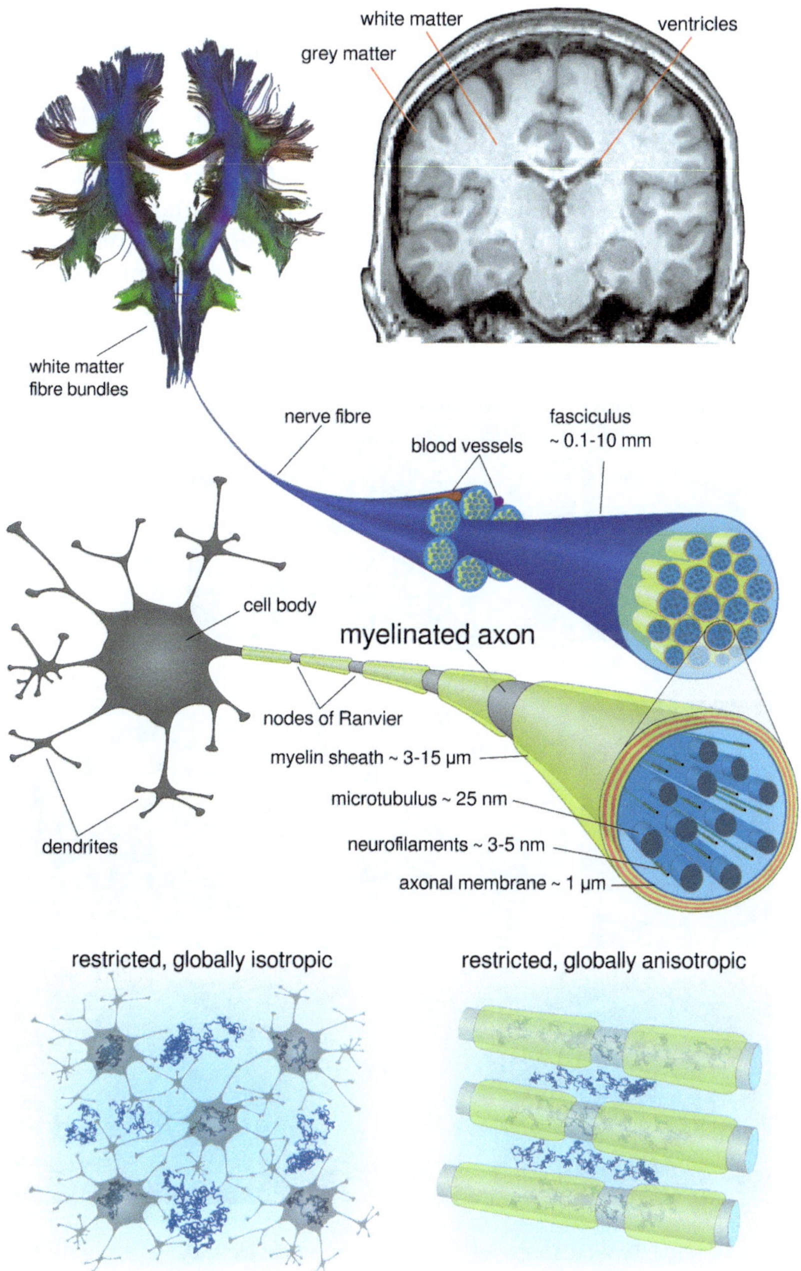

Figure 3.9 A schematic representation of the complexity of the central nervous system on various length scales: a coronal longitudinal-relaxation-weighted image showing GM, WM and CSF regions (top-right); WM fibre tracts (generated with the help of the ExploreDTI toolkit.[153]) (top-left); a geometrical description of a neuron and of a nerve fibre (middle); a schematic representation of restricted, globally isotropic and globally anisotropic diffusion in brain tissue (bottom).

Based on the tensor eigenvalues, several rotational invariants[87] can be defined and used to produce parameter maps, providing information on water mobility and directionality in space. Typical maps of mean diffusivity (MD) (the mean of tensor eigenvalues), fractional anisotropy (FA) (a normalised variance of the eigenvalues), and (RGB) colour-coded FA can be seen in Figure 3.10a. DTI metrics are sensitive biomarkers of various neurodegenerative diseases such as multiple sclerosis,[88] epilepsy,[89] Alzheimer's disease[90,91] and various other disorders (schizophrenia, dementia).[92–94] Among the clinical applications of DTI, the most prominent example is its use in the early diagnosis and assessment of stroke.[66,95] Ischaemic lesions

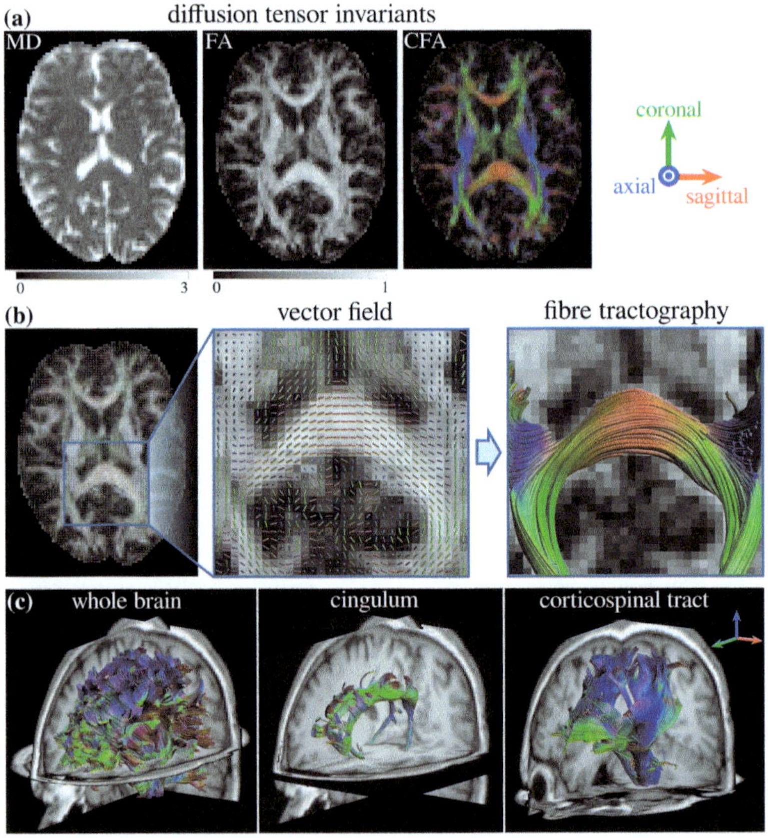

Figure 3.10 (a) Typical maps of DTI metrics, such as MD, FA and colour-coded FA (CFA), in *in-vivo* human brain in a slice perpendicular to the axial direction. (b) The RGB-colour-coded field of the eigenvector (linked to the tensor largest eigenvalue) overlaid on the FA map. In oriented fibre bundles, eigenvectors are assumed to be parallel to the main fibre direction, thus allowing for the fibre reconstruction shown on the right. (c) Examples of DTI-based streamline fibre tractography: whole brain (left), cingulum (middle), and corticospinal tract (right). Maps, vector fields, and fibre tracts were generated with the help of the ExploreDTI toolkit.[153]

tend to manifest themselves within the first 30 min after the onset of infarction in the DW images prior to their visualisation on other conventional MR images. Diffusion changes will occur in the event of a failure of the sodium/potassium pump, giving rise to an inter-compartmental water shift and cell swelling (cytotoxic oedema).[66,95] DTI has also greatly improved our understanding of the link between microstructure and function in healthy development and ageing: cognitive changes in early maturation have been associated with changes in brain integrity and circuitry,[96,97] whereas the decline of microstructural properties in later life has been correlated with cognitive changes/deficits such as processing speed, memory, and executive functioning.[98,99]

A unique feature of dMRI, in general, and DTI, in particular, is that it allows one to access the patterns of WM global connections in the *in-vivo* brain.[74,82,100] Due to the global orientational ordering of axonal fibres across the brain, diffusion of water appears anisotropic on the macroscopic length scales and enables reconstruction of fibre pathways and global anatomical brain connectivity.[101–104] This technique, *i.e.* virtual *in-vivo* reconstruction of the brain circuitry, Figure 3.10b, is known as 'fibre tractography'.[86,105] Examples of some fibre tracts are shown in Figure 3.10. Recently, fibre tractography has advanced rapidly, although in parallel with the recognition of its limitations[106] and is currently emerging as a new field of its own, referred to as 'human connectomics'.[107–109]

In spite of its success though, DTI suffers from intrinsic limitations. First, because of the underlying model of Gaussian diffusion, the applicability of DTI is restricted to a range of relatively low *b*-factors, see Figure 3.8. Second, DTI fails to properly capture diffusional anisotropy and fibre orientations in regions with complex fibre configurations (crossing, kissing, bending, and twisting), where FA appears artificially 'reduced' and can be erroneously interpreted as microstructural decline.

3.3.4 Non-Gaussian Diffusion Methods

Beyond the range of diffusion weightings valid for DTI, Figure 3.8, clear deviations from the mono-exponential behaviour have been reported.[110–113] Although the underlying mechanisms are not well understood,[73] numerous methods[112–126] have been proposed to describe the patterns of non-Gaussian diffusion in brain tissue and to quantify diffusion non-Gaussianity. These include biexponential fitting,[112,113,120–122] diffusion kurtosis imaging (DKI),[114,115] stretched-exponential function,[118,123,124] CHARMED,[119] AxCaliber[127] or NODDI.[128] Some of the suggested methods are empirical in nature, such as DKI,[114,115] allowing one to describe the dMRI signal using phenomenological parameters, such as mean kurtosis (MK). Other approaches, such as CHARMED[119] or NODDI,[128] use biophysical modelling in order to estimate important microstructural parameters (axonal water fraction, axonal radii, neurite orientation dispersion, *etc.*) directly from experimental data. The majority of the suggested models comprise the so-called multi-compartment models (see ref. 129 for an overview), which assume that water is located in spatially distinct compartments (for example, intra-axonal and extra-axonal spaces), whereas an alternative interpretation[130] proposes the existence of two differently structured water pools.

Non-Gaussian methods provide a more accurate approximation of the DW signal than conventional DTI in the moderate to high *b*-value ranges[131] and enrich information related to brain tissue microstructure.[132,133] Regarding clinical applications, these methods are still at an early stage, which is mainly due to the severe decrease in the SNR with increasing *b*-factors, leading to impractically long acquisition times. Recently, a significant breakthrough was achieved with DKI that demonstrated promising biomarkers in development,[134,135] healthy aging,[136,137] stroke,[116,138–141] neurodegenerative diseases, such as Parkinson's,[142] Huntington's,[143] and Alzheimer's diseases,[144] and in glioma grading.[145,146] As an example, Figure 3.11 demonstrates the greatly enhanced contrast of stroke lesions in MK maps compared to the

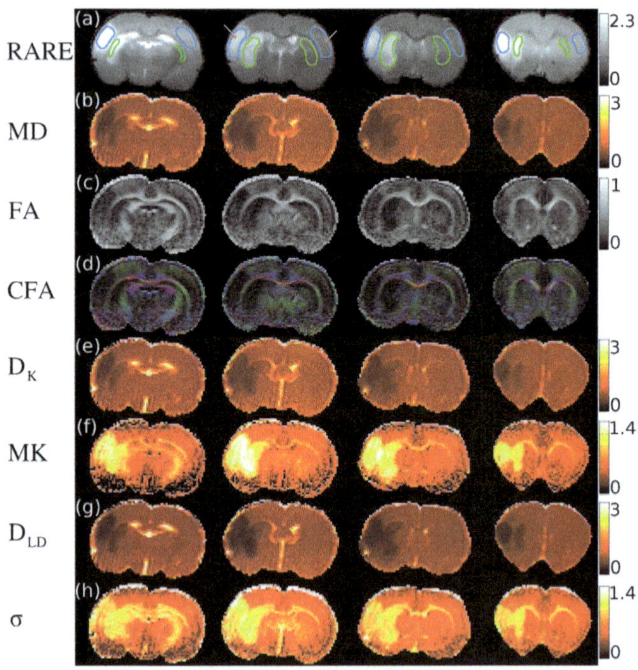

Figure 3.11 (a) Anatomical transverse-relaxation-weighted images (RARE) and the maps of (b) MD, (c) FA, (d) CFA, (e) D_K (diffusion coefficient evaluated in DKI), (f) MK, (g) D_{LD} (diffusion coefficient evaluated in 'log-normal distribution function imaging') and (h) σ (the width of the diffusion distribution function) of the rat brain affected by stroke. MD, D_K and D_{LD} are decreased in the ischaemic region, as expected, but both MK and σ are strongly enhanced, providing greater contrast. The numbers on the scale bars that refer to MD, D_K and D_{LD} are to be multiplied by 10^{-3} mm^2 s^{-1}. Reproduced from ref. 139 with permission from John Wiley and Sons, Copyright © 2012 John Wiley & Sons, Ltd.

gold standard MD. However, the field has not been sufficiently investigated and challenging efforts are required in order to increase the applicability of non-Gaussian methods in the clinical environment. Fortunately, the state-of-the-art ultra-high magnetic field[147] and ultra-high magnetic field gradients,[148,149] combined with cutting-edge methodological approaches and large-scale efforts of international consortia,[150–152] promise to achieve a significant breakthrough in our understanding of the human connectome in the foreseeable future.

3.3.5 Conclusions

Due to the sensitivity of molecular propagation to the local geometrical and physiological environment, dMRI has established itself as a powerful quantitative method, providing access to the sub-voxel microstructural properties of tissue and its global macroscopic organisation. It has contributed to the development of novel diffusion-based neuromarkers and has improved our understanding of relationships between brain anatomy and function. The recent development of a variety of non-Gaussian methods has helped to enhance the information obtained from DW data and to provide new tools in clinical diagnostics. Further elucidation of the human connectome has evolved to become a challenging aim for large-scale international studies and has opened up great opportunities for the future of neuroscience.

3.4 Quantitative Imaging

MRI is based on one of the most powerful analytical techniques (NMR) and as such is fully capable of measuring different tissue parameters quantitatively.

However, MRI has been so successful in producing high contrast images of the brain that its quantitative aspect has been largely neglected for quite some time. This is especially so in clinical settings where scan time is at a premium.

The one notable exception over the years had been diffusion imaging (see Section 3.3), mainly due to its power in characterising areas affected by stroke.

Fortunately, in recent years, there has been an increased interest in using quantitative imaging, ignited by the increased availability of high-field systems providing the high SNR most quantitative techniques demand. In addition, an increase in the number of large cohort studies worldwide and the need to characterise them by centre-independent measures has demonstrated the advantage of using quantitative, as opposed to mixed-contrast, *qualitative* imaging.

The general scheme of quantitative imaging is to acquire a number of images that are solely, or largely, weighted by a given parameter. Using an appropriate model, which is parameter dependent, fitting can be carried out on a voxel-by-voxel basis to derive the desired parameter. The process becomes somewhat more complicated when accuracy and precision need to be taken into account, which is necessary for quantitative methods to be at all useful. A number of corrections are usually required to be applied to the images before fitting, or included in the fit model, in order to achieve the desired precision and accuracy. These correction terms can include such basic attributes as the flip angle;[154] B_0 homogeneity;[155] inversion pulse efficiency;[156] temperature;[157,158] receive B_1 inhomogeneities[159] *etc.*, or more complicated aspects such as effects of unspoiled magnetisation[160] or deriving many parameters simultaneously as in fingerprinting.[161] In the end, although much of quantitative imaging methodology is straightforward, including the measurements for the required corrections can make the acquisition time prohibitive, at least in a clinical environment. In general, progress in quantitative imaging revolves around (a) accelerated imaging methods without sacrificing SNR; (b) minimising the number of correction measurements; and (c) appropriate models that are amenable to rapid and robust fitting. There are a considerable number of quantitative parameters that have been investigated in the MRI literature. In the following, emphasis is placed upon the most basic and ubiquitous of these parameters which influence contrast in all sequences: magnetisation density/water content, longitudinal relaxation T_1, transverse relaxation T_2 and effective transverse relaxation T_2^*.

A number of methods for mapping of tissue relaxation parameters and water content have been published.[156–158,162–171] An important characteristic of relaxation times *in-vivo* is that they depend on field strength,[172,173] whereas water content—as a true physiological parameter—is field independent.

3.4.1 T_2 and T_2^* Mapping

The transverse relaxation time T_2 is one of the two principal relaxation times in MRI. Its quantification—the technique of T_2 relaxometry—provides important information about the mobility and chemical environment of the tissue of interest.

T_2 is changed in a number of pathologies. Most notably, it is increased in stroke, brain tumour accompanying oedema, and multiple sclerosis (MS) lesions.

The associated parameter T_2^* describes the effective transverse relaxation and includes the effect of magnetic field inhomogeneities; the effect of magnetic field inhomogeneities can be, to a very large extent, refocused by the use of spin echoes.

The theoretical basis of T_2 relaxometry techniques is the classical Hahn spin echo induced by 90° (excitation) and 180° (refocusing) RF pulses. The T_2 decay constant can be measured by sequentially acquiring a number of spin-echo images each with a different echo time, TE. However, the long duration of such an acquisition makes it impractical for most clinical applications. Furthermore, the use of long echo times makes the signal susceptible to an additional dephasing in internal field gradients mediated by diffusion. The alternative multiple spin echo approach uses multiple refocusing pulses in each pass so that the acquisition length is shortened. The effect of diffusion is minimised by this acquisition scheme, but the quantitative accuracy of the sequence is, however, compromised by imperfect refocusing. The Carr-Purcell-Meiboom-Gill (CPMG) sequence[174,175] is an enhancement of this approach in which the phase of the RF pulses is modulated to compensate for small deviations in the refocusing efficiency. However, this implementation has still been found to be insufficient for accurate multi-slice T_2 determination[176] due to the residual sensitivity to refocusing imperfections. For multi-slice imaging, which relies on slice selective pulses, this is exacerbated as a result of non-uniformities in the refocusing flip angle across the slice, especially at the slice edges. This results in the formation of stimulated (secondary) echoes that disrupt the T_2 decay of the spin (primary) echoes.

T_2^* relaxometry has multiple important applications in MRI, including blood oxygenation level dependent functional imaging,[177] detection and tracking of super-paramagnetic iron oxides[178] and assessment of iron content in brain.[179] T_2^* maps can be obtained from rapid acquisitions, such as spoiled gradient-echo multi-echo sequences. At high fields, the effect of microscopic field inhomogeneities, such as those created by myelinated fibres, gives rise to exquisite local T_2^* contrast. At lower fields and lower resolution, the T_2^* contrast between healthy grey and white matter is small.

T_2^* mapping is affected by several confounding factors, most importantly by macroscopic B_0 field variations and therefore only partly characterises the properties of the tissue itself. These effects can be corrected to a large extent at excitation[160,180] or included in the signal equation.[155,181] The signal obtained with both T_2 and T_2^* multi-echo mapping techniques can be used to investigate the presence of different water pools in tissue by performing a multi-component analysis.[182,183]

3.4.2 T_1 Mapping

The longitudinal relaxation time (T_1) has a particular role in the investigation of brain tissue, because of an increased sensitivity in pathology detection,[184] such as brain cancer, demyelinating diseases, ischemia, hepatic encephalopathy or chronic alcoholism.[157,158,165,166] Moreover, quantitative T_1 maps can help clinicians to evaluate contrast agent uptake, iron overload, and blood perfusion and volume.[167] They also provide a more robust template for morphometry studies and a more specific marker of disease progression in comparison to conventional T_1 weighted images.[168]

The gold standard method for T_1 mapping makes use of inversion-recovery spin-echo (IR-SE) sequence.[169,170] However, IR-SE requires an extremely long scan time, which is impractical for clinical use and can lead to inaccurate measurements due to patient motion. Many efforts have been made to shorten the T_1 measurement time. We mention here two extremes in terms of relaxation curve sampling: Look-Locker methods such as TAPIR[156] or IR-EPI,[185] which typically describe the signal evolution with many points; and two-point methods such as MP2RAGE[186] or based on the steady state equation.[158,169,171] The utility of mapping T_1 becomes evident when one notes that the high T_1 contrast between grey and white matter in the brain has been explained as being due to the presence of cholesterol in myelin.[187,188]

3.4.3 Water Content

Water content is a conceptually simple quantity, highly relevant to biological systems, which can be measured by MRI. Water concentration is highly regulated in the healthy human brain and changes only slightly with age and sex.[157,166,189] Several pathologies such as alcoholism, haemodialysis, stroke, tumour, hepatic encephalopathy and multiple sclerosis have been shown to lead to either local or global disturbances in water distribution and content.[157,190–194] A remarkable correlation between water content measured by non-NMR methods and T_1 in biological tissue was noticed early on[195–197] and has been used by several groups to derive surrogate water content values based on T_1.[198–202] However, this is only an inferred quantity and changes in nuclear magnetic resonance (NMR) relaxation times represent a composite of both water content and water structuring in the tissue under study.[203]

Whereas changes in water content are highly relevant for the characterisation of disease, these changes are usually in the low percentage range. Any useful method for water content measurement must cope with this challenge of very high accuracy and precision.

Here, we present a method which is conceptually simple and easy to implement, both on the scanner and in postprocessing. It is based on a long TR (10 s) multiple-echo gradient-echo (GRE) measurement with ~7 min acquisition time.[164] At TR = 10 s, the signal intensity in brain tissue is determined largely by M_0, the proton density of free water; signal from macromolecules and other fast-relaxing species is difficult to capture. For $T_1 < 2$ s and TR = 10 s (resulting in TR/$T_1 > 5$), the steady state correction is negligible for any flip angle, such that the maximum SNR for $\alpha = 90°$ can be obtained. The signal equation (eqn (3.12)) is therefore

$$M = M_0 \sin\alpha_{\text{eff}} \exp\left(-\text{TE}/T_2^{*}\right) B_1^{-} \tag{3.12}$$

with $\alpha_{\text{eff}} = \alpha_{\text{nom}} B_1^{+}$ the effective and nominal flip angle; B_1^{+} and B_1^{-} are corrections reflecting deviations of the transmit and the receive field from a homogeneous distribution of the nominal flip angle and from unity, respectively.

Since α_{eff} is close to 90°, the intensity inhomogeneity due to B_1^{+} is minimised ($dM/d\alpha \sim 0$). Fitting the exponential decay characterised by the relaxation time constant T_2^{*} allows one to extrapolate the signal to TE = 0.

Compared to other variants of water mapping, this method has the advantage that all B_1 corrections, which are noticeable even at 1.5 T and become large at fields of 3 T and above, are multiplicative. They can be approached heuristically, using a bias field correction, as implemented in several software packages used in neuroscientific applications. Water content can be derived by normalisation to the CSF signal from the lateral ventricles, considered to consist of 100% water. Given the long T_1 of CSF, of ~4.3 s,[172,204,205] its signal is reduced to ~0.94 of the maximum.

Figure 3.12 shows water content maps obtained from the same healthy volunteer at 1.5 T and 3 T and its histogram over the whole brain tissue. A clear separation between the distribution in WM (~68%) and that in GM (~82%) is visible. Additionally, the beneficial effect of using an acquisition with high SNR is apparent in the lack of influence of increasing field strength on the water content distribution. This indicates that the influence of noise on this parameter is already small at 1.5 T and water content values can be compared across fields.

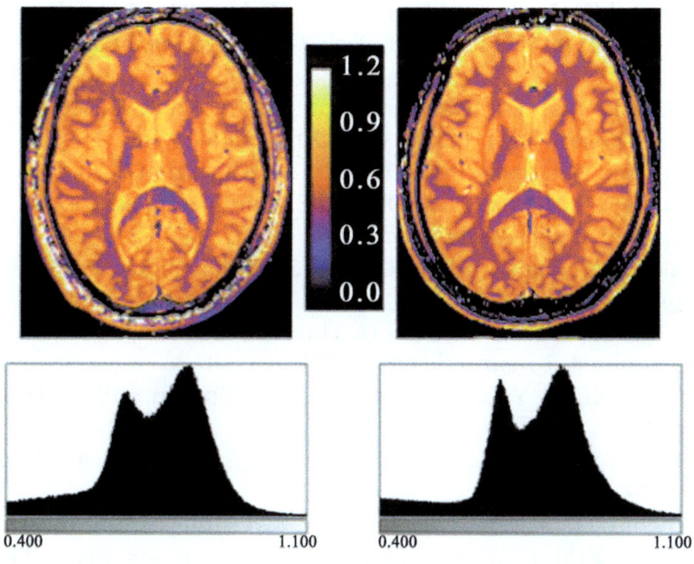

Figure 3.12 Water content maps obtained from the same healthy volunteer at 1.5 T and 3 T and the histograms of water content over the whole brain for each field strength. A clear separation between the distribution in WM (~68%) and that in GM (~82%) is visible at both fields. Increasing the field strength by a factor of two, and thus doubling the SNR, has only a small effect on the water content distribution, slightly decreasing its width, while the mean values per tissue type are practically unchanged. This demonstrates the beneficial effect of using an acquisition with high SNR, indicating that the influence of noise on the MRI-measured water content is already small at 1.5 T and water content values can be compared across fields.

References

1. K. Brodmann, *Vergleichende Lokalisationslehre der Grosshirnrinde*, Johann Ambrosius Barth, Leipzig, 1909.
2. J. C. Mazziotta, S. C. Huang, M. E. Phelps, R. E. Carson, N. S. MacDonald and K. Mahoney, *J. Cereb. Blood Flow Metab.*, 1985, **5**(1), 70.
3. P. T. Fox, M. A. Mintun, E. M. Reiman and M. E. Raichle, *J. Cereb. Blood Flow Metab.*, 1988, **8**(5), 642.
4. M. E. Raichle, *Res. Publ. Assoc. Res. Nerv. Ment. Dis.*, 1990, **67**, 257.
5. P. E. Roland, C. J. Graufelds, J. Wohlin, L. Ingelman, M. Andersson, A. Ledberg, J. Pedersen, S. Akerman, A. Dabringhaus and K. Zilles, *Hum. Brain Mapp.*, 1994, **1**(3), 173.
6. G. McGibney, M. R. Smith, S. T. Nichols and A. Crawley, *Magn. Reson. Med.*, 1993, **30**(1), 51.
7. M. A. Griswold, P. M. Jakob, R. M. Heidemann, M. Nittka, V. Jellus, J. Wang, B. Kiefer and A. Haase, *Magn. Reson. Med.*, 2002, **47**(6), 1202.
8. K. Setsompop, B. A. Gagoski, J. R. Polimeni, T. Witzel, V. J. Wedeen and L. L. Wald, *Magn. Reson. Med.*, 2012, **67**(5), 1210.
9. Y. Hu and G. H. Glover, *Magn. Reson. Med.*, 2007, **58**(5), 947.
10. S. Yun, S. S. Oh, Y. Han and H. Park, *J. Magn. Reson. Imaging*, 2009, **29**(4), 924.
11. R. M. Heidemann, D. Ivanov, R. Trampel, F. Fasano, H. Meyer, J. Pfeuffer and R. Turner, *Magn. Reson. Med.*, 2012, **68**(5), 1506.
12. S. Yun and N. J. Shah, *PLoS One*, 2017, **12**(9), e0184759.
13. S. Moeller, E. Yacoub, C. A. Olman, E. Auerbach, J. Strupp, N. Harel and K. Uğurbil, *Magn. Reson. Med.*, 2010, **63**(5), 1144.
14. J. Budde, G. Shajan, M. Zaitsev, K. Scheffler and R. Pohmann, *Magn. Reson. Med.*, 2014, **74**, 209.
15. C. J. Hodgetts, N. L. Voets, A. G. Thomas, S. Stuart, A. D. Lawrence and K. S. Graham, *J. Neurosci.*, 2017, **37**(12), 3150.
16. A. Sengupta, R. Yakupov, O. Speck, S. Pollmann and M. Hanke, *Data Brief*, 2017, **24**, 219.
17. S. Ogawa, T. M. Lee, A. R. Kay and D. W. Tank, *Proc. Natl. Acad. Sci. U. S. A.*, 1990, **87**, 9868.
18. E. Yacoub, T. Q. Duong, P. F. van De Moortele, M. Lindquist, G. Adriany, S. G. Kim, K. Uğurbil and X. Hu, *Magn. Reson. Med.*, 2003, **49**(4), 655.
19. T. Q. Duong, E. Yacoub, G. Adriany, X. Hu, K. Ugurbil, J. T. Vaughan, H. Merkle and S. G. Kim, *Magn. Reson. Med.*, 2002, **48**(4), 589.
20. A. D. Halai, S. R. Welbourne, K. Embleton and L. M. Parkes, *Hum. Brain Mapp.*, 2014, **35**(8), 4118.
21. E. Yacoub, A. Shmuel, J. Pfeuffer, P. F. van De Moortele, G. Adriany, P. Andersen, J. T. Vaughan, H. Merkle, K. Ugurbil and X. Hu, *Magn. Reson. Med.*, 2001, **45**(4), 588.
22. F. Pedregosa, M. Eickenberg, P. Ciuciu, B. Thirion and A. Gramfort, *Neuroimage*, 2015, **104**, 209.

23. R. Srikanth and A. G. Ramakrishnan, *Int. J. Neural Syst.*, 2006, **16**(2), 125.
24. J. Wang, H. Zhu, J. Fan, K. Giovanello and W. Lin, *Med. Image Comput. Comput. Assisted Intervention*, 2011, **14**, 269.
25. D. A. Handwerker, J. M. Ollinger and M. D'Esposito, *Neuroimage*, 2004, **21**(4), 1639.
26. S. E. Petersen and J. W. Dubis, *Neuroimage*, 2012, **62**(2), 1177.
27. M. W. Chee, V. Venkatraman, C. Westphal and S. C. Siong, *Hum. Brain Mapp.*, 2003, **18**(3), 186.
28. K. J. Friston, A. P. Holmes, C. J. Price, C. Büchel and K. J. Worsley, *Neuroimage*, 1999, **10**(4), 385.
29. M. D'Esposito, E. Zarahn and G. K. Aguirre, *Psychol. Bull.*, 1999, **125**(1), 155.
30. D. L. Schacter, R. L. Buckner, W. Koutstaal, A. M. Dale and B. R. Rosen, *Neuroimage*, 1997, **6**(4), 259.
31. M. H. Lee, C. D. Smyser and J. S. Shimony, *AJNR Am. J. Neuroradiol.*, 2013, **34**(10), 1866.
32. M. E. Raichle, A. M. MacLeod, A. Z. Snyder, W. J. Powers, D. A. Gusnard and G. L. Shulman, *Proc. Natl. Acad. Sci. U. S. A.*, 2001, **98**(2), 676.
33. M. M. Martin, *Front. Hum. Neurosci.*, 2011, **5**, 28.
34. J. B. Poline, F. Kherif and W. D. Penny, in *Human Brain Function*, ed. R. S. Frackowiak, K. J. Friston, C. D. Frith, R. J. Dolan, C. J. Price, S. Zeki, J. Ashburner and W. D. Penny, Elsevier, San Diego, 2nd edn, 2007, p. 126.
35. D. Saloner, *Radiographics*, 1995, **15**(2), 453.
36. M. Alkhorayef, E. Babikir, A. Alrushoud, H. Al-Mohammed and A. Sulieman, *Saudi J. Biol. Sci.*, 2017, **24**, 235.
37. H. Lusic and M. W. Grinstaff, *Chem. Rev.*, 2013, **113**(3), 1641.
38. G. C. Kane, A. W. Stanson, D. Kalnicka, D. W. Rosenthal, C. U. Lee, S. C. Textor and V. D. Garovic, *Nephrol. Dial. Transplant.*, 2008, **23**(4), 1233.
39. M. Miyazaki and V. S. Lee, *Radiology*, 2008, **248**(1), 20.
40. J. H. Maki, T. L. Chenevert and M. Prince, *Abdom. Imaging*, 1998, **23**, 469.
41. F. R. Korosec, R. Frayne, T. M. Grist and C. A. Mistretta, *Magn. Reson. Med.*, 1996, **36**(3), 345.
42. T. K. Foo, M. Saranathan, M. R. Prince and T. L. Chenevert, *Radiology*, 1997, **203**(1), 275.
43. J. W. Goldfarb, M. G. Hochman, D. S. Kim and R. R. Edelman, *AJR Am. J. Roentgenol.*, 2001, **177**(5), 1177.
44. G. H. Jahng, K. L. Li, L. Ostergaard and F. Calamante, *Korean J. Radiol.*, 2014, **15**(5), 554.
45. L. Willats and F. Calamante, *NMR Biomed.*, 2013, **26**(8), 913.
46. F. Calamante, *Prog. Nucl. Magn. Reson. Spectrosc.*, 2013, **74**, 1.
47. F. Calamante, D. L. Thomas, G. S. Pell, J. Wiersma and R. Turner, *J. Cereb. Blood Flow Metab.*, 1999, **19**(7), 701.
48. P. Meier and K. L. Zierler, *J. Appl. Physiol.*, 1954, **6**(12), 731.
49. P. S. Tofts and A. G. Kermode, *Magn. Reson. Med.*, 1991, **17**(2), 357.

50. P. S. Tofts, G. Brix, D. L. Buckley, J. L. Evelhoch, E. Henderson, M. V. Knopp, H. B. Larsson, T. Y. Lee, N. A. Mayr, G. J. Parker, R. E. Port, J. Taylor and R. M. Weisskoff, *J. Magn. Reson. Imaging*, 1999, **10**(3), 223.
51. M. Essig, M. S. Shiroishi, T. B. Nguyen, M. Saake, J. M. Provenzale, D. Enterline, N. Anzalone, A. Dörfler, A. Rovira, M. Wintermark and M. Law, *AJR Am. J. Roentgenol.*, 2013, **200**(1), 24.
52. X. H. Zhu, S. G. Kim, P. Andersen, S. Ogawa, K. Uğurbil and W. Chen, *Magn. Reson. Med.*, 1998, **40**(5), 703.
53. T. Grobner, *Nephrol. Dial. Transplant.*, 2006, **21**(4), 1104.
54. X. Golay, J. Hendrikse and T. C. Lim, *Top. Magn. Reson. Imaging*, 2004, **15**(1), 10.
55. D. S. Williams, J. A. Detre, J. S. Leigh and A. P. Koretsky, *Proc. Natl. Acad. Sci. U. S. A.*, 1992, **89**(1), 212.
56. R. R. Edelman, B. Siewert, D. G. Darby, V. Thangaraj, A. C. Nobre, M. M. Mesulam and S. Warach, *Radiology*, 1994, **192**(2), 513.
57. R. R. Edelman and Q. Chen, *Magn. Reson. Med.*, 1998, **40**(6), 800.
58. S. G. Kim, *Magn. Reson. Med.*, 1995, **34**(3), 293.
59. R. B. Buxton, L. R. Frank, E. C. Wong, B. Siewert, S. Warach and R. R. Edelman, *Magn. Reson. Med.*, 1998, **40**(3), 383.
60. P. Mansfield and P. G. Morris, *Adv. Magn. Reson.*, 1982, 1.
61. D. Le Bihan and E. Breton, *C. R. Acad. Sci.*, 1985, **T.301**, 1109.
62. G. E. Wesbey, M. E. Moseley and R. L. Ehman, *Invest. Radiol.*, 1984, **19**, 491.
63. D. Le Bihan, E. Breton, D. Lallemand, P. Grenier, E. Cabanis and M. Lavaljeantet, *Radiology*, 1986, **161**, 401.
64. D. Jones, *Diffusion MRI*, Oxford University Press, Oxford, New York, 2011.
65. H. Johansen-Berg and T. E. Behrens, *Diffusion MRI: From Quantitative Measurement to In Vivo Measurement*, Academic Press, Amsterdam, Boston, 2009.
66. M. E. Moseley, Y. Cohen, J. Mintorovitch, L. Chileuitt, H. Shimizu, J. Kucharczyk, M. F. Wendland and P. R. Weinstein, *Magn. Reson. Med.*, 1990, **14**, 330.
67. S. E. Maier, Y. Sun and R. V. Mulkern, *NMR Biomed.*, 2010, **23**, 849.
68. O. Ciccarelli, M. Catani, H. Johansen-Berg, C. Clark and A. Thompson, *Lancet Neurol.*, 2008, **7**, 715.
69. R. Brown, *Philos. Mag.*, 1828, **4**, 161.
70. A. Fick, *Poggendorff's Ann. Phys. Chem.*, 1855, **94**, 59.
71. A. Einstein, *Ann. Phys.*, 1905, **17**, 549.
72. D. Le Bihan, *Nat. Rev. Neurosci.*, 2003, **4**, 469.
73. D. Le Bihan and H. Johansen-Berg, *Neuroimage*, 2012, **61**, 324.
74. S. Wakana, H. Jiang, L. M. Nagae-Poetscher, P. C. van Zijl and S. Mori, *Radiology*, 2004, **230**, 77.
75. C. Beaulieu, *NMR Biomed.*, 2002, **15**, 435.
76. P. J. Basser, J. Mattiello and D. LeBihan, *Biophys. J.*, 1994, **66**, 259.
77. P. J. Basser, J. Mattiello and D. LeBihan, *J. Magn. Reson. B*, 1994, **103**, 247.

78. P. B. Kingsley, *Concept Magn. Reson. A*, 2006, **28a**, 101.
79. P. B. Kingsley, *Concept Magn. Reson. A*, 2006, **28a**, 155.
80. P. B. Kingsley, *Concept Magn. Reson. A*, 2006, **28a**, 123.
81. Y. Assaf and O. Pasternak, *J. Mol. Neurosci.*, 2008, **34**, 51.
82. B. J. Jellison, A. S. Field, J. Medow, M. Lazar, M. S. Salamat and A. L. Alexander, *AJNR Am. J. Neuroradiol.*, 2004, **25**, 356.
83. D. K. Jones and A. Leemans, *Methods Mol. Biol.*, 2011, **711**, 127.
84. J. Mattiello, P. J. Basser and D. Le Bihan, *Magn. Reson. Med.*, 1997, **37**, 292.
85. C. Pierpaoli, P. Jezzard, P. J. Basser, A. Barnett and G. Di Chiro, *Radiology*, 1996, **201**, 637.
86. D. Le Bihan, J. F. Mangin, C. Poupon, C. A. Clark, S. Pappata, N. Molko and H. Chabriat, *J. Magn. Reson. Imaging*, 2001, **13**, 534.
87. P. J. Basser and D. K. Jones, *NMR Biomed.*, 2002, **15**, 456.
88. S. D. Roosendaal, J. J. Geurts, H. Vrenken, H. E. Hulst, K. S. Cover, J. A. Castelijns, P. J. Pouwels and F. Barkhof, *Neuroimage*, 2009, **44**, 1397.
89. D. W. Gross, *Epilepsia*, 2011, **52**(Suppl 4), 32.
90. Y. Assaf, *Bioessays*, 2008, **30**, 1235.
91. D. A. Medina and M. Gaviria, *Neuropsychiatr. Dis. Treat.*, 2008, **4**, 737.
92. M. Naik, A. Lundervold, H. Nygaard and J. T. Geitung, *Acta Radiol.*, 2010, **51**, 662.
93. Z. Nedelska, C. G. Schwarz, B. F. Boeve, V. J. Lowe, R. I. Reid, S. A. Przybelski, T. G. Lesnick, J. L. Gunter, M. L. Senjem, T. J. Ferman, G. E. Smith, Y. E. Geda, D. S. Knopman, R. C. Petersen, C. R. Jack Jr and K. Kantarci, *Neurobiol. Aging*, 2015, **36**, 2010.
94. Y. W. Shin, J. S. Kwon, T. H. Ha, H. J. Park, D. J. Kim, S. B. Hong, W. J. Moon, J. M. Lee, I. Y. Kim, S. I. Kim and E. C. Chung, *Neuroimage*, 2006, **30**, 1285.
95. C. H. Sotak, *NMR Biomed.*, 2002, **15**, 561.
96. J. Dubois, G. Dehaene-Lambertz, J. F. Mangin, D. Le Bihan, P. S. Huppi and L. Hertz-Pannier, *Neurophysiol. Clin.*, 2012, **42**, 1.
97. P. S. Huppi and J. Dubois, *Semin. Fetal Neonat. Med.*, 2006, **11**, 489.
98. D. J. Madden, I. J. Bennett, A. Burzynska, G. G. Potter, N. K. Chen and A. W. Song, *Biochim. Biophys. Acta*, 2012, **1822**, 386.
99. A. N. Voineskos, T. K. Rajji, N. J. Lobaugh, D. Miranda, M. E. Shenton, J. L. Kennedy, B. G. Pollock and B. H. Mulsant, *Neurobiol. Aging*, 2012, **33**, 21.
100. O. Sporns, *Ann. N. Y. Acad. Sci.*, 2011, **1224**, 109.
101. F. Grinberg, I. I. Maximov, E. Farrher and N. J. Shah, *Magn. Reson. Imaging*, 2017, **45**, 7.
102. M. Lazar, *NMR Biomed.*, 2010, **23**, 821.
103. M. Bastiani, N. J. Shah, R. Goebel and A. Roebroeck, *Neuroimage*, 2012, **62**, 1732.
104. M. Catani, M. Thiebaut de Schotten, D. Slater and F. Dell'Acqua, *Neuroimage*, 2013, **80**, 2.
105. S. Mori and P. C. van Zijl, *NMR Biomed.*, 2002, **15**, 468.

106. S. Jbabdi and H. Johansen-Berg, *Brain Connect.*, 2011, **1**, 169.
107. T. E. Behrens and O. Sporns, *Curr. Opin. Neurobiol.*, 2012, **22**, 144.
108. R. C. Craddock, R. L. Tungaraza and M. P. Milham, *Gigascience*, 2015, **4**, 13.
109. H. Johansen-Berg, *Neuroimage*, 2013, **80**, 541.
110. C. A. Clark and D. Le Bihan, *Magn. Reson. Med.*, 2000, **44**, 852.
111. Y. Cohen and Y. Assaf, *NMR Biomed.*, 2002, **15**, 516.
112. R. V. Mulkern, S. J. Haker and S. E. Maier, *Magn. Reson. Imaging*, 2009, **27**, 1151.
113. F. Grinberg, E. Farrher, J. Kaffanke, A. M. Oros-Peusquens and N. J. Shah, *Neuroimage*, 2011, **57**, 1087.
114. J. H. Jensen, J. A. Helpern, A. Ramani, H. Lu and K. Kaczynski, *Magn. Reson. Med.*, 2005, **53**, 1432.
115. J. H. Jensen and J. A. Helpern, *NMR Biomed.*, 2010, **23**, 698.
116. F. Grinberg, E. Farrher, L. Ciobanu, F. Geffroy, D. Le Bihan and N. J. Shah, *PLoS One*, 2014, **9**, e89225.
117. T. Niendorf, R. M. Dijkhuizen, D. G. Norris, M. van Lookeren Campagne and K. Nicolay, *Magn. Reson. Med.*, 1996, **36**, 847.
118. M. G. Hall and T. R. Barrick, *Magn. Reson. Med.*, 2008, **59**, 447.
119. Y. Assaf and P. J. Basser, *Neuroimage*, 2005, **27**, 48.
120. C. A. Clark, M. Hedehus and M. E. Moseley, *Magn. Reson. Med.*, 2002, **47**, 623.
121. V. G. Kiselev and K. A. Il'yasov, *Magn. Reson. Med.*, 2007, **57**, 464.
122. S. E. Maier, S. Vajapeyam, H. Mamata, C. F. Westin, F. A. Jolesz and R. V. Mulkern, *Magn. Reson. Med.*, 2004, **51**, 321.
123. K. M. Bennett, K. M. Schmainda, R. T. Bennett, D. B. Rowe, H. Lu and J. S. Hyde, *Magn. Reson. Med.*, 2003, **50**, 727.
124. S. De Santis, A. Gabrielli, M. Bozzali, B. Maraviglia, E. Macaluso and S. Capuani, *Magn. Reson. Med.*, 2011, **65**, 1043.
125. D. A. Yablonskiy, G. L. Bretthorst and J. J. Ackerman, *Magn. Reson. Med.*, 2003, **50**, 664.
126. P. N. Sen and P. J. Basser, *Magn. Reson. Imaging*, 2005, **23**, 215.
127. Y. Assaf, T. Blumenfeld-Katzir, Y. Yovel and P. J. Basser, *Magn. Reson. Med.*, 2008, **59**, 1347.
128. H. Zhang, T. Schneider, C. A. Wheeler-Kingshott and D. C. Alexander, *Neuroimage*, 2012, **61**, 1000.
129. E. Panagiotaki, T. Schneider, B. Siow, M. G. Hall, M. F. Lythgoe and D. C. Alexander, *Neuroimage*, 2012, **59**, 2241.
130. D. Le Bihan, *Phys. Med. Biol.*, 2007, **52**, R57.
131. A. Kristoffersen, *Magn. Reson. Med.*, 2011, **66**, 1639.
132. E. S. Hui, M. M. Cheung, L. Qi and E. X. Wu, *Neuroimage*, 2008, **42**, 122.
133. M. M. Cheung, E. S. Hui, K. C. Chan, J. A. Helpern, L. Qi and E. X. Wu, *Neuroimage*, 2009, **45**, 386.
134. F. Grinberg, I. I. Maximov, E. Farrher, I. Neuner, L. Amort, H. Thonnessen, E. Oberwelland, K. Konrad and N. J. Shah, *Neuroimage*, 2017, **144**, 12.

135. A. Paydar, E. Fieremans, J. I. Nwankwo, M. Lazar, H. D. Sheth, V. Adisetiyo, J. A. Helpern, J. H. Jensen and S. S. Milla, *AJNR Am. J. Neuroradiol.*, 2014, **35**, 808.
136. M. F. Falangola, J. H. Jensen, J. S. Babb, C. Hu, F. X. Castellanos, A. Di Martino, S. H. Ferris and J. A. Helpern, *J. Magn. Reson. Imaging*, 2008, **28**, 1345.
137. J. Latt, M. Nilsson, R. Wirestam, F. Stahlberg, N. Karlsson, M. Johansson, P. C. Sundgren and D. van Westen, *J. Magn. Reson. Imaging*, 2013, **37**, 610.
138. J. H. Jensen, M. F. Falangola, C. Hu, A. Tabesh, O. Rapalino, C. Lo and J. A. Helpern, *NMR Biomed.*, 2011, **24**, 452.
139. F. Grinberg, L. Ciobanu, E. Farrher and N. J. Shah, *NMR Biomed.*, 2012, **25**, 1295.
140. J. S. Cheung, E. Wang, E. H. Lo and P. Z. Sun, *Stroke*, 2012, **43**, 2252.
141. E. S. Hui, E. Fieremans, J. H. Jensen, A. Tabesh, W. Feng, L. Bonilha, M. V. Spampinato, R. Adams and J. A. Helpern, *Stroke*, 2012, **43**, 2968.
142. J. J. Wang, W. Y. Lin, C. S. Lu, Y. H. Weng, S. H. Ng, C. H. Wang, H. L. Liu, R. H. Hsieh, Y. L. Wan and Y. Y. Wai, *Radiology*, 2011, **261**, 210.
143. I. Blockx, G. De Groof, M. Verhoye, J. Van Audekerke, K. Raber, D. Poot, J. Sijbers, A. P. Osmand, S. Von Horsten and A. Van der Linden, *Neuroimage*, 2012, **59**, 957.
144. N. J. Gong, C. S. Wong, C. C. Chan, L. M. Leung and Y. C. Chu, *Magn. Reson. Imaging*, 2013, **31**, 688.
145. S. Van Cauter, J. Veraart, J. Sijbers, R. R. Peeters, U. Himmelreich, F. De Keyzer, S. W. Van Gool, F. Van Calenbergh, S. De Vleeschouwer, W. Van Hecke and S. Sunaert, *Radiology*, 2012, **263**, 492.
146. P. Raab, E. Hattingen, K. Franz, F. E. Zanella and H. Lanfermann, *Radiology*, 2010, **254**, 876.
147. N. J. Shah, *Brain Struct. Funct.*, 2015, **220**, 1867.
148. Q. Fan, T. Witzel, A. Nummenmaa, K. R. Van Dijk, J. D. Van Horn, M. K. Drews, L. H. Somerville, M. A. Sheridan, R. M. Santillana, J. Snyder, T. Hedden, E. E. Shaw, M. O. Hollinshead, V. Renvall, R. Zanzonico, B. Keil, S. Cauley, J. R. Polimeni, D. Tisdall, R. L. Buckner, V. J. Wedeen, L. L. Wald, A. W. Toga and B. R. Rosen, *Neuroimage*, 2016, **124**, 1108.
149. J. A. McNab, B. L. Edlow, T. Witzel, S. Y. Huang, H. Bhat, K. Heberlein, T. Feiweier, K. Liu, B. Keil, J. Cohen-Adad, M. D. Tisdall, R. D. Folkerth, H. C. Kinney and L. L. Wald, *Neuroimage*, 2013, **80**, 234.
150. S. N. Sotiropoulos, S. Jbabdi, J. Xu, J. L. Andersson, S. Moeller, E. J. Auerbach, M. F. Glasser, M. Hernandez, G. Sapiro, M. Jenkinson, D. A. Feinberg, E. Yacoub, C. Lenglet, D. C. Van Essen, K. Ugurbil, T. E. Behrens and WU-Minn HCP Consortium, *Neuroimage*, 2013, **80**, 125.
151. P. Kochunov, N. Jahanshad, D. Marcus, A. Winkler, E. Sprooten, T. E. Nichols, S. N. Wright, L. E. Hong, B. Patel, T. Behrens, S. Jbabdi, J. Andersson, C. Lenglet, E. Yacoub, S. Moeller, E. Auerbach, K. Ugurbil, S. N. Sotiropoulos, R. M. Brouwer, B. Landman, H. Lemaitre, A. den Braber, M. P. Zwiers, S. Ritchie, K. van Hulzen, L. Almasy, J. Curran, G.

I. deZubicaray, R. Duggirala, P. Fox, N. G. Martin, K. L. McMahon, B. Mitchell, R. L. Olvera, C. Peterson, J. Starr, J. Sussmann, J. Wardlaw, M. Wright, D. I. Boomsma, R. Kahn, E. J. de Geus, D. E. Williamson, A. Hariri, D. van 't Ent, M. E. Bastin, A. McIntosh, I. J. Deary, H. E. Hulshoff Pol, J. Blangero, P. M. Thompson, D. C. Glahn and D. C. Van Essen, *Neuroimage*, 2015, **111**, 300.
152. D. C. Van Essen, S. M. Smith, D. M. Barch, T. E. Behrens, E. Yacoub, K. Ugurbil and WU-Minn HCP Consortium, *Neuroimage*, 2013, **80**, 62.
153. A. Leemans, B. Jeurissen, J. Sijbers and D. K. Jones, in Proc. Intl. Soc. Mag. Reson. Med., 2009, p. 3537.
154. V. L. Yarnykh, *Magn. Reson. Med.*, 2007, **57**, 192.
155. M. Fernández-Seara and F. W. Wehrli, *Magn. Reson. Med.*, 2000, **44**, 358.
156. N. J. Shah, M. Zaitsev, S. Steinhoff and K. Zilles, *Neuroimage*, 2001, **14**, 1175.
157. H. Neeb and N. J. Shah, *Magn. Reson. Med.*, 2006, **56**, 224.
158. H. Neeb, K. Zilles and N. J. Shah, *Neuroimage*, 2006, **31**, 1156.
159. R. Deichmann, C. D. Good and R. Turner, *Magn. Reson. Med.*, 2002, **47**, 398.
160. C. Preibisch and R. Deichmann, *Magn. Reson. Med.*, 2009, **61**, 125.
161. D. Ma, V. Gulani, N. Seiberlich, K. Liu, J. L. Sunshine, J. L. Duerk and M. A. Griswold, *Nature*, 2013, **495**, 187.
162. J. B. M. Warntjes, O. Dahlqvist Leinhard, J. West and P. Lundberg, *Magn. Reson. Med.*, 2008, **60**, 320.
163. A. M. Oros-Peusquens, F. Keil, K. J. Langen, H. Herzog, G. Stoffels, C. Weiss and N. J. Shah, *Nucl. Instruments Methods Phys. Res. Sect. A*, 2014, **734**, 185.
164. A. M. Oros-Peusquens, R. Loução, M. Zimmermann, K. J. Langen and N. J. Shah, *Methods*, 2017, **130**, 135.
165. S. Steinhoff, M. Zaitsev, K. Zilles and N. J. Shah, *Magn. Reson. Med.*, 2001, **46**, 131.
166. H. Neeb, K. Zilles and N. J. Shah, *Neuroimage*, 2006, **29**, 910.
167. H. Neeb, V. Ermer, T. Stocker and N. J. Shah, *Neuroimage*, 2008, **42**, 1094.
168. P. Tofts, *Quantitative MRI of the Brain: Measuring Changes Caused by Disease*, John Wiley & Sons, 2005.
169. S. C. L. Deoni, B. K. Rutt and T. M. Peters, *Magn. Reson. Med.*, 2003, **49**, 515.
170. S. C. L. Deoni, T. M. Peters and B. K. Rutt, *Magn. Reson. Med.*, 2005, **53**, 237.
171. S. C. L. Deoni, B. K. Rutt, T. Arun, C. Pierpaoli and D. K. Jones, *Magn. Reson. Med.*, 2008, **60**, 1372.
172. W. D. Rooney, G. Johnson, X. Li, E. R. Cohen, S. G. Kim, K. Ugurbil and C. S. Springer, *Magn. Reson. Med.*, 2007, **57**, 308.
173. A. M. Oros-Peusquens, M. Laurila and N. J. Shah, *Magn. Reson. Mater. Phys., Biol. Med.*, 2008, **21**, 131.
174. H. Y. Carr and E. M. Purcell, *Phys. Rev.*, 1954, **94**, 630.

175. S. Meiboom and D. Gill, *Rev. Sci. Instrum.*, 1958, **29**, 688.
176. S. Majumdar, S. C. Orphanoudakis, A. Gmitro, M. O'Donnell and J. C. Gore, *Magn. Reson. Med.*, 1986, **3**, 562.
177. S. Ogawa, R. S. Menon, D. W. Tank, S. G. Kim, H. Merkle, J. M. Ellermann, K. Ugurbil, K. Ugurbilt and K. Ugurbil, *Biophys. J.*, 1993, **64**, 803.
178. H. Dahnke and T. Schaeffter, *Magn. Reson. Med.*, 2005, **53**, 1202.
179. R. J. Ordidge, J. M. Gorell, J. C. Deniau, R. A. Knight and J. A. Helpern, *Magn. Reson. Med.*, 1994, **32**, 335.
180. S. Baudrexel, S. Volz, C. Preibisch, J. C. Klein, H. Steinmetz, R. Hilker and R. Deichmann, *Magn. Reson. Med.*, 2009, **62**, 263.
181. J. R. Reichenbach, R. Venkatesan, D. A. Yablonskiy, M. R. Thompson, S. Lai and E. M. Haacke, *J. Magn. Reson. Imaging*, 1997, **7**, 266.
182. A. MacKay, K. Whittall, J. Adler, D. Li, D. Paty and D. Graeb, *Magn. Reson. Med.*, 1994, **31**, 673.
183. P. Sati, P. van Gelderen, A. C. Silva, D. S. Reich, H. Merkle, J. A. de Zwart and J. H. Duyn, *Neuroimage*, 2013, **77**, 268.
184. R. Damadian, *Science*, 1971, **171**, 1151.
185. R. Deichmann, D. Hahn and A. Haase, *Magn. Reson. Med.*, 1999, **42**, 206.
186. J. P. Marques, T. Kober, G. Krueger, W. van der Zwaag, P. F. Van de Moortele and R. Gruetter, *Neuroimage*, 2010, **49**, 1271.
187. S. H. Koenig, R. D. Brown, M. Spiller and N. Lundbom, *Magn. Reson. Med.*, 1990, **14**, 482.
188. S. H. Koenig, *Magn. Reson. Med.*, 1991, **20**, 285.
189. P. Christiansen, P. B. Toft, P. Gideon, E. R. Danielsen, P. Ring and O. Henriksen, *Magn. Reson. Imaging*, 1994, **12**, 1237.
190. M. A. Smith, J. Chick, D. M. Kean, R. H. B. Douglas, A. Singer, R. E. Kendell and J. J. K. Best, *Lancet*, 1985, **325**, 1273.
191. R. J. Winney, D. M. Kean, J. J. K. Best and M. A. Smith, *Lancet*, 1986, **328**, 1107.
192. P. Gideon, S. Rosenbaum, B. Sperling and P. Petersen, *Magn. Reson. Imaging*, 1999, **17**, 301.
193. C. Laule, I. M. Vavasour, G. R. W. Moore, J. Oger, D. K. B. Li, D. W. Paty and A. L. MacKay, *J. Neurol.*, 2004, **251**, 284.
194. N. J. Shah, H. Neeb, G. Kircheis, P. Engels, D. Häussinger and K. Zilles, *Neuroimage*, 2008, **41**, 706.
195. I. C. Kiricuta and V. Simplăceanu, *Cancer Res.*, 1975, **35**, 1164.
196. P. A. Bottomley, T. H. Foster, R. R. Argersinger and L. M. Pfeifer, *Med. Phys.*, 1984, **11**, 425.
197. I. L. Cameron, V. A. Ord and G. D. Fullerton, *Magn. Reson. Imaging*, 1984, **2**, 97.
198. S. Naruse, Y. Horikawa, C. Tanaka, K. Hirakawa, H. Nishikawa and K. Yoshizaki, *J. Neurosurg.*, 1982, **56**, 747.
199. B. A. Bell, D. M. Kean, H. L. Macdonald, G. H. Barnett, R. H. B. Douglas, M. A. Smith, C. N. J. Mcghee, J. D. Miller, J. L. Tocher and J. J. K. Best, *Lancet*, 1987, **329**, 66.

200. B. A. Bell, *Ann. R. Coll. Surg. Engl.*, 1989, **71**, 375.
201. P. P. Fatouros, A. Marmarou, K. A. Kraft, S. Inao and F. P. Schwarz, *Magn. Reson. Med.*, 1991, **17**, 402.
202. P. P. Fatouros and A. Marmarou, *J. Neurosurg.*, 1999, **90**, 109.
203. R. Mathur-De Vré, *Br. J. Radiol.*, 1984, **57**, 955.
204. A. L. Hopkins, H. N. Yeung and C. B. Bratton, *Magn. Reson. Med.*, 1986, **3**, 303.
205. P. S. Tofts, J. S. Jackson, D. J. Tozer, M. Cercignani, G. Keir, D. G. MacManus, G. R. Ridgway, B. H. Ridha, K. Schmierer, D. Siddique, J. S. Thornton, S. J. Wroe and N. C. Fox, *Magn. Reson. Med.*, 2008, **59**, 190.

CHAPTER 4

Ultra-high Field Imaging

W. A. WORTHOFF[a], J. MAULER[a], A. M. OROS-PEUSQUENS[a] AND N. J. SHAH*[a,b,c]

[a]Forschungszentrum Jülich, Institute of Neuroscience and Medicine – 4, Wilhelm-Johnen-Straße 1, 52428 Jülich, Germany; [b]JARA-BRAIN – Translational Medicine, Aachen, Germany; [c]Department of Neurology, RWTH Aachen University, Pauwelsstr. 30, 52074 Aachen, Germany
*E-mail: n.j.shah@fz-juelich.de

4.1 Introduction

Until recently, standard clinical MRI systems have been operating at magnetic field strengths of up to 3 T. However, improvements in the availability of clinical MRI systems with greater field strengths have begun to open up the possibility of significantly enhanced image quality, contrast and resolution, both in anatomical and functional imaging, as well as in spectroscopy.[1–3] In addition to improvements in the acquired images, increased field strengths also enable the *in vivo* use of non-proton target nuclei, the so-called X-nuclei, making the analysis of metabolism and biochemical processes through spectroscopic and imaging methods possible. This feature greatly enhances the applicability and universality of MR as a tool with which to analyse all manner of pathology and developments in living beings.

However, despite the clear benefits of high field and ultra-high field MRI, the use of higher magnetic field strength brings with it complex technological

New Developments in NMR No. 19
Hybrid MR-PET Imaging: Systems, Methods and Applications
Edited by N. Jon Shah
© The Royal Society of Chemistry 2019
Published by the Royal Society of Chemistry, www.rsc.org

demands. This is largely due to several important factors. First, the magnet needs to be capable of generating the strong magnetic field while maintaining sufficient homogeneity over the sampling region. Second, the increased Larmor frequency, in particular of protons, causes the performance of the radiofrequency (RF) hardware to be affected. Wave effects, which are less pronounced at lower magnetic field strengths, make homogeneous excitation and reception difficult as the RF wavelength is shortened. In addition, the specific absorption rate (SAR) induced by the RF pulses is higher once the RF increases. This implies that the RF hardware and coil design require careful consideration. Finally, the quality of acquired images and spectra suffer from field variations caused by the distribution of magnetic susceptibility within the subject, which is more pronounced in ultra-high field MRI. Solutions to this are more advanced and require the use of higher-order shim methods and potentially the use of additional shim hardware, such as shim inserts. The special issues associated with ultra-high field MRI are, in part, mitigated by careful pulse sequence design, the use of appropriate hardware, and by advanced means of post-processing. Nevertheless, the imaging process is still considerably more challenging and prone to artefacts compared to conventional MRI at more conservative field strengths. However, despite these difficulties, the benefits of increased image quality and the versatility derived from these methods offers a new level of insight into the understanding of human anatomy and function.

4.2 High-resolution Spectroscopy

4.2.1 Properties of Magnetic Resonance Spectroscopy (MRS)

In contrast to conventional MR imaging, which transfers the physical properties of protons, such as density and relaxation rate, into intensity values, while neglecting the interconnecting chemical relations; MRS elucidates information relating to the molecular structures underlying the MR signal. Therefore, MRS can be used to identify and quantify MRS-visible metabolites in the volume under investigation. Conversely, if the molecular formula is known, the chemical bonds can be characterised. Unlike the positron emission tomography (PET) technique, MRS offers the opportunity to analyse multiple metabolites in one measurement with no requirement for special probes or radioactivity. However, its drawback compared to PET is its orders of magnitude lower sensitivity and inferior specificity. This introductive chapter focuses on proton MRS.

4.2.2 Physical Basis of MRS: Chemical Shift and Scalar Coupling

In a simplified model of magnetic resonance, the effect of electrons is neglected, and nuclei of the same kind resonate at the Larmor frequency $\omega_0 = \gamma B_0$. This model is frequently referred to when the basics of MR imaging are introduced, but it is insufficient when it comes to understanding MR spectroscopy. In a more realistic approach, the electrons in the *chemical* environment shield the nuclei, which affects the resonance frequency; this is the so-called chemical shift. The electrons in the vicinity of the nuclei under investigation rotate about the external magnetic field $\vec{B}_0$ in the opposite direction to the proton spin precession. At the same time, the rotary motion of the negatively charged electrons in the magnetic field induces a magnetic moment $\vec{\mu}_e$ opposite to the orientation of the external field $\vec{B}_0$. Consequently, the nuclei experience a lower effective field strength $\vec{B} = \vec{B}_0(1 - \sigma)$, which causes the resonance frequency

$$\omega_0 = \gamma B_0(1 - \sigma) \tag{4.1}$$

of the nuclei to be reduced (eqn (4.1)). The shielding constant, σ, depends on the chemical environment and has no unit.[4]

In order to express/quantify the chemical shift, δ, which would correspond to the use of absolute units of Hertz in a way that is independent from the strength of the magnetic field B_0, a reference substance with chemical shift ν_{ref} is used to calculate the fraction (eqn (4.2))

$$\delta = \frac{\nu - \nu_{ref}}{\nu_{ref}} \times 10^6 \tag{4.2}$$

yielding relative units in the form of parts per million (ppm).[4,5]

In addition to the effect of the external magnetic field B_0, which is partially shielded by the electron cloud, the nuclear spins experience spin–spin coupling, *i.e.* coupling with each other. Because the impact of direct spin–spin coupling averages out to zero for molecules in a solution, this chapter focuses on indirect spin–spin coupling (scalar coupling, *J*-coupling), which is much more relevant for spectroscopy applications in humans. Indirect spin–spin coupling is transmitted by the bonding electrons. The more bonds there are between two coupled nuclei, the lower the magnitude of the *J*-coupling.

The distribution of the resonance frequencies of a nucleus A, that is coupled to a nucleus X, shows a symmetrically split spectrum, which is centred at the chemical shift frequency υ_A.[4]

4.2.3 Basic Excitation–Acquisition Experiment and Data Processing

A very basic MR spectroscopy experiment comprises a 90° pulse that excites the spin system. The signal then decays in a free induction decay (FID) and is sampled within an acquisition interval ACQ (Figure 4.1). The 90° pulse needs to have a bandwidth broad enough, with an amplitude that is strong

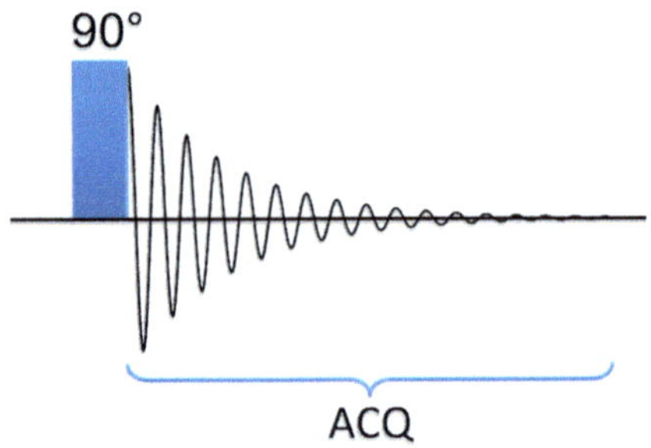

Figure 4.1 Pulse-FID, ACQ.

sufficient, to excite the range of offset frequencies given by the spin system under investigation.

If the pulse is applied along the x-axis, the corresponding magnetisation vectors are flipped from the equilibrium state to the $-y$-direction. Therefore, N magnetisation vectors precess at their individual precession frequency, Ω, about the z-axis and build up the total magnetisation $\boldsymbol{M} = (M_x, M_y)$ with (eqn (4.3))

$$M_x = \sum_{i=1}^{N} M_{0,i} \sin(\Omega_i t) \text{ and } M_y = -\sum_{i=1}^{N} M_{0,i} \cos(\Omega_i t) \tag{4.3}$$

where $M_{0,\mathrm{i}}$ denotes the equilibrium magnetisation of the vector i. The Fourier transformation of the FID yields a spectrum comprising the frequencies from Ω_1 to Ω_N.

Generally, the detectable signal $\boldsymbol{S}$ is proportional to the magnetisation vector $\boldsymbol{M}$. For the sake of convenience, both components can be regarded as the real and imaginary part of a complex signal $S(t) = S_x(t) + iS_y(t)$. If the signal undergoes transverse relaxation with a time constant T_2, it is necessary to extend the signal equation by introducing an exponential relaxation term such that (eqn (4.4))

$$S(t) = \left(S_0 \cos(\Omega t) + iS_0 \sin(\Omega t)\right) \mathrm{e}^{-\frac{t}{T_2}} = S_0 \mathrm{e}^{i\Omega t} \mathrm{e}^{-\frac{t}{T_2}} \tag{4.4}$$

The Fourier transform transfers data from the time domain into the frequency domain. Therefore, the complex signal can be decomposed into its frequency compounds by using the Fourier transform, which consequently yields the real and the imaginary part of the spectrum. The lines of the real part are termed the absorption mode, whereas the imaginary part corresponds to the dispersion mode. The exponential decay of the signal imprints the so-called Lorentzian shape to the spectral lines, which are characterised by the line width $1/(\pi T_2)$. The area under the peak is proportional to the number of contributing spins and so larger T_2 relaxation constants result in concomitantly narrower peak widths and increased peak heights. Depending on the time when data acquisition begins, the signal vector may not necessarily be aligned along the *x*-axis. However, only this orientation would result in the above-mentioned positive absorption and dispersion mode lines in both the x- and y- directions. Different time offsets result in varying mixtures of the pure modes or even the occurrence of completely swapped modes between the *x*- and *y*-axis. This property is referred to as phase shift and can be modelled by an additional phase factor $\exp(i\varphi)$ in the signal equation. For analysis, it is desirable to obtain the real part of the spectrum as pure absorption mode lines. This ensures that no negative spectral lines occur which may cancel out superimposed positive lines from other neighbouring spins. The output of a pure absorption mode spectrum can be accomplished by the multiplication of the signal $S(t)$ with the phase correction term $\exp(i\tilde{\varphi})$ with $\varphi + \tilde{\varphi} = 0$ (eqn (4.5))

$$\exp(i\tilde{\varphi})S(t) = \exp(i\tilde{\varphi})S_0\exp(i\varphi)\exp\left(i\Omega t - \frac{t}{T_2}\right) = S_0\exp\left(i\Omega t - \frac{t}{T_2}\right) \tag{4.5}$$

Phase correction terms in the above-mentioned form are independent of the frequency, and as such, they are called zero-order correction terms. In contrast to this, first-order phase shifts depend linearly on the offset frequency. Such phase errors arise if the RF field strength of the pulse is on the order of magnitude of the range of the offset frequencies.[6] Thus, manual phase corrections are performed by adjusting the zero-order phase correction with respect to the main peak of the spectrum and by setting the linear correction term to additionally correct all neighbouring peaks stretching out to the borders of the total spectrum.

In addition to the essential processing steps described above, real spectroscopy measurements usually require further processing steps that can only be briefly listed here. Thermal noise from the receive coil, and to a lesser extent noise originating from the spectrometer electronics, give a constant contribution to the decaying MR signal. In order to reduce the impact of this noise on the spectrum, the acquisition time should be restricted to the time frame in which the MR signal has not completely decayed. Moreover, the multiplication of the decaying signal with a decreasing function, *i.e.* with an exponential function, cancels out the noise rather than real signal contributions and results in a sensitivity enhancement. This procedure, which is referred to as apodisation, is performed at the expense of a somewhat broadened line width as a consequence of the procedure. This drawback is overcome by the additional multiplication with an increasing function, which partially compensates for the decay of the signal and leads to an improvement in resolution. The effect of both functions together can be replaced by the multiplication with a single weighting function, for instance with a Gaussian function. For all these operations, the underlying data are discrete points that are displayed as a continuous line. As it is frequently desirable to gain a visually smoother spectrum, zero filling is employed, which means appending zeroes to the end of the sampled signal. Subsequently, the Fourier transform yields exactly the same resolution of the peaks as without zero filling, because the information content of the data has not changed, but the course of the points is more densely populated.[6]

Spectroscopy measurements on human subjects may yield up to about 20 metabolites, which can only be efficiently and accurately analysed by using numerical fitting algorithms. These algorithms can operate on both time domain or on frequency domain data; however, this overview chapter focuses only on the latter. The respective programs (*i.e.* LCModel,[7] Tarquin,[8] jMRUI[9]) decompose the measured spectra into a linear combination of the spectra of the pure compounds. These basis sets need to be complete in the respect that all measured substances are represented by their individual basis spectra.[4] Basis sets can be gained in two ways: either one uses the exact same pulse sequence that is going to be used in the later measurements to measure high concentration watery solutions of all pure metabolites, or one numerically computes the basis sets by exploiting the density matrix formalism in conjunction with the properties of the pulse sequence to be used. Due to having the advantages of being noiseless, artefact-free, cheap and easily adaptable to different pulse sequences, the latter approach is frequently used. The fitting of human MRS data is complicated by unspecific macromolecular baseline contributions that stretch over a large range of the ppm scale. Their contribution can be captured using polynomial functions.[10]

4.2.4 *In vivo* Measurements with Spatial Localisation and Water Suppression

Spatial localisation techniques are required to assign spectroscopic signals to the tissue of their origin. Moreover, such techniques are beneficial for improved spectral resolution and for reduced contamination from signals from irrelevant anatomical compartments, *i.e.* from extra-cranial lipids. This is significant because they can restrict the signal acquisition to smaller volumes with consequently better B_0 homogeneity and therefore with a narrower spectral line width. Furthermore, they can restrict data acquisition to volumes that are sufficiently far away from sources of unwanted lipid signal contamination. However, the latter may be addressed by dedicated lipid suppression techniques. In order to omit signal contributions from outside of a single volume-of-interest, the formation of echoes can be exploited to selectively refocus the magnetisation from the voxel under investigation; this is implemented in the Stimulated Echo Acquisition Mode (STEAM) and Point Resolved Spectroscopy (PRESS) techniques. Multi-voxel spectroscopic data can be acquired with spatial resolution more efficiently by using chemical shift imaging (CSI).

The STEAM sequence[11] comprises three 90° pulses (Figure 4.2(a)), which, in addition to three FIDs and four spin echoes, finally evoke the build-up

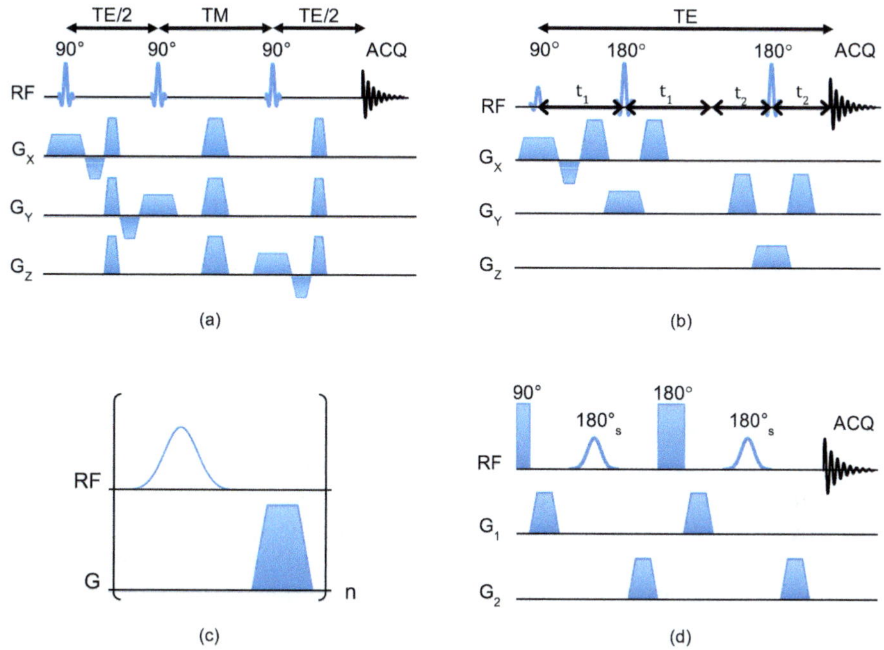

Figure 4.2 Pulse sequences for spatial localisation and water signal suppression. (a) STEAM sequence; (b) PRESS sequence; (c) CHESS water suppression; and (d) MEGA water suppression integrated into a spin echo pulse sequence.

of the so-called stimulated echo.[12] The first pulse excites the magnetisation in the transverse plane and, after the time interval TE/2 has passed, the second pulse flips it in the longitudinal direction. The subsequent mixing time period, TM, does not add to the echo time because the magnetisation is 'stored' along the *z*-direction and no T_2 relaxation can occur. Water suppression steps can be carried out during the TM time interval. In particular, the unwanted spin echoes can be dephased with the application of crusher gradients. The third 90° pulse rotates the magnetisation back to the transverse plane, where the stimulated echo can be acquired after another TE/2 interval. Spatial selection is accomplished with the use of frequency selective RF pulses and magnetic field gradients. The signal intensity of a stimulated echo only reaches 50% of the amplitude of a spin echo, because the second 90° pulse only flips 50% of the magnetisation back in the longitudinal direction, whereas the other half undergoes dephasing caused by the crusher gradients. This is a disadvantage of the STEAM method.

The higher signal intensity of spin echoes can be exploited using the PRESS sequence,[13] which evokes a spin echo from the magnetisation located in the cuboid formed by the intersection of three slices. The magnetisation in the first plane is excited by a slice-selective 90° pulse and refocused by two 180° pulses applied in the other two mutually perpendicular planes.

This induces the formation of two spin echoes. The first one occurs after the time interval $2t_1$ and the second one appears after an additional time. Non-ideal RF pulses give rise to additional coherences, which can be dephased by spoiler gradients. Compared to STEAM, the minimum echo time of PRESS is longer.

MRS imaging (MRSI) methods acquire spectroscopic data from a sufficiently large number of voxels so that the assessment of tissue heterogeneity becomes possible. For spatial encoding, the frequency data must be preserved, as it contains the spectroscopic information. Thus, spatial encoding is accomplished by pure phase encoding and not frequency encoding. Chemical shift imaging (CSI)[14] is an example of such a technique. In the case of 1D CSI, N time domain signals are sampled under incrementing phase gradient strength over subsequent acquisitions. The Fourier transform results in N individual MR spectra.

Because the most abundant fraction of molecules in mammalian bodies is water, its dominating signal contribution from the two H atoms at 4.7 ppm partially interferes with the metabolite peaks in the spectrum and gives rise to baseline distortions and spurious signals due to vibrations evoked by the gradient system.[4] In order to prevent the occurrence of the water peak, the differences in properties such as T_1 or T_2 relaxation, chemical shift, and scalar coupling between water and metabolites can be exploited. This chapter highlights the principles of two important methods: frequency selective excitation and refocussing.

The basic components of frequency selective excitation (chemical shift selective (CHESS) water suppression[15]) consist of an RF pulse, which is designed in such a way that it selectively excites the frequency band spanning only the water resonance,[16] and a subsequently applied magnetic field gradient that dephases all the coherences (Figure 4.2(c)). If the B_0 and B_1 fields are homogeneous enough, one CHESS step would be sufficient. However, in real applications with imperfect shim results, up to six steps are usually required to achieve satisfying water suppression. Additional cycles extend the frequency range around the water resonance so that on the occurrence of B_0 inhomogeneities, even spins of water molecules with frequency offsets become excited. Moreover, the impact of B_1 inhomogeneities when the nutation angle deviates from 90° is lowered. In the STEAM sequence, CHESS water suppression can be integrated in the preparation period and in the TM time interval just before the spins are flipped again for signal acquisition.[17]

By using 180° frequency selective pulses, the vulnerability of excitation-based water suppression to reappearing magnetisation due to T_1 relaxation can be traded against a longer minimum echo time, required by the introduction of these 180° pulses (Figure 4.2(d)). However, this comes with the penalty of enhanced signal relaxation due to T_2^*. The requirement for longer echo times means that these methods are only applicable to long echo time MRS, such as spectral editing sequences. The common principle of such methods is the exclusive refocusing of the metabolite signals during the dephasing of the water signal, which is exemplified by MEGA water suppression[18,19] integrated into a spin echo sequence (Figure 4.2(d)). The total magnetisation (from water and metabolites) is flipped to the transverse plane by applying a non-selective 90° pulse. The signal arising from the metabolites under investigation dephases when exposed to the gradients G_1 and G_2 and is only refocused by the subsequent non-selective 180° pulse. In addition, the water signal experiences two selective refocusing pulses so that it dephases under the influence of the gradient pairs.

4.3 X-nuclei

4.3.1 Overview

Conventional MRI targets the hydrogen nucleus, which is highly abundant in the human body and offers high sensitivity. In contrast, as the use of ultra-high field scanners has, to some extent, alleviated the problems of sensitivity and abundance associated with the use of non-proton nuclei (or X-nuclei), the *in vivo* use of nuclei such as sodium (^{23}Na) or phosphorous (^{31}P) has recently gained prominence. X-nuclei participate in many metabolic processes and are of substantial interest to clinicians and researchers. Although metabolic information can be obtained using other technologies, for example PET in particular, with emerging hybrid technology this can be seen as a complementary means to analyse various aspects of metabolic processes.[20] By itself, X-nuclei MRI is, in most cases, non-invasive, allows measurements at concentrations from natural abundance and can track entire chains of biochemical processes. Consequently, this makes X-nuclei MRI a very useful aid in observing, understanding and analysing metabolism and its disease.

The most common X-nuclei investigated *in vivo* are ^{2}H, ^{7}Li, ^{13}C, ^{17}O, ^{19}F, ^{23}Na, ^{31}P, ^{35}Cl and ^{39}K, the use of which is, however, penalised by lower abundances and reduced gyromagnetic ratios (see Table 4.1). It should be noted

Table 4.1 Common X-nuclei in biological tissue with non-zero spin (data from de Graaf,[4] Stone[103] and Cercignani, Dowell and Tofts[104]).

Nuclide	Spin	Gyromagnetic ratio $\gamma/2\pi$/MHz T^{-1}	Magnetic dipole moment μ_z/μ_N	Natural abundance (%)	Relative sensitivity (%)
^{1}H	1/2	42.58	2.7928	99.9885	100
^{2}H	1	6.54	0.8574	0.0115	0.97
^{7}Li	3/2	16.55	3.2565	92.41	29.4
^{13}C	1/2	1.07	0.7024	1.07	1.59
^{17}O	5/2	−5.77	−1.8938	0.038	2.91
^{19}F	1/2	40.05	2.6269	100	83.2
^{23}Na	3/2	11.27	2.2177	100	9.27
^{31}P	1/2	17.25	1.1316	100	6.65
^{35}Cl	3/2	4.18	0.8219	75.76	0.47
^{39}K	3/2	1.83	0.3915	93.258	0.05

that many, but not all, of these nuclei exhibit an intricate spin dynamic compared to hydrogen. In other words, they possess a nuclear spin greater than ½ and consequently offer unique features. Although these remarkable properties can be exploited to gather additional information about the current state and environment of the nuclei, they bring significant technical implications, rendering the measurement process methodologically more challenging. For example, the relaxation processes of X-nuclei often occur on a significantly faster timescale than those of hydrogen nuclei. Furthermore, the relaxation processes are also phenomenologically more sophisticated and deviate from the relatively slow monoexponential decay exhibited by proton spin.

4.3.2 X-nuclei Hardware

Measuring X-nuclei is possible if the scanner is equipped with the hardware necessary to excite and measure at the appropriate frequency of the nuclei in question. In general, this requires RF coils to be tuned to the X-nuclei Larmor frequencies and the corresponding transmit and receive chains to be installed. Double-tuned coils (see Chapter 2) are commonly used, which allow anatomical imaging and shimming to take advantage of the high signal levels offered by protons. In most cases, these coils are optimised for the acquisition of the X-nucleus signal. The proton signal offers a plentiful MR signal, which can be sacrificed to some extent in favour of the best possible X-nucleus image.

4.3.3 Fast Image Acquisition

Fast relaxation rates, as exhibited by *e.g.* ^{39}K, ^{35}Cl, ^{23}Na, ^{17}O, and low thermal equilibrium magnetisation make signal acquisition necessary as early as possible after excitation. Regular (Cartesian) imaging techniques require additional time between the RF pulse and readout to encode spatial information in the image (see Chapter 1). As the acquisition is not carried out from FID data, but from recalled echoes at a later time in the acquisition, a loss in the X-nuclei signal occurs. For nuclei exhibiting rapid relaxation, ultra-short TE (UTE) imaging sequences can be used to maximise the detected signal. These sequences do not require additional echoes for the imaging process; instead, image acquisition begins shortly after the excitation pulse, and thus sampling commences almost at the start of the FID signal and originates from the centre of the *k*-space (or close to it).[21–23]

A simple UTE imaging process is based on radial trajectories. Immediately after the RF pulse, the sampling of spokes takes places to obtain spatially encoded imaging data directly from the FID. Projections are oriented in such a way that the *k*-space is covered as efficiently as possible. Consequently, the sampling of the *k*-space is not uniform and is prone to artefacts. This is, in particular, due to under sampling, which leads to aliasing or streaking and requires more sophisticated reconstruction methods to obtain the images. Often it is necessary to resample the data and apply density compensation, making the reconstruction more computationally demanding. In order to cover a FOV similar to that of a Cartesian sequence, more projections are, in general, necessary compared to the number of phase encode steps in the corresponding Cartesian acquisition.[24,25] If sampling commences during the ramp-up phase of the acquisition gradients, it is referred to as ramped sampling. This is done so measurements can be as free of relaxation as possible, to further reduce TE, and be taken in close proximity to the centre of the *k*-space. Alternatively, it is possible to compensate for the gradient ramp by dephasing outwards in the opposite direction of the readout *via* a gradient blip. To do this, a gradient moment similar to the gradient moment covered during the ramp-up of the acquisition gradient, but opposite in polarity, is applied; the readout begins at the centre of the *k*-space once the acquisition gradient has settled. It is important that the imaging process commences at the centre of the *k*-space, albeit at the expense of a slightly increased echo time. Recently, acquisition of what is referred to as density adapted radial imaging has been successfully applied,[26] where the *k*-space coverage along the projections is optimised by means of changing the *k*-space distance between individual data points. Although (3D) radial imaging has successfully provided images of fast decaying nuclei,[27–30] alternative sampling schemes have been developed in an attempt to make the sampling of *k*-space even more efficient, *e.g.* spiral,[31] cone[32] and twisted projection imaging.[33,34]

4.3.4 Multiple Quantum Filtering

For nuclei with a spin greater than ½, higher order coherences can be accessed *via* a sequence of multiple RF pulses.[35,36] A common method with which to achieve this is with the application of three 90° RF pulses.[37,38] The first pulse creates transverse magnetisation from first order coherences, after which the system is left to evolve for a preparatory time, τ. The second pulse populates third-order coherences. After a mixing time, δ, the third RF pulse transfers higher-order coherence to the first order and thus makes detection *via* MR methods possible. In order to isolate the signal only to the desired coherences, additional measures, such as phase cycling, must be taken.[39] Multiple quantum filters have the potential to induce weightings towards intra/extracellular space, exploiting the fact that quadrupolar nuclei, such as sodium, exhibit an energy structure that is strongly dependent on the environment of the nuclei. Unfortunately, multiple quantum filters suffer from significant signal-to-noise ratio (SNR) penalties, making their application demanding and time-consuming. However, emerging ultra-high field technologies offer significant benefits in terms of SNR and make high-resolution multiple quantum filtering feasible in certain clinical protocols.

4.3.5 Expose of Important X-nuclei

4.3.5.1 Phosphorus

Phosphorus is found in various molecular compounds, such as deoxyribonucleic acid (DNA) and ribonucleic acid (RNA), adenosine triphosphate (ATP), phosphocreatine (PCr), phosphodiester (PDE), inorganic phosphate (Pi) and phosphomonoester (PME). Phosphorus is involved in a multitude of metabolic processes including, for example, as part of the body's energy transport and storage system. Magnetic resonance spectroscopy is able to detect phosphorus in ATP, PCr, PDE, Pi and PME. In phosphorous spectroscopic methods, the area under the phosphorus spectral peaks representing the concentration of these compounds is a marker of metabolic function and disease.

The natural abundance of phosphorus is rather low and its MR sensitivity is significantly lower than that of hydrogen. Additionally, T_1 is long while T_2 is short,[40,41] making the data acquisition process lengthy and fairly inefficient. Increased magnetic field strength,[42] sophisticated sampling schemes[43] or the nuclear Overhauser effect (NOE)[41,44] can be used to improve the acquisition process.

4.3.5.2 Carbon

Carbon is a key component in the biochemistry of life and is found in a multitude of molecules and compounds. It participates in a wide range of metabolic processes, fluxes and conversions. Unfortunately, the majority of carbon is attributed to the MR-inactive C-12 isotope. The C-13 isotope, which possesses a spin of ½, is penalised with a low natural abundance. The low sensitivity motivates the use of sophisticated means to boost MR signal creation *via* NOE,[45] the application of polarisation transfer,[46] hyperpolarisation[47,48] or with the external administration of C-13.[49] The Larmor frequency of C-13 is roughly a quarter of the hydrogen Larmor frequency, thus appropriately tuned coils and matching transmit and receive chains are necessary for its detection. Due to the large chemical shift of C-13, analysing C-13 spectra is often easier than proton spectra, nevertheless, the area under the MR peaks is not a reliable measure of the number of nuclei contributing to the resonance, as attached hydrogen elevates the carbon signal or induces splitting through coupling mechanisms. NOE, polarisation transfer and proton decoupling might help mitigate these issues to some extent and boost SNR significantly.[50,51]

4.3.5.3 Fluorine

Fluorine is a nucleus with a high NMR sensitivity, second only to the proton. Unfortunately, fluorine is not present in noteworthy quantities in living organisms. Short relaxation times and therefore a broad spectrum from nuclei in solid structures, such as bone, hinder natural abundance measurements. However, good contrast can be achieved if fluorine is administered as a contrast agent because there is almost no endogenous background signal. For example, with the use of appropriate hardware, the intake and distribution of pharmaceuticals containing fluorine can be monitored, while cellular hypoxia, extracellular pH and intracellular metal ions can be observed using fluorine MRI.[52–56] Even though the range of chemical shift allows for high spectral resolution, spectroscopy of fluorine is challenging. This is due to the fact that the chemical shift range is so large that a sufficiently broadband, powerful transmitter and an equally broadband and sensitive receiver are necessary. Similarly, spin-lock techniques also suffer from the large chemical shift range. Thus, fluorine is commonly used as a target for imaging, rather than for spectroscopy. Usually, fluorine imaging is complemented with the use of additional proton-based scans to obtain anatomical images.

4.3.5.4 Sodium

Sodium is the second most abundant element in the human body and is a promising biomarker.[57,58] It exists in both intra- and extra-cellular space, participates in numerous metabolic processes and its concentrations are strictly regulated in healthy tissue *via* Na/K ATPase. The sodium–potassium exchange maintains a constant concentration gradient across the cell membrane and consumes energy from ATP to do this. Resting potentials arise from different intra- and extra-cellular ion concentrations. Intracellular sodium concentrations are usually between 10–15 mmol L^{-1}, while the extracellular concentration is about 140 mmol L^{-1}.[59] Sodium concentration, in particular intracellular sodium, can serve as a biomarker for existing or emerging pathology.[60,61] Often, when sodium concentration increases are detected, they have been caused either by an increased intracellular sodium concentration, an increase of the volume of the extracellular space or by an increase of vascularisation.[62] Although the concentration of intracellular sodium is lower than that of sodium in the extracellular space, it is an attractive parameter relating to metabolic function and cell viability. Therefore, the ability to extract the intracellular signal and observe its changes, which might otherwise be concealed by changes in extracellular space, is of great interest. Methods to reliably discriminate between MRI signals from both compartments are based on principles exploiting the fact that sodium, being quadrupolar in nature, exhibits different spectral properties and relaxation dynamics, depending on the environment of the nuclei. In addition to invasive methods, which enrich the extracellular space with paramagnetic shift reagents to separate the intra- and extra-cellular sodium peaks,[63–65] non-invasive techniques based on relaxometry[66,67] and multiple quantum filters[68–70] have been proposed to induce weightings towards compartmental sodium. However, multiple quantum filtered methods are currently the subject of on-going discussion in the scientific community; the origin of the triple quantum filtered sodium signal is not yet fully understood. Nevertheless, it has been observed in experiment, that whereas the sodium density weighted signal commonly originates from cerebrospinal fluid (CSF) in ventricles, central canal, vitreous humour (VH), and subarachnoid space, the multiple quantum weighted signal is elevated in tissue, *i.e.* grey and white matter.[37,68,71]

4.3.5.5 *Lithium*

Although lithium does not play a major role in human metabolism *per se*, it does affect neurological processes and is used as a treatment for mental disorders.[72] In the human brain it is detectable *via* spectroscopy.[73] Thus, lithium MRS offers a means to observe lithium treatments of acute mania and bipolar disorders.[74] Lithium MR methods offer a relatively high sensitivity and the nucleus is sufficiently abundant in the human brain, nevertheless its chemical shift is fairly low.[75] Applications are mainly based on localised MRS. Lithium relaxometry has been shown to be able to discriminate between intra- and extra-cellular lithium.[76]

4.3.5.6 *Potassium and Chlorine*

Potassium is mainly present in cells, while chlorine is mostly present in extracellular space. Synaptic efficacy is strongly dependent on the presence of transmembrane ionic gradients, in which potassium and chlorine are involved. Rapid relaxation processes and low abundances, and in the case of potassium a large number of MR inactive nuclei in tissue, make measurement and quantification extremely challenging. To date, only a few *in vivo* spectroscopy[77–79] and imaging[80–83] results using potassium and chlorine have been reported in humans.

4.3.5.7 *Oxygen*

Oxygen plays a vital role in the metabolism. Measuring dynamic changes in the distribution of oxygen in the human body can be used to observe functional activity[84] and to observe oxygen metabolism.[85,86] For example, by direct detection of O-17, the cerebral rate of oxygen consumption ($CMRO_2$) can be probed[85,87–92] as opposed to using indirect fMRI *via* the blood oxygen level dependent (BOLD) effect, oxygen MRI could, in principle, be used as a direct measure of brain function. However, the only MR-active oxygen isotope (O-17) is low in natural abundance and in sensitivity. Commonly, O-17 is administered externally *via* the inhalation of O-17 enriched oxygen. As the relaxation times of O-17 are largely field independent, the use of ultra-high fields is especially beneficial, bringing a four-fold increase in sensitivity upon going from 4.7 T to 9.4 T.[93] The indirect detection of O-17 is also possible *via* scalar coupling between hydrogen and oxygen in water molecules, where the T_2 and rotating frame spin lattice relaxation time $T_1\rho$ of protons is altered. Commonly, RF saturation and spin-lock techniques are used to reduce J coupling and, thus, to amplify sensitivity.[94–98]

4.3.5.8 Deuterium

Deuterium is a spin-1 nucleus. It is quadrupolar and has low natural abundance. Compared to the nucleus of 'normal' hydrogen, the proton, its T_1 time is significantly shorter.[99,100] Additionally, the prolonged T_1 experienced by coupled protons, due to lack of dipolar interaction, makes indirect detection possible. Deuterium is a promising tracer, as it is stable and can be easily distributed within the body.[101] This makes it a useful tool with which to observe various effects such as processes in the nervous system, mitosis and membrane function.[102]

4.4 Anatomical and Functional Imaging

MRI has revolutionised modern medicine with its ability to produce images of the brain, non-invasively and with exquisite soft tissue contrast.

Furthermore, this is increasingly feasible at ultra-high fields, where additional contrast mechanisms, such as phase contrast, become increasingly relevant. With SAR limits being an important constraint in high-field imaging, gradient-echo imaging, which deposits little power in the brain, is the method of choice for many MRI applications. Not only is it fast and versatile in contrast, but also offers two, simultaneously acquired and largely complementary, contrasts—magnitude and phase. The human brain can thus be investigated *in vivo*, with a resolution of a few hundred microns, with adequate SNR and with multiple contrasts in a reasonable scanning time.

At high fields, it becomes possible to characterise anatomical landmarks which are clinically important but too small to be accurately characterised at lower fields. Among others, the visualisation of the substantia nigra (SN) and the subthalamic nucleus (STN) can be achieved using high resolution T_2^* imaging[105] and quantitative susceptibility contrast (Figure 4.3). These regions are particularly relevant in Parkinson's disease, most crucially for localising the STN for precise implantation of electrodes for deep brain stimulation.

Early concerns regarding high field imaging in humans included the issue of diminishing T_1 contrast between white and grey matter with increasing field strength.[106] T_1 is currently the most widely used contrast for healthy brain imaging and is mainly acquired with inversion-recovery prepared and rapid read-out sequences (*e.g.* MP-RAGE). Fortunately, concerns regarding contrast were unfounded and T_1 contrast *in vivo* proved to be a powerful tool for investigating brain structure at high fields.[107,108] It is now routinely used in ultra-high field anatomical imaging as a quantitative contrast, based on an extension of the lower field imaging method, MP2RAGE.[109]

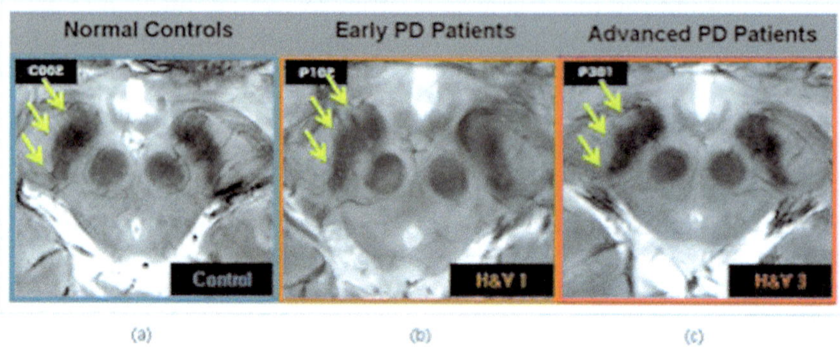

Figure 4.3 Typical selected axial view images of normal control and PD patients obtained from 7.0 T T_2^*-weighted imaging. Note the boundaries between the SN and CP, which show the difference between normal and PD patients. Reproduced from ref. 105 with permission from John Wiley and Sons, Copyright © 2008 Wiley Periodicals, Inc.

One of the probable targets of anatomical and functional research at ultra-high field is the cerebellum. Although it contains more neurons than the cerebrum, its complicated, filigrane anatomical and fibre structure has largely prevented a detailed MRI, diffusion and fMRI study of its properties at lower fields. This becomes feasible at 7 T and Marques *et al.*,[110] report identification of cerebellar layer structure, *i.e.* the granule and molecular layer. An example of high-contrast imaging of the cerebellum at 9.4 T (own data) is shown in Figure 4.4.

A major goal of imaging neuroscience is the precise, routine and objective identification of the different brain areas associated with specific observable operations for individual human subjects. This enables questions regarding the computational role of specific cortical architectures to be empirically addressed. This goal would be easier to achieve if MRI-based structural parcellation of the living human brain could be automatically performed. The identification of cortical features, and *in vivo* parcellation based on them, is one of the main aims of anatomical imaging at ultra-high fields. Variations in myelin distribution can be well visualised, especially at high fields. Myelin is the major source of contrast in T_1-weighted[111] and, to some extent, also in proton density images, and is responsible for T_2 and T_2^* contrast in most of the brain.

The identification of anatomical cortical landmarks based on their T_2 or T_2^* contrast, but also most notably, phase/susceptibility contrast, has been demonstrated several times. This is also possible at clinical fields,[112,113] especially if extensive averaging is performed,[114] but becomes much more feasible, even routinely, at ultra-high fields such as 7 T and beyond.[115,116]

However, for ultra-high-resolution imaging *in vivo*, subject motion presents severe difficulty. The effects of head motion on a typical structural scan,

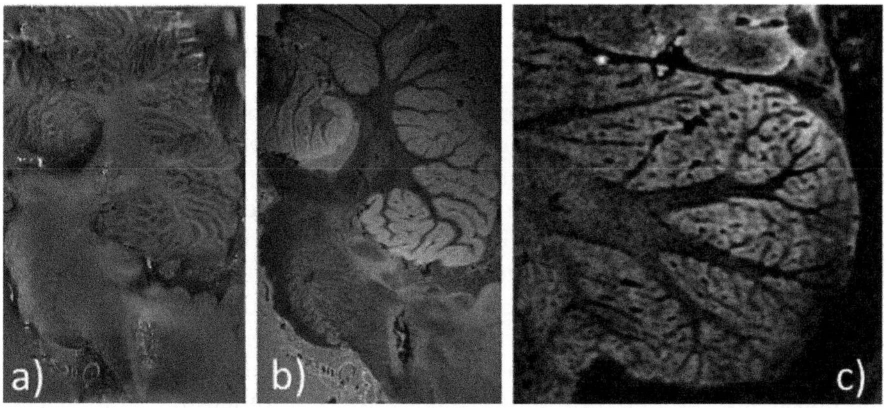

Figure 4.4 High-resolution images of the cerebellum at 9.4 T with slices crossing the dentate nucleus. (a) and (b) are images obtained from a whole, fixed post-mortem brain; (c) image obtained from a healthy volunteer. A multiple-echo 3D GRE sequence was used in both cases with isotropic acquisitions having a linear voxel dimension of 150 μm for post-mortem imaging and 300 μm for *in vivo*. Phase (a) and magnitude (b) images show largely complementary contrast. The very high WM-GM contrast in (c) could allow for segmentation of tissue classes and enable accurate characterisation of cerebellar GM volume *in vivo*.

acquired using multiple spin excitations, are blurring and the formation of multiple spatially translated images. These are often accompanied by an artefact distributed across the image, which results from discontinuities in *k*-space data arising from motion between successively acquired lines in *k*-space. *Post hoc* methods for the correction of such artefacts[117] have had little success.

Effective measures to overcome these problems depend mainly on the prospective adjustment of the parameters determining slice position. Measured head movement parameters are passed to the scanner and converted into a corresponding adjustment of the imaging gradients. The imaged volume, therefore, always moves with the subject's head, and the resulting images require far less post-acquisition correction for head motion artefacts. Methods include the use of navigator echoes and several strategies for independent tracking of the head position.

4.5 Emerging Applications

4.5.1 Phase and Susceptibility Weighted Imaging

Up until approximately a decade ago, phase information was still largely discarded from the reconstructed images offered by manufacturers on clinical scanners.

Phase images, however, contain unique information about local susceptibility changes between blood and tissues or between tissue types,[118,119] which can be useful in measuring iron content,[120] calcifications,[121] or other substances which change the local field, *e.g.* deoxygenated blood or contrast agents.

This tissue-specific information is, unfortunately, obscured by the inhomogeneities of the applied field, especially those created by the head itself by its geometry and rough susceptibility distribution (air spaces, bone, tissue). These effects only increase with field strength. In addition to the development of better shim systems for ultra-high field imaging, methods for phase filtering have also been developed.

The simplest one, homodyne filtering,[118] opened the way to inspecting phase information on line. To make the information more accessible to radiologists, in addition to using phase and magnitude images separately, the phase and the magnitude information was combined in a new susceptibility-weighted magnitude image, which is now referred to as SWI.[122] The combination of magnitude and phase data is obtained by binarising the phase information (*e.g.* by setting paramagnetic values to 1 and all the rest to 0) and multiplying it a number of times into the magnitude image.[122] An example of the resulting contrast, strongly enhancing visualisation of the venous system, is shown in Figure 4.5.

A major problem of using phase information, even after removing the background field, is that its precise distribution shows dependence on the geometry of the object creating it and its orientation with respect to the applied field.[123] This is circumvented to some extent in SWI venography by choosing the transverse orientation of the slices and an optimised voxel aspect ratio.[122]

4.5.2 Quantitative Susceptibility Mapping

In order to make the phase information quantitative, and, in principle, independent of the orientation of the object in the magnetic field, the relation between the measured local field and susceptibility sources needs to be inverted.[124]

An initial issue relating to this is finding a more realistic way to eliminate the effect of the background phase than homodyne filtering, which quite severely reduces tissue contrast. Several algorithms for background phase filtering have been proposed and reviewed.[125]

The magnetic field at a given point is the sum of contributions from all surrounding magnetic susceptibility sources, each contribution depending on the distance to the observation point and orientation in the applied field. The sources can be modelled as a distribution of dipoles. This gives rise to a complicated relationship between the resulting magnetic field and the underlying susceptibility distribution in the R^3 space. However, the inversion of this relationship is easily feasible in Fourier space,[126] where the convolution of the susceptibility distribution with the dipole field becomes multiplicative. Unfortunately, in the Fourier domain, the dipole kernel that has values close to zero on two cone surfaces centred at $\pm 54.7°$ angle with

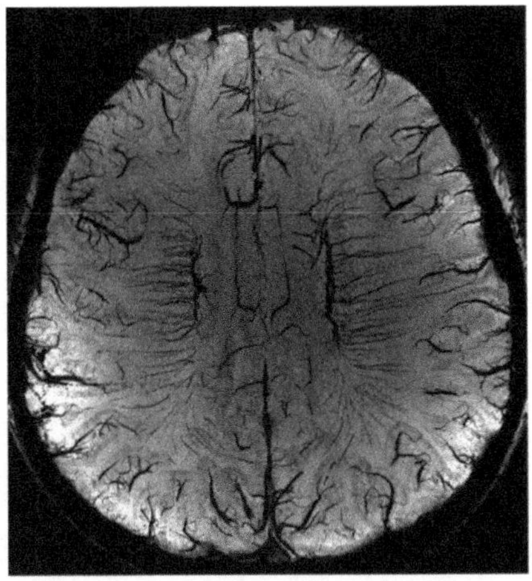

Figure 4.5 A sample minimum Intensity Projection (mIP) from 7 T data, with a resolution of 215 × 215 × 1000 μ, TE = 16 ms, TR = 45 ms, FA = 25° and mIP over 8 sections.[121]

the direction of the main magnetic field and a naive inversion gives rise to severe streaking artefacts.[125,127] This problem can be circumvented, for example by acquiring phase data at different orientations with respect to the field and using a subset of the information from each data set. Thus, 'missing data' relating to susceptibility in one orientation can be recovered by reorienting the subject in a fixed magnet and resampling the MRI signal (COSMOS).[128] Several techniques for accurate susceptibility reconstruction from data acquired at a single orientation have been developed and reviewed.[125]

Thus, it is feasible to use QSM to measure physiological parameters that are sources of susceptibility, such as ferritin content, which is relevant to *e.g.* most neurodegenerative diseases, or venous deoxyhaemoglobin used for quantitative fMRI, or—in combination with perfusion information—to quantify the cerebral metabolic consumption rate CMRO2.

The phase information required for QSM inversion can be obtained at any field with equivalent contrast for B_0 × TE = constant. However, the increased phase SNR obtainable at ultra-high fields, in combination with the availability of mature inversion algorithms, opens the way to routinely using QSM as an anatomical (Figure 4.6) as well as a quantitative method for clinical applications.

4.5.3 Chemical Exchange Saturation Transfer (CEST)

Chemical exchange saturation transfer (CEST) is a novel MR technique that enables the imaging of certain compounds at concentrations that are too low to be detected directly in MR spectroscopy. For this to be possible, the target compound must be capable of exchanging protons with the surrounding

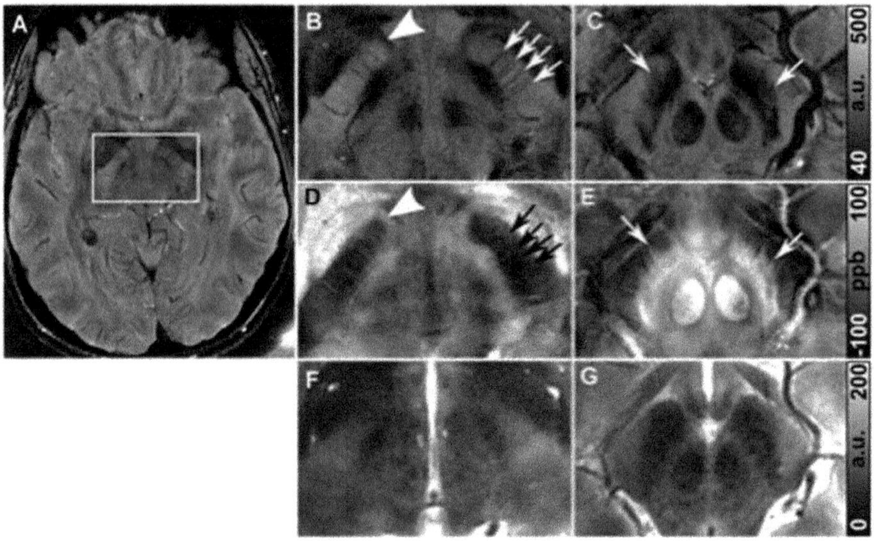

Figure 4.6 Gradient-echo imaging of a 29 year-old subject at 7 T showing the midbrain with 0.4 mm in-plane resolution. An axial magnitude image at the level of the subthalamic nucleus is shown in (A). The white box indicates the anatomic region that is depicted enlarged in (B), (D), and (F), displaying a susceptibility weighted image, a susceptibility map, and a T_2-weighted image, respectively. Sections of a corresponding susceptibility weighted image, susceptibility map, and T_2-weighted image at the level of the red nucleus are presented in (C), (E), and (G), respectively. Arrows in (B) and (D) indicate tubular structures running from the STN to the GPi that may represent pallidofugal fibres made up by the lenticular and subthalamic fasciculus. The arrowhead points to fibre tracts that correspond to the anatomic course of the ansa lenticularis. The arrows in (C) and (E) mark featherlike structures starting in the substantia nigra and running through the cortical spinal tract along a course that matches the anatomic course of striatonigral fibres. Reproduced from ref. 137 with permission from Oxford University Press.

water molecules. This property can be exploited to cause a continuous increase in the saturation of water, leading to greatly enhanced sensitivity. Figure 4.7 depicts this effect schematically.

In addition to exogenous contrast agents, all endogenous biomolecules can be used as 'contrast agents' if they have a proton which exchanges with those from a water molecule. This opens the way to the availability of molecular information for the diagnosis of pathologies, or for the detection of molecular response to treatment—true molecular imaging with MRI. The potential of CEST is exemplified with applications to stroke and brain tumours.

An important CEST application is amide proton transfer (APT) MRI. Mobile proteins and peptides have amide groups that exhibit a relatively narrow range of MR frequencies centred at ~3.5 ppm and a chemical

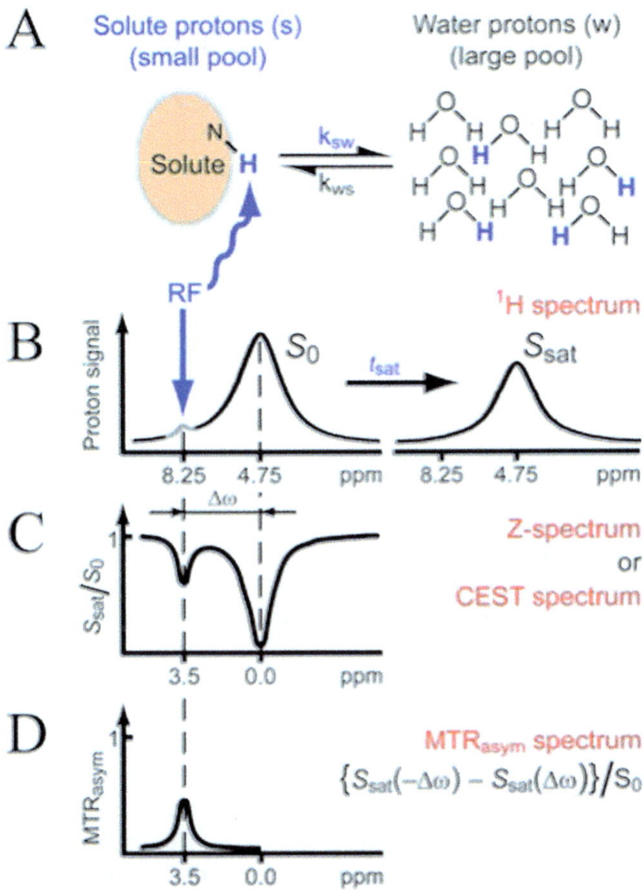

Figure 4.7 Chemical Exchange Saturation Transfer (CEST): principles and measurement approach for pure exchange effects (A, B) solute protons (blue) are saturated at their specific resonance frequency in the proton spectrum (here 8.25 ppm for amide protons). This saturation is transferred to water (4.75 ppm) at exchange rate k_{sw} and non-saturated protons (black) return. After a period (t_{sat}), this effect becomes visible on the water signal (B, right). (C) Measurement of normalised water saturation (S_{sat}/S_0) as a function of irradiation frequency, generating a so-called Z-spectrum (or CEST spectrum or MT spectrum). When irradiating the water protons at 4.75 ppm, the signal disappears due to direct (water) saturation (DS). This frequency is assigned to 0 ppm in Z-spectra. At short saturation times, only this direct saturation is apparent. At longer t_{sat} the CEST effect becomes visible at the frequency of the low-concentration exchangeable solute protons, now assigned to 8.25 − 4.75 = 3.5 ppm in the Z-spectrum. (D) Result of Magnetisation Transfer Ratio (MTR = 1 − S_{sat}/S_0) asymmetry analysis of the Z-spectrum with respect to the water frequency to remove the effect of direct saturation. We note that the standard NMR chemical shift assignment for water in ^{1}H spectra is at 4.75 ppm, while the 0 ppm referencing is used in Z-spectra. Reproduced from ref. 129 with permission from John Wiley and Sons, Copyright © 2011 Wiley-Liss, Inc.

exchange rate of around 30 Hz.[130] The concentration of amide groups in endogenous mobile peptides and proteins is in the range of 5–8 mM in most tissues.[130] These three conditions are appropriate for the detection of amide protons with CEST MRI. APT MRI was successful in identifying ischemic regions of brain tissue in patients who had suffered a cerebrovascular stroke. Decreased cerebral blood flow during a stroke causes the intracellular pH to decrease.[131] This decreased pH causes a decrease in the exchange rate of amide protons because the chemical exchange of amides in proteins is base-catalysed. The lower exchange rates of amide protons cause ischemic tissue to have decreased APT MRI contrast. Very importantly, the ischemic core can show a significant decrease in APT compared to the hypoperfused region surrounding the infarcted tissue that ultimately survives.[132,133]

Similar to APT MRI arising from endogenous amide protons, also the amine protons of endogenous mobile proteins, peptides and amino acids can generate CEST MRI contrast. As with amide protons, the chemical exchange of amine protons with water is base-catalysed. The rate of chemical exchange of amine protons with water is faster than that displayed by amide protons. It can be even too fast to generate CEST at a pH greater than 7.0. An acidic tumour microenvironment of less than pH 7.0 slows the chemical exchange of amide protons, causing an increase in CEST contrast.

APT MRI has been used in the investigation of tumours, although the precise source of the imaging contrast is still being discussed. The concentrations of mobile proteins and peptides were determined to be higher in tumours[134] thus generating higher APT MRI contrast. Metabolically active tumours produce excess lactic acid in the extracellular tumour microenvironment as a consequence of enhanced aerobic glycolysis, known as the Warburg effect.[135] Because the chemical exchange of protons from amides to water is base-catalysed, this decrease in tumour pH should generate lower APT MRI contrast due to the slower exchange rates from amide groups. These two counteracting conditions can cause the APT MRI contrast from tumours to be only slightly higher or unchanged, relative to normal tissues, especially for low-grade tumours.

A combination of APT MRI and amine CEST MRI can be used by comparing the ratio of CEST contrasts at 3.5 and 2.75 ppm to exploit the different pH dependencies of CEST signal amplitudes on the chemical exchange rates of amide and amine protons. This is known as amine–amide concentration independent detection (AACID)[136] and shows potential for monitoring tumour response to treatment.

References

1. K. Ugurbil, *IEEE Trans. Biomed. Eng.*, 2014, **61**, 1364.
2. J. H. Duyn, *NeuroImage*, 2012, **62**, 1241.

3. P. Balchandani and T. P. Naidich, *AJNR Am. J. Neuroradiol.*, 2015, **36**, 1204.
4. R. A. de Graaf, *In vivo NMR Spectroscopy: Principles and Techniques*, Wiley-Interscience, 2nd edn, 2007.
5. C. Juchem and D. L. Rothman, in *Magnetic Resonance Spectroscopy*, ed. C. Stagg and D. Rothman, Academic Press, San Diego, 2014, p. 3.
6. J. Keeler, *Understanding NMR Spectroscopy*, John Wiley and Sons Ltd, 2nd edn, 2010.
7. S. W. Provencher, *Magn. Reson. Med.*, 1993, **30**(6), 672.
8. M. Wilson, G. Reynolds, R. A. Kauppinen, T. N. Arvanitis and A. C. Peet, *Magn. Reson. Med.*, 2011, **65**(1), 1.
9. D. Stefan, F. Di Cesare, A. Andrasescu, E. Popa, A. Lazariev, E. Vescovo, O. Strbak, S. Williams, Z. Starcuk, M. Cabanas and others, *Meas. Sci. Technol.*, 2009, **20**(10), 104035.
10. J.-B. Poullet, D. M. Sima and S. Van Huffel, *J. Magn. Reson.*, 2008, **195**(2), 134.
11. J. Frahm, K.-D. Merboldt and W. Hänicke, *J. Magn. Reson.*, 1987, **72**(3), 502.
12. E. L. Hahn, *Phys. Rev.*, 1950, **80**(4), 580.
13. P. A. Bottomley, *Ann. N. Y. Acad. Sci.*, 1987, **508**, 333.
14. T. R. Brown, B. M. Kincaid and K. Ugurbil, *Proc. Natl. Acad. Sci. U. S. A.*, 1982, **79**(11), 3523.
15. A. Haase, J. Frahm, W. Hänicke and D. Matthaei, *Phys. Med. Biol.*, 1985, **30**(4), 341.
16. M. A. Bernstein, K. F. King and X. J. Zhou, in *Handbook of MRI Pulse Sequences*, ed. M. A. Bernstein, K. F. King and X. J. Zhou, Academic Press, Burlington, 2004, p. 96.
17. C. T. W. Moonen and P. C. M. Van Zijl, *J. Magn. Reson.*, 1990, **88**(1), 28.
18. M. Mescher, A. Tannus, M. Johnson and M. Garwood, *J. Magn. Reson., Ser. A*, 1996, **123**(2), 226.
19. M. Mescher, H. Merkle, J. Kirsch, M. Garwood and R. Gruetter, *NMR Biomed.*, 1998, **11**(6), 266.
20. N. J. Shah, A. M. Oros-Peusquens, J. Arrubla, K. Zhang, T. Warbrick, J. Mauler, K. Vahedipour, S. Romanzetti, J. Felder, A. Celik, E. Rota-Kops, H. Iida, K. J. Langen, H. Herzog and I. Neuner, *J. Magn. Reson.*, 2013, **229**, 101.
21. E. M. Haacke, R. W. Brown, M. R. Thompson and R. Venkatesan, *Magnetic Resonance Imaging: Physical Principles and Sequence Design*, Wiley, 1999.
22. M. A. Bernstein, K. F. King and X. J. Zhou, *Handbook of MRI Pulse Sequences*, Elsevier Science, 2004.
23. D. J. Tyler, M. D. Robson, R. M. Henkelman, I. R. Young and G. M. Bydder, *J. Magn. Reson. Imaging*, 2007, **25**(2), 279.
24. M. L. Lauzon and B. K. Rutt, *Magn. Reson. Med.*, 1996, **36**(6), 940.
25. M. L. Lauzon and B. K. Rutt, *Magn. Reson. Med.*, 1998, **40**(5), 769.

26. A. M. Nagel, F. B. Laun, M. A. Weber, C. Matthies, W. Semmler and L. R. Schad, *Magn. Reson. Med.*, 2009, **62**(6), 1565.
27. R. Jerecic, M. Bock, C. Wacker, W. Bauer and L. R. Schad, *Biomed. Tech.*, 2002, **47**(Suppl 1 Pt 1), 458.
28. S. Nielles-Vallespin, M. A. Weber, M. Bock, A. Bongers, P. Speier, S. E. Combs, J. Wohrle, F. Lehmann-Horn, M. Essig and L. R. Schad, *Magn. Reson. Med.*, 2007, **57**(1), 74.
29. S. Anumula, J. Magland, S. L. Wehrli, H. Zhang, H. Ong, H. K. Song and F. W. Wehrli, *Magn. Reson. Med.*, 2006, **56**(5), 946.
30. Y. Liu, Y. Zhang, C. Wu, J. Zhu, C. Wang, N. Tomko, M. D. Linetsky, R. G. Salomon, C. Ramos-Estebanez, Y. Wang and X. Yu, *Magn. Reson. Med.*, 2018, **79**(1), 256.
31. N. D. Gai, C. Rochitte, M. S. Nacif and D. A. Bluemke, *Magn. Reson. Med.*, 2015, **73**(2), 623.
32. F. Riemer, B. S. Solanky, C. Stehning, M. Clemence, C. A. Wheeler-Kingshott and X. Golay, *MAGMA*, 2014, **27**(1), 35.
33. F. E. Boada, G. X. Shen, S. Y. Chang and K. R. Thulborn, *Magn. Reson. Med.*, 1997, **38**(6), 1022.
34. A. Borthakur, I. Hancu, F. E. Boada, G. X. Shen, E. M. Shapiro and R. Reddy, *J. Magn. Reson.*, 1999, **141**(2), 286.
35. J. R. Keltner, S. T. Wong and M. S. Roos, *J. Magn. Reson., Ser. B*, 1994, **104**(3), 219.
36. D. E. Woessner and N. Bansal, *J. Magn. Reson.*, 1998, **133**(1), 21.
37. I. Hancu, F. E. Boada and G. X. Shen, *Magn. Reson. Med.*, 1999, **42**(6), 1146.
38. D. P. Fiege, S. Romanzetti, C. C. Mirkes, D. Brenner and N. J. Shah, *Magn. Reson. Med.*, 2013, **69**(6), 1691.
39. N. Ivchenko, C. E. Hughes and M. H. Levitt, *J. Magn. Reson.*, 2003, **160**(1), 52.
40. J. Ren, A. D. Sherry and C. R. Malloy, *NMR Biomed.*, 2015, **28**(11), 1455.
41. H. Lei, X. H. Zhu, X. L. Zhang, K. Ugurbil and W. Chen, *Magn. Reson. Med.*, 2003, **49**(2), 199.
42. H. Qiao, X. Zhang, X. H. Zhu, F. Du and W. Chen, *Magn. Reson. Imaging*, 2006, **24**(10), 1281.
43. F. Niess, G. B. Fiedler, A. I. Schmid, S. Goluch, R. Kriegl, M. Wolzt, E. Moser and M. Meyerspeer, *Magn. Reson. Med.*, 2017, **77**(3), 921.
44. M. W. Lagemaat, B. L. van de Bank, P. Sati, S. Li, M. C. Maas and T. W. Scheenen, *NMR Biomed.*, 2016, **29**(3), 256.
45. D. M. Doddrell, D. T. Pegg and M. R. Bendall, *J. Magn. Reson.*, 1982, **48**, 323.
46. D. W. J. Klomp, W. K. Renema, M. van der Graaf, B. E. de Galan, A. P. Kentgens and A. Heerschap, *Magn. Reson. Med.*, 2006, **55**, 271.
47. K. Golman, L. E. Olsson, O. Axelsson, S. Mansson, M. Karlsson and J. S. Petersson, *Br. J. Radiol.*, 2003, **76**, 118.
48. K. P. Golman and J. S. Petersson, *Acad. Radiol.*, 2006, **13**, 932.
49. A. Moreno, B. D. Ross and S. J. Blüml, *J. Neurochem.*, 2001, **77**, 347.

50. R. Gruetter, G. Adriany, I. Y. Choi, P. G. Henry, H. Lei and G. Oz, *NMR Biomed.*, 2003, **16**(6–7), 313.
51. D. W. Klomp, W. K. Renema, M. van der Graaf, B. E. de Galan, A. P. Kentgens and A. Heerschap, *Magn. Reson. Med.*, 2006, **55**(2), 271.
52. G. Brix, M. E. Bellemann, U. Haberkorn, L. Gerlach and W. J. Lorenz, *Nucl. Med. Biol.*, 1996, **23**(7), 897.
53. J. S. van Gorp, P. R. Seevinck, A. Andreychenko, A. J. Raaijmakers, P. R. Luijten, M. A. Viergever, M. Koopman, V. O. Boer and D. W. Klomp, *NMR Biomed.*, 2015, **28**(11), 1433.
54. H. W. van Laarhoven, D. W. Klomp, M. Rijpkema, Y. L. Kamm, D. J. Wagener, J. O. Barentsz, C. J. Punt and A. Heerschap, *NMR Biomed.*, 2007, **20**(2), 128.
55. A. S. Ojugo, P. M. McSheehy, D. J. McIntyre, C. McCoy, M. Stubbs, M. O. Leach, I. R. Judson and J. R. Griffiths, *NMR Biomed.*, 1999, **12**(8), 495.
56. Y. Aoki, K. Akagi, Y. Tanaka, J. Kawai and M. Takahashi, *Invest. Radiol.*, 1996, **31**(11), 680.
57. N. J. Shah, W. A. Worthoff and K. J. Langen, *NMR Biomed.*, 2016, **29**(2), 162.
58. S. Konstandin and L. R. Schad, *MAGMA*, 2014, **27**(1), 1.
59. M. Modo and J. W. Bulte, *Methods Mol. Biol.*, 2011, **711**, 3–14.
60. S. S. Winkler, *Neuroradiology*, 1990, **32**(5), 416.
61. R. Sharma, *Informatica Medica Slovenica*, 2005, **10**(1), 56.
62. G. Madelin, J. S. Lee, R. R. Regatte and A. Jerschow, *Prog. Nucl. Magn. Reson. Spectrosc.*, 2014, **79**, 14.
63. N. Bansal, M. J. Germann, V. Seshan, G. T. Shires 3rd, C. R. Malloy and A. D. Sherry, *Biochemistry*, 1993, **32**(21), 5638.
64. R. K. Gupta and P. Gupta, *Biophys. J.*, 1982, **37**(2), A76.
65. H. Naritomi, M. Kanashiro, M. Sasaki, Y. Kuribayashi and T. Sawada, *Biophys. J.*, 1987, **52**(4), 611.
66. A. M. Nagel, E. Amarteifio, F. Lehmann-Horn, K. Jurkat-Rott, W. Semmler, L. R. Schad and M. A. Weber, *Invest. Radiol.*, 2011, **46**(12), 759.
67. G. Madelin, R. Kline, R. Walvick and R. R. Regatte, *Sci. Rep.*, 2014, **4**, 4763.
68. A. Tsang, R. W. Stobbe and C. Beaulieu, *Magn. Reson. Med.*, 2012, **67**(6), 1633.
69. R. H. Griffey, B. V. Griffey and N. A. Matwiyoff, *Magn. Reson. Med.*, 1990, **13**(2), 305.
70. F. Boada, G. Laverde, C. Jungreis, E. Nemoto and C. Tanase, *Conf. Proc. IEEE Eng. Med. Biol. Soc.*, 2005, **1**, 731.
71. W. A. Worthoff, A. Shymanskaya and N. J. Shah, *Magn. Reson. Med.*, 2018, DOI: 10.1002/mrm.27387.
72. D. P. Devanand, J. G. Strickler, E. D. Huey, E. Crocco, B. P. Forester, M. M. Husain, I. V. Vahia, H. Andrews, M. M. Wall and G. H. Pelton, *Contemp. Clin. Trials*, 2018, **71**, 33.
73. R. A. Komoroski, *NMR Biomed.*, 2005, **18**(2), 67.
74. G. F. Mason and J. H. Krystal, *NMR Biomed.*, 2006, **19**(6), 690.

75. P. F. Renshaw, J. C. Haselgrove, J. S. Leigh and B. Chance, *Magn. Reson. Med.*, 1985, **2**(5), 512.
76. R. A. Komoroski, D. M. Lindquist and J. M. Pearce, *NMR Biomed.*, 2013, **26**(9), 1152.
77. W. R. Adam, A. P. Koretsky and M. W. Weiner, *Biophys. J.*, 1987, **51**(2), 265.
78. A. Stevens, P. Paschalis and T. Schleich, *Biophys. J.*, 1992, **61**(5), 1061.
79. R. M. Wellard, B. P. Shehan, W. R. Adam and D. J. Craik, *Magn. Reson. Med.*, 1993, **29**(1), 68.
80. I. C. Atkinson, T. C. Claiborne and K. R. Thulborn, *Magn. Reson. Med.*, 2014, **71**(5), 1819.
81. A. M. Nagel, F. Lehmann-Horn, M. A. Weber, K. Jurkat-Rott, M. B. Wolf, A. Radbruch, R. Umathum and W. Semmler, *Radiology*, 2014, **271**(2), 585.
82. R. Umathum, M. B. Rosler and A. M. Nagel, *Radiology*, 2013, **269**(2), 569.
83. M. A. Weber, A. M. Nagel, A. M. Marschar, P. Glemser, K. Jurkat-Rott, M. B. Wolf, M. E. Ladd, H. P. Schlemmer, H. U. Kauczor and F. Lehmann-Horn, *Radiology*, 2016, **280**(3), 848.
84. G. D. Mateescu, *Adv. Exp. Med. Biol.*, 2003, **510**, 213.
85. X. H. Zhu and W. Chen, *Prog. Nucl. Magn. Reson. Spectrosc.*, 2011, **59**(4), 319.
86. A. M. Oros-Peusquens, F. Keil, K. J. Langen, H. Herzog, G. Stoffels, C. Weiss and N. J. Shah, *Nucl. Instrum. Methods Phys. Res., Sect. A*, 2014, **734**, 185.
87. J. Pekar, L. Ligeti, Z. Ruttner, R. C. Lyon, T. M. Sinnwell, P. van Gelderen, D. Fiat, C. T. Moonen and A. C. McLaughlin, *Magn. Reson. Med.*, 1991, **21**(2), 313.
88. G. D. Mateescu, in *Adv. Exp. Med. Biol.*, ed. D. F. Wilson, S. M. Evans, J. Biaglow and A. Pastuszko, Springer US, Boston, MA, 2003, p. 213.
89. T. Arai, S. Nakao, K. Mori, K. Ishimori, I. Morishima, T. Miyazawa and B. Fritz-Zieroth, *Biochem. Biophys. Res. Commun.*, 1990, **169**(1), 153.
90. T. Arai, K. Mori, S. Nakao, K. Watanabe, K. Kito, M. Aoki, H. Mori, S. Morikawa and T. Inubushi, *Biochem. Biophys. Res. Commun.*, 1991, **179**(2), 954.
91. X. H. Zhu and W. Chen, *Anal. Biochem.*, 2017, **529**, 171.
92. A. Gordji-Nejad, K. Mollenhoff, A. M. Oros-Peusquens, D. R. Pillai and N. J. Shah, *MAGMA*, 2014, **27**(1), 81.
93. X. H. Zhu, H. Merkle, J. H. Kwag, K. Ugurbil and W. Chen, *Magn. Reson. Med.*, 2001, **45**(4), 543.
94. A. H. Stolpen, R. Reddy and J. S. Leigh, *J. Magn. Reson.*, 1997, **125**(1), 1.
95. I. Ronen, J. H. Lee, H. Merkle, K. Ugurbil and G. Navon, *NMR Biomed.*, 1997, **10**(7), 333.
96. R. Reddy, A. H. Stolpen, S. R. Charagundla, E. K. Insko and J. S. Leigh, *Magn. Reson. Imaging*, 1996, **14**(9), 1073.
97. E. A. Mellon, R. S. Beesam, J. E. Baumgardner, A. Borthakur, W. R. Witschey 2nd and R. Reddy, *J. Neurosci. Methods*, 2009, **179**(1), 29.

98. A. J. de Crespigny, H. E. D'Arceuil, T. Engelhorn and M. E. Moseley, *Magn. Reson. Med.*, 2000, **43**(6), 876.
99. H. H. Mantsch, H. Saito and I. C. P. Smith, *Prog. Nucl. Magn. Reson. Spectrosc.*, 1977, **11**(4), 211.
100. S. H. Koenig, K. Hallenga and M. Shporer, *Proc. Natl. Acad. Sci. U. S. A.*, 1975, **72**(7), 2667.
101. V. E. Kinsey, M. Grant and D. G. Cogan, *Arch. Ophthalmol.*, 1942, **27**(2), 242.
102. D. J. Kushner, A. Baker and T. G. Dunstall, *Can. J. Physiol. Pharmacol.*, 1999, **77**(2), 79.
103. N. J. Stone, *At. Data Nucl. Data Tables*, 2005, **90**(1), 75.
104. M. Cercignani, N. G. Dowell and P. S. Tofts, *Quantitative MRI of the Brain: Principles of Physical Measurement*, CRC Press, New York, 2nd edn, 2018.
105. Z. H. Cho, Y. B. o Kim, J. Y. Han, H. K. Min, K. N. Kim, S. H. Choi, E. Veklerov and L. A. Shepp, *Int. J. Imaging Syst. Technol.*, 2008, **18**, 2.
106. H. W. Fischer, P. A. Rinck, Y. Van Haverbeke and R. N. Muller, *Magn. Reson. Med.*, 1990, **16**, 317.
107. W. D. Rooney, G. Johnson, X. Li, E. R. Cohen, S. Kim, K. Ugurbil and C. S. Springer, *Magn. Reson. Med.*, 2007, **57**, 308.
108. A. M. Oros-Peusquens, M. Laurila and N. J. Shah, *Magn. Reson. Mater. Phys., Biol. Med.*, 2008, **21**, 131.
109. J. P. Marques, T. Kober, G. Krueger, W. van der Zwaag, P. F. Van de Moortele and R. Gruetter, *NeuroImage*, 2010, **49**, 1271.
110. J. Marques, W. van der Zwaag, C. Granziera, G. Krueger and R. Gruetter, *Radiology*, 2010, **254**, 942.
111. S. H. Koenig, *Magn. Reson. Med.*, 1991, **20**(2), 285.
112. E. L. Barbier, S. Marrett, A. Danek, A. Vortmeyer, P. van Gelderen, J. Duyn, P. Bandettini, J. Grafman and A. P. Koretsky, *Magn. Reson. Med.*, 2002, **48**, 735.
113. R. Turner, A. M. Oros-Peusquens, S. Romanzetti, K. Zilles and N. J. Shah, *Magn. Reson. Imaging*, 2008, **26**(7), 935.
114. A. M. Oros-Peusquens, T. Stoecker, K. Amunts, K. Zilles and N. J. Shah, *Magn. Reson. Imaging*, 2010, **28**(3), 329.
115. R. Trampel, D. V. M. Ott and R. Turner, *Cereb. Cortex*, 2011, **21**, 2075.
116. N. J. Shah, A. M. Oros-Peusquens, J. Arrubla, K. Zhang, T. Warbrick, J. Mauler, K. Vahedipour, S. Romanzetti, J. Felder, A. Celik, E. Rota-Kops, H. Iida, K. J. Langen, H. Herzog and I. Neuner, *J. Magn. Reson.*, 2013, **229**, 101.
117. M. E. Bourgeois, F. T. Wajer, M. Roth, A. Briguet, M. Décorps, D. van Ormondt, C. Segebarth and D. Graveron-Demilly, *J. Magn. Reson.*, 2003, **163**(2), 277.
118. J. R. Reichenbach, R. Venkatesan, D. J. Schillinger, D. K. Kido and E. M. Haacke, *Radiology*, 1997, **204**, 272.
119. A. M. Abduljalil, P. Schmalbrock, V. Novak and D. W. Chakeres, *J. Magn. Reson. Imaging*, 2003, **18**, 284.

120. E. M. Haacke, N. Y. Cheng, M. J. House, Q. Liu, J. Neelavalli, R. J. Ogg, A. Khan, M. Ayaz, W. Kirsch and A. Obenaus, *Magn. Reson. Imaging*, 2005, **23**, 1.
121. J. Berberat, R. Grobholz, L. Boxheimer and S. Rogers, *AJR, Am. J. Roentgenol.*, 2014, **202**, 847.
122. E. M. Haacke, S. Mittal, Z. Wu, J. Neelavalli and Y. C. N. Cheng, *AJNR Am. J. Neuroradiol.*, 2009, **30**(1), 19.
123. J. D. Jackson, *Classical Electrodynamics*, John Wiley and Son, Hoboken, NJ, 3rd edn, 1999, p. 184.
124. J. Li, S. Chang, T. Liu, Q. Wang, D. Cui, X. Chen, M. Jin, B. Wang, M. Pei, C. Wisnieff, P. Spincemaille, M. Zhang and Y. Wang, *Magn. Reson. Med.*, 2012, **68**(5), 1563.
125. Y. Wang and T. Liu, *Magn. Reson. Med.*, 2015, **73**(1), 82.
126. G. Deville, M. Bernier and J. Delrieux, *Phys. Rev. B: Condens. Matter Mater. Phys.*, 1979, **19**, 5666.
127. J. P. Marques and R. Bowtell, *Concepts Magn. Reson., Part B*, 2005, **25**, 65.
128. T. Liu, P. Spincemaille, L. de Rochefort, B. Kressler and Y. Wang, *Magn. Reson. Med.*, 2009, **61**(1), 196.
129. P. C. M. Van Zijl and N. N. Yadav, *Magn. Reson. Med.*, 2011, **65**(4), 927.
130. J. Zhou, J. F. Payen, D. A. Wilson, R. J. Traystman and P. C. van Zijl, *Nat. Med.*, 2003, **9**, 1085.
131. P. Z. Sun, J. S. Cheung, E. Wang and E. H. Lo, *J. Cereb. Blood Flow Metab.*, 2011, **31**, 1743.
132. G. W. Harston, Y. K. Tee, N. Blockley, T. W. Okell, S. Thandeswaran, G. Shaya, F. Sheerin, M. Cellerini, S. Payne, P. Jezzard, M. Chappell and J. Kennedy, *Brain*, 2015, **138**, 36.
133. A. Tietze, J. Blicher, I. K. Mikkelsen, L. Østergaard, M. K. Strother, S. A. Smith and M. J. Donahue, *NMR Biomed.*, 2014, **27**, 163.
134. K. Yan, Z. Fu, C. Yang, K. Zhang, S. Jiang, D. H. Lee, H. Y. Heo, Y. Zhang, R. N. Cole, J. E. Van Eyk and J. Zhou, *Mol. Imaging Biol.*, 2015, **17**, 479.
135. O. Warburg, *Science*, 1956, **123**, 309.
136. N. McVicar, A. X. Li, D. F. Goncalves, M. Bellyou, S. O. Meakin, M. A. Prado and R. Bartha, *J. Cereb. Blood Flow Metab.*, 2014, **34**, 690.
137. T. M. Schneider, A. Deistung, U. Biedermann, C. Matthies, R.-I. Ernestus, J. Volkmann, S. Heiland, M. Bendszus and J. R. Reichenbach, *Operative Neurosurgery*, 2016, **12**(4), 330–338.

Section II

Positron Emission Tomography

CHAPTER 5

Introduction to PET

H. HERZOG*[a] AND P. ALMEIDA[b]

[a]Forschungszentrum Jülich, Institute of Neuroscience and Medicine – 4, 52425 Jülich, Germany; [b]Institute of Biophysics and Biomedical Engineering, University of Lisbon, 1749-016 Lisbon, Portugal
*E-mail: h.herzog@fz-juelich.de

5.1 Introduction

Physicians can choose between different imaging procedures to support their diagnostic decisions. These procedures rely on methods that differ in their underlying physical basis, the achievable spatial and temporal resolution of the acquired images and the kind of information they deliver. Whereas structural and anatomical information is primarily obtained by computed tomography (CT), magnetic resonance imaging (MRI) and ultrasound (US), the attractive characteristics of positron emission tomography (PET) are its ability to visualise a variety of metabolic and physiological processes in the living body and to do that quantitatively.

For this purpose, molecules that take part in the metabolic and physiological processes to be investigated are labelled with radioactive nuclides that primarily emit positrons. The most frequently used positron emitters are ^{18}F, ^{11}C, ^{15}O and ^{13}N. The cold sister isotopes of ^{11}C, ^{15}O and ^{13}N are natural elements of organic molecules, which play numerous roles in metabolism. Therefore, when molecules such as glucose or water are labelled with the positron emitting isotopes ^{11}C and ^{15}O, respectively, they remain

New Developments in NMR No. 19
Hybrid MR-PET Imaging: Systems, Methods and Applications
Edited by N. Jon Shah
© The Royal Society of Chemistry 2019
Published by the Royal Society of Chemistry, www.rsc.org

chemically unchanged, but become measurable by external radiation detectors. Most molecules labelled with ^{18}F show a slight change in biochemical behaviour (often in a specifically desired way) since this positron emitter usually substitutes a hydroxyl or methyl group in a naturally existing molecule. A specific feature of PET is that the mass of the radiopharmaceutical injected into the patient is so small that the tracer principle, which is generally valid in nuclear medicine, is not violated. This means that the metabolic processes underlying the investigation are not disturbed by the PET measurement and that the injection does not produce any pharmacological effects.

The measurement of PET radiopharmaceuticals is achieved by positioning radiation detectors, typically as a sequence of rings, around the patient (Figure 5.1). These detectors primarily measure the radiation resulting from the annihilation of the emitted positrons with electrons in tissue. Since the first prototypes PET scanners in the 1970s, the development of better and more sophisticated hardware and software has continuously improved the imaging capabilities of PET. This chapter deals with the physical and technological basics of PET.

5.2 The Physics of Positron Emitters

From the physical point-of-view, PET is based on the emission of positrons from radioactive nuclei having an excess of positive charge. In water or soft body tissue, emitted positrons can travel from several mm to less than

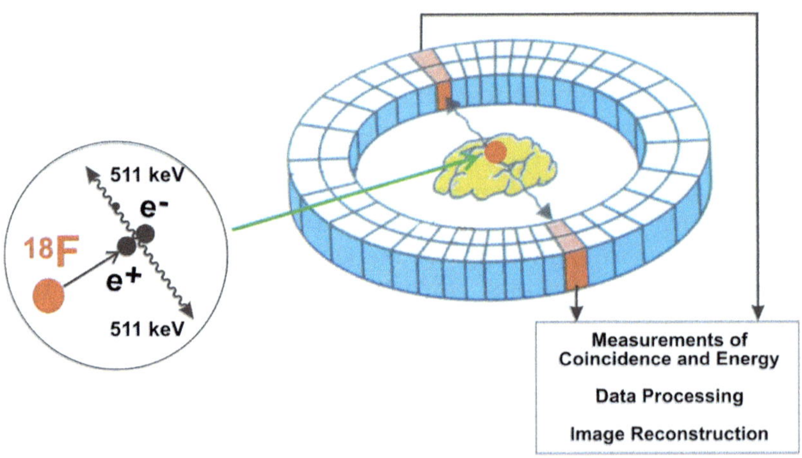

Figure 5.1 A pair of 511 keV photons result from the annihilation of a positron originating from a positron emitting radionuclide (here ^{18}F). The photons are registered by opposing elements of the detector ring, which surrounds the patient. In this way, the line of response (LOR), *i.e.* the line on which the positron emitter is located, can be determined. During a PET measurement thousands of LORs are detected, processed and used for image reconstruction.

Table 5.1 Characteristic data of commonly used positron emitters.

Isotope	Mean positron energy/MeV	Maximum positron energy/MeV	Positron range in water/FWHM in mm	Half-life/min
^{11}C	0.39	0.96	1.1	20.3
^{13}N	0.49	1.19	1.4	9.97
^{15}O	0.74	1.70	1.5	2.03
^{18}F	0.25	0.64	1.0	109.8

1 mm depending on their kinetic energy upon leaving the nucleus (Table 5.1). This range is a fundamental limit to the spatial resolution of PET. Positron emitters differ not only in respect to their range but also in respect to their physical half-life. Positron emitters with a short half-life of only a few minutes (^{15}O, ^{13}N, ^{11}C) must be produced by a cyclotron close to the PET scanner or by a generator system, as in the case of ^{82}Rb. For ^{18}F, distribution systems have been established worldwide so that this positron emitter, with a half-life of about 2 h, is often produced in a remote cyclotron with an attached radiochemical facility, where the final ^{18}F-labelled radiopharmaceutical is synthesised and thereafter transported to the PET scanner.

When the positron has lost its energy to the surrounding material and comes very close to rest, it combines with an electron to become an ultra-short-lived positronium. At the annihilation of the positron–electron pair, two photons, each with an energy of 511 keV, are produced and emitted in nearly opposite directions to ensure the preservation of linear momentum (Figure 5.1). The energy of the two photons corresponds to the mass energy of the positronium so that the preservation of energy is ensured. Since the process of annihilation and the emission of the photon pair are equal for all positron emitters, a detector system can be optimised according to this constant physical property and can be used to detect all types of PET radiopharmaceuticals.

The emission of a pair of 511 keV photons in opposite directions and its registration by two detector elements of the PET ring within a certain time interval, the so-called coincidence time window, can be regarded as the most important feature of PET. If two detector elements of the PET ring each record a 511 keV photon within the coincidence time window of, *e.g.* 5 ns, it is assumed that the two photons originate from the same positron–electron annihilation and that the position of the radiolabelled molecule is on the line connecting the two detector elements. This line is called the line of response (LOR), however, this assumption neglects the positron range since the spatial position of the positron emitter is not the same as that at which annihilation took place. Since the LOR is determined by the coincidence measurement, PET does not need collimators and only applies an electronic 'collimation'. This is in contrast to planar scintigraphy and single photon emission tomography (SPECT), where lead collimators are used to define the direction of the incoming radiation. Like all radioactive events, the positron emission, and thereby the occurrence of the annihilation photons, as well as their recording by the PET detector, varies according to Poisson statistics. Consequently, if a detector measures N annihilation photons in a certain time interval, the standard deviation σ of the measurement is $N^{1/2}$ and the coefficient-of-variance (COV) is $1/N^{1/2}$ For example, if 100 events are measured, COV is 10% and 10 000 measured events lead to a COV of 1%. A PET detector with better sensitivity, *i.e.* with more recorded events or counts per time unit leads to less variance of the measurement. This fact is one motivation to look for improvements to be made in PET detectors, especially with respect to the scintillation crystals, the so-called frontend, but also in terms of faster processing electronics.

5.3 The Detector System

The primary element of a typical PET detector system is the scintillation detector. Such a detector consists of a scintillation crystal and a photodetector, *e.g.* photomultiplier tube (PMT) (see also Section 6.2.2). The annihilation photon excites the scintillator material so that it emits light photons, a part of which hits the photocathode of the PMT where they cause a photoelectric effect, producing electrons. These electrons are amplified within the PMT tube by a factor of approximately 10^6, resulting in a voltage pulse which is proportional to the number of light photons produced within the scintillation crystal and, in this way, proportional to the energy of the incoming annihilation photon. Although this photon has an original energy of 511 keV, it may lose part of this energy on its way through the patient's body to the surface of the scintillation crystal due to Compton scattering. Furthermore, it can happen that the annihilation photon does not lose all its energy in the primarily hit crystal and its energy is scattered to a neighbouring crystal where it is absorbed. This results in each of the two crystals receiving only a part of the original energy of 511 keV (modern systems can rebuild this information and obtain increased sensitivity by doing this). More details about different types of photodetectors can be found in Chapter 6.

Figure 5.2 shows and compares the typical energy spectra of a scintillation detector irradiated with a positron emitter and equipped with either bismuth germanate (BGO) or lutetium orthosilicate (LSO) as scintillation material (see also Table 6.2). Each energy spectrum consists of two parts, the photopeak and the Compton spectrum. The photopeak has its maximum energy value at 511 keV and is Gaussian in shape, due to the statistical processes of the light emission in the crystal and the amplification of the electrons within the PMT. If a crystal material emits more light after being hit by the incoming annihilation photon, the statistical fluctuation becomes less important, resulting in a narrower photopeak. This is the case of LSO when compared to, for example, BGO (Figure 5.2). The Compton spectrum represents the energy distribution of those 511 keV-photons that have lost part of their energy by Compton scattering within the body. As explained below, the overlap between the photopeak and the Compton spectrum should be as small

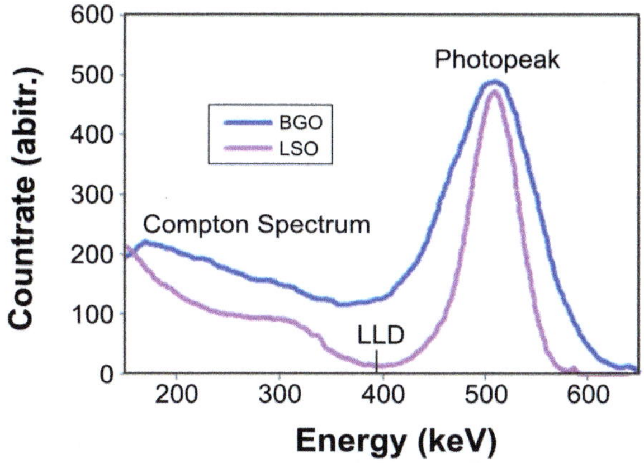

Figure 5.2 Energy spectra of two crystals used in PET. In current systems, BGO has been replaced by LSO because of its narrower photopeak, which results from its better light yield. The lower energy-level discriminator (LLD) is to separate photons which undergo Compton scattering from unscattered photons.

as possible. In this respect, LSO is preferable to BGO, which is one major reason why contemporary PET detectors are nearly exclusively equipped with LSO or lutetium yttrium orthosilicate (LYSO), which has similar behaviour. To separate non-scattered from scattered annihilation photons, an energy window is defined around the maximum of the photopeak, *e.g.* from 450 keV to 650 keV. Only annihilation photons which have an energy compatible with the defined energy window are tested for belonging to a coincidence event. Photons with an energy below the lower limit of the energy window are regarded as scattered and are excluded from further processing. Crystals with a narrow photopeak show a smaller overlap between the photopeak and the Compton spectrum so that scattered and non-scattered coincidences can be better separated.

A PET detector system consists of many scintillation detectors which are commonly arranged into adjacent rings surrounding the patient (Figure 5.3). The PET system installed in the current version of the Siemens PET/CT scanner mCT, for example, consists of 4 rings with 48 detector blocks each leading to an axial field-of-view (FOV) of 21.6 cm and a diameter of 70 cm. In commercial PET systems, the typical detector unit (known as a detector block) consists of a matrix of scintillation crystals with a size of 4 × 4 × 20 mm^3 each, and 4 PMTs, for example. The light emitted from the crystal matrix is distributed within the matrix of the PMTs so that, by appropriate weighing of the voltage pulses of the PMTs, the crystal initially hit by the annihilation photon can be identified.

The spatial resolution of PET is essentially determined by the area of the crystal perpendicular to the radius of the ring. In the example above this area is 4 × 4 mm (16 mm^2). One of the first commercial PET scanners, the

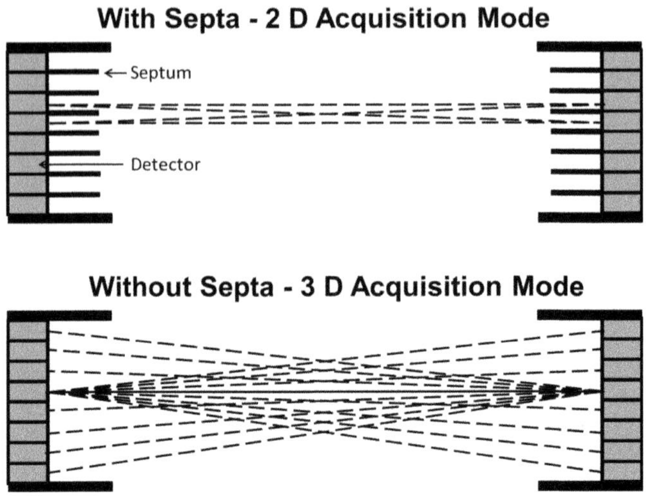

Figure 5.3 Sections of detectors rings with two- and three-dimensional acquisitions modes. In the three-dimensional mode, the septa between the detector rings are retracted or removed so that much more pairs of annihilation photons can be recorded for a given period of time.

ECAT II, had circular crystals with a diameter of 1 inch resulting in an area of 507 mm^2). During the development of PET, this area became smaller and smaller. The PET module of the hybrid 3T MR-BrainPET operated in Jülich has a crystal area of 2.5 × 2.5 mm (6.25 mm^2) with a length of 20 mm. Preclinical, small-animal PET scanners have even smaller crystals, as in the case of the MEDISO PET 122S, which has a crystal size of 1.12 × 1.12 (1.25 mm^2) with a length of 13 mm. As a general rule of thumb, the resolution of a PET system is similar to the side length of the frontal crystal area. The PET modules installed in the whole-body PET/CT scanners of the main manufactures have an intrinsic (*i.e.* not influenced by specific reconstruction features) resolution of 4 to 5 mm (full-width-at-half-maximum = FWHM) based on crystal areas with a side length of about 4 mm.

The sensitivity of a PET system has the unit cpm MBq^{-1} mL^{-1} and represents the number of coincidences (counts per minute = cpm) recorded in a specific measurement arrangement when the underlying radioactivity has a concentration of 1 MBq mL^{-1}. Sensitivity is primarily determined by the crystal material, the diameter and axial length of the PET system, and the length of the scintillation crystal. The early, and some later (due to cost reasons), PET scanners were equipped with NaI crystals.

As shown in Table 6.1, this type of crystal has a low density and was replaced by BGO and more recently by LSO and LYSO, because the higher density of these latter crystals results in a better sensitivity. A PET scanner with a small diameter and a long axial FOV has a greater solid angle, so that annihilation photons emitted from a positron source distribution located at the middle of the scanner hit more scintillation crystals. Finally, the sensitivity can be increased by using longer crystals, so that the chance of a photon being stopped on its way through the crystal is greater. However, despite this benefit, using long crystals results in an increase of the so-called parallax error (see Section 5.5). The first generations of PET scanners measured in a two-dimensional (2D) way, *i.e.* coincidences were only accepted between detector elements belonging to the same ring or the directly neighbouring ring (Figure 5.3). The single rings of the PET scanner were separated by septa made of lead or tungsten. Nowadays, all PET systems measure in a three-dimensional (3D) way without any septa, so that coincidences between detectors belonging to non-adjacent rings are accepted.[1] This development has resulted in a dramatic increase in the sensitivity of PET.

5.4 Time-of-Flight PET

Coincidences are counted independently of the place along the LOR where the annihilations occurred. If the time points when the two annihilation photons hit the opposing detector crystals, and their difference Δt, *i.e.* the difference in time-of-flight (TOF) between hitting the crystals, can be measured exactly, the location of the annihilation can be calculated using eqn (5.1)

$$\Delta x = \Delta t\, c/2 \tag{5.1}$$

where Δx is the distance between the centre of the LOR and the point of annihilation, and c is the speed of light. In this case, the PET image can be constructed by assembling the annihilation events into an image matrix.

However, Δt cannot be measured exactly, only with an uncertainty of some hundreds of picoseconds, depending on the scintillation material used. The scintillator must have a fast decay time (on the order of 500 ps or less), be properly sized regarding low resolution and high sensitivity, and be coupled with fast photodetectors (see also Chapter 6). With the introduction of LSO or LYSO, TOF measurements with an uncertainty of, *e.g.* 500 ps became possible. Therefore, the location of the annihilation event can be restricted to an interval of 7.5 cm instead of to the total length of the LOR.[2] This restriction can then be used in analytical and algebraic image reconstruction algorithms to reduce noise. Furthermore, the excellent timing resolution of TOF PET systems leads to a reduction of random events. Thus, the TOF method effectively reduces the variance in the reconstructed images, allowing a reduction of the injected doses or, alternatively, a reduction in the examination time.

In order to achieve a sub-centimetre spatial resolution from a TOF measurement, a timing resolution of less than 50 ps is necessary, which is not yet achievable in commercial PET systems. However, recently, timing resolutions smaller than 200 ps were achieved in experimental setups (see also Section 6.3.2), offering a great deal of promise for future high-performance, TOF PET systems.[3]

5.5 Depth of Interaction

Ideally, an annihilation photon enters a crystal perpendicularly to its smaller area face and is stopped by interacting with the crystal material on its path through the 2–3 cm long crystal. In this way, one end of a LOR can be assigned to one crystal. However, most annihilation photons hit the frontal crystal surface in an oblique way, especially if the positron emission is located away from the central axis of the detector ring. The photon path through the crystal may, therefore, be short, resulting in the photon being absorbed by a neighbouring crystal. In this case, the LOR linked with this photon is assigned to the neighbouring crystal and not to the crystal originally hit (crossed) by the photon. This results in the so-called parallax error, or depth-of-interaction (DOI) effect and leads to a loss of image resolution due to a broadening of its tangential component (see also Chapter 6). The DOI effect can be reduced by using larger detector rings and shorter crystals. However, both a larger detector ring and a shortened crystal lead to a loss of sensitivity. Therefore, crystals with two or more layers have been suggested[4] and successfully used in experimental PET scanners but have not yet been applied in commercial PET detectors for human studies. Although more relevant for brain PET, the DOI effect is also present in whole-body PET: in the case of a PET scanner with a diameter of 80 cm and 4 mm-wide crystals, the DOI effect leads to a 40% degradation of resolution at a radius of 10 cm from the central axis.[5]

5.6 True, False and Lost Coincidences

The information available in reconstructed PET images should be linear to the underlying activity, *i.e.* a two-fold increase in activity concentration in a certain tissue must lead to a two-fold increase in the number of pairs of annihilation photons emitted from this tissue and recorded by the detectors. Then, the image reconstruction based on these—true—coincidences must lead to a two-fold increase of the pixel value in the respective area of the reconstructed image. However, the coincidences recorded by the PET detectors not only contain true coincidences, whose two annihilation photons reach opposite detectors in an undisturbed way, but also random, scattered and even lost coincidences (Figure 5.4). These not-true coincidences distort the reconstructed image if they are not taken into account by the inclusion of an appropriate correction procedure.

As shown in Figure 5.4, a random coincidence occurs if two single photons belonging to two different positron emissions are registered by two opposing detectors within the coincidence time window and are falsely regarded as a coincidence event. The random coincidences cause a non-structured and nearly flat background that requires an appropriate correction (see Section 7.2.4).

Consequently, random coincidences can be diminished by a shorter coincidence time window, which becomes possible using faster scintillations

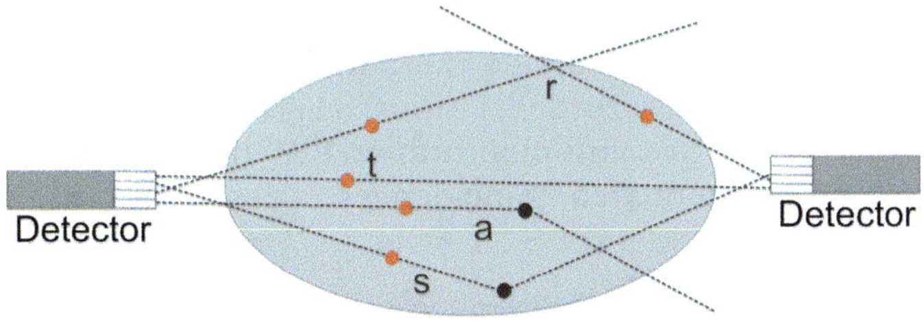

Figure 5.4 Different 'kinds' of coincidences. The primary aim of the PET measurement is obtaining true coincidences (t). Random coincidences (r) coming from two different simultaneously occurring positron emissions and scattered coincidences (s) with at least photon deviated by Compton scattering disturb the reconstructed image and must be corrected appropriately. Another correction concerns attenuated coincidences (a), which are not recorded because a photon is lost by Compton scattering. Red dots indicate the points of positron emission or positron–electron annihilation, respectively, and black dots represent atoms at which the annihilation photons are scattered.

crystals such as modern, commonly employed crystals LSO or LYSO and faster electronics.

A scattered coincidence results from the Compton scattering of one or both annihilation photons in the subject lying in the PET scanner (Figure 5.4). Scattered coincidences cause a low-frequency background and decrease image contrast. The primary tool to eliminate scattered coincidences is the lower energy-level discriminator (LLD), which is used to separate photons that undergo Compton scattering from unscattered photons (Figure 5.2). The definition of the LLD results in a compromise because of the overlap between the Compton spectrum and photopeak. The lower energy threshold must not be set too high in order to avoid loss of too many unchanged photons and sensitivity. The width of the photopeak is dependent on the light yield of the scintillation crystal. Here LSO and LYSO are more favourable than BGO, which was used in older PET scanners. Although many scattered coincidences are rejected by the LLD, there are still scattered events with just a small scattering angle and consequently a small loss of energy. In this case, their energy overlaps with the lower energy range of the photopeak. The remaining scattered coincidences must be corrected by dedicated software (see Chapter 7, Section 7.2.2). This correction became even more essential with the advent of 3D PET scanners without septa, as in 3D PET the chance of scattered annihilation photons hitting the photodetectors increases considerably. In whole-body PET the scatter fraction is 40–50%. In brain PET it is about 30%.

In general, radiation is attenuated in material. This attenuation is caused by the photoelectric effect, where the radiation interacts with an electron of the material and is completely absorbed, or by Compton scattering. In the case of 511 keV photons, the predominant effect is Compton scattering. If one or two photons coming from a positron annihilation are scattered within the patient's body or by the material of the patient bed, so that as their way leads outside the detector ring, the corresponding coincidence event is not recorded, *i.e.* 'lost' (Figure 5.4). To achieve undistorted, highly qualitative and quantitative PET imaging, an appropriate correction for the attenuated coincidences is essential (see Chapter 11, Section 11.2).

5.7 Performance Characteristics

The aim of each and every user of an imaging system is to ensure that the system has optimum performance. In most cases, however, a unique performance parameter does not exist, but rather a number of parameters assess different aspects of performance. Spatial resolution, count rate behaviour, sensitivity, accuracy and image quality are the commonly used items which characterise the performance of a PET scanner. In order to measure these parameters in a generally accepted and repeatable way, measurement standards have been defined by a dedicated task force of the (North American) National Electrical Manufacturers Association (NEMA). All major PET manufacturers worldwide were represented in this task force. The document containing the present standard for PET is the 'NEMA NU 2-2012, Performance Measurements of Positron Emission Tomographs'.[6]

5.7.1 Spatial Resolution

Within the context of hybrid MR-PET, it is particularly important to know how the spatial resolution of PET is defined. One should also be aware that spatial resolution in MRI has a different meaning. The spatial resolution of MRI most often, but not always, corresponds to the size of an image voxel resulting from the MR sequence. In MRI, the size of a voxel is related to the sampling *k*-space, which contains the originally acquired frequency information, from which the image is obtained by the discrete inverse Fourier transform (DIFT). The spatial resolution of PET is defined by the NEMA standard as the ability of an imaging system 'to distinguish between two (radioactive) points after image reconstruction'. To measure the spatial resolution, a point source positioned in the free bore at certain locations is imaged (Figure 5.5(A) and (B)). For image reconstruction, filtered back-projection without any smoothing is applied. Thereafter, straight lines crossing the centre of the point image are defined in x, y, or z direction and profiles of voxel values along these lines are obtained (Figure 5.5(C) and (D)). The outcome parameter of this measurement is the FWHM, *i.e.* the width of the profile at 50% of its maximum. In the example shown in Figure 5.5, the FWHM is 3.1 mm, resulting from five voxel bins at 50% of the profile maximum and a voxel size of 0.625 mm.

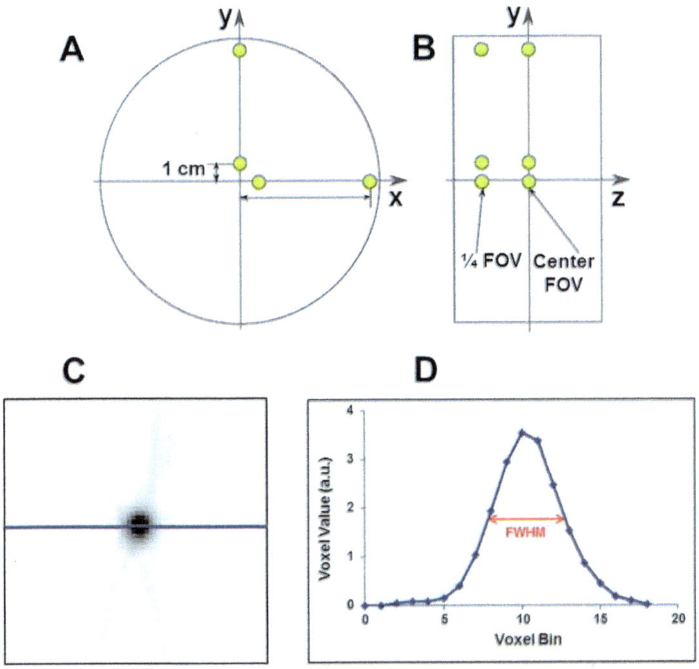

Figure 5.5 (A) Point sources in a transversal plane for measuring the radial (*e.g.* in x-direction at $y = 0$ cm) and tangential (*e.g.* in y-direction at $x = 10$ cm) resolution. (B) Point sources in a sagittal plane for measuring the axial (in z-direction) resolution at $z = 0$ cm and at ¼ of the axial FOV. (C) Image of a point source at $x = 1$ cm and $y = 0$ cm and horizontal line crossing the point. (D) Profile of image voxel values along the horizontal line marked in (C) and double-arrowed red line at 50% of the profile maximum representing the full width at half maximum (FWHM) with a length of five voxel bins.

5.7.2 Count Rate Behaviour

As discussed in Section 5.4, PET detectors do not just acquire all true coincidences, but also random and scattered coincidences. These additional false coincidences influence the dead time of the detectors, which increases with the total number of incoming events. Figure 5.6 shows the dependence of different kinds of coincidences on the activity concentration to be measured. The total count rate R_{tot}—also called prompt count rate—is the sum of true (R_{true}), random (R_r), and scattered (R_s) coincidences. In particular, the random coincidences, with their quadratic dependence on the activity concentration influence, lead to a saturation of the detectors at about 30 kBq in this example. In order to describe a balance between true and false coincidences Strother *et al.* introduced the so-called noise equivalent count rate (R_{NEC}) (eqn (5.2))[7]

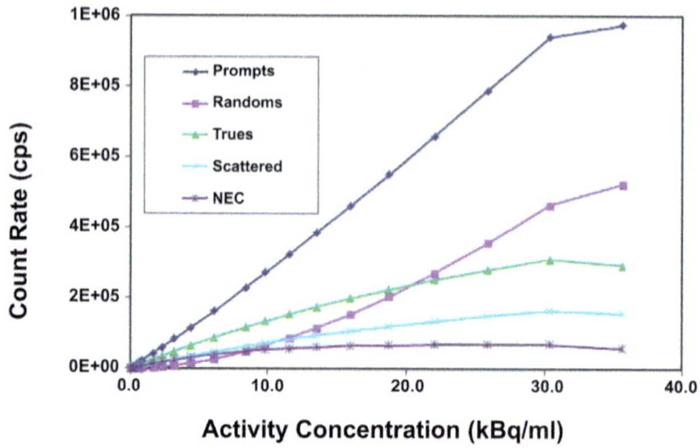

Figure 5.6 Plots of all measured (prompts), true, random and, scattered coincidences and of the NEC rate as functions of the activity concentrations in the phantom used for this measurement.

$$R_{\text{NEC}} = \frac{{R_{\text{true}}}^2}{R_{\text{true}} + R_{\text{s}} + kR_{\text{r}}} \tag{5.2}$$

The NEMA standard proposes a dedicated procedure to measure the total count rate and to determine R_{true}, R_{S}, and R_{r} from this measurement. k depends on the way of correcting for the random coincidences (see Chapter 7, Section 7.2.4): if it equals 1, the correction employs the delayed window technique, and if it equals 2, the subtraction is based on eqn (7.11), using the knowledge of the single rates of the detectors. R_{NEC} has become a well-accepted parameter to describe the count rate behaviour of a PET scanner. When comparing the count rate capability of different scanners, the maximum R_{NEC} and its corresponding activity concentrations are especially important. An impressive increase of this parameter can be observed as the development of PET scanners has progressed: The Siemens ECAT Exact HR+ PET-only scanner based on BGO crystals and operated in our institute since 1996 has a maximum R_{NEC} of 26 000 cps at 8.3 kBq mL^{-1},[8] whereas the present LSO/LYSO-based PET systems of the integrated MR-PET devices of Siemens and General Electric offer 184 000 cps at 23 kBq mL^{-1},[9] and 218 000 cps at 18 kBq mL^{-1},[10] respectively.

5.7.3 Sensitivity

As already described in Section 5.3, the sensitivity of a PET scanner depends on the crystal material and various design parameters. The removal of septa from between the detector rings has led to a considerable increase of sensitivity and further increases have been obtained with a longer axial field of view, longer crystals and a smaller ring diameter. However, the actual number of annihilation photons registered by the PET detector depends heavily on the shape and attenuation properties of the tissue of the investigated patient. Therefore, a standardised procedure is necessary to measure and compare the sensitivity of different PET scanners. Based on the original suggestion of Bailey *et al.*,[11] the NEMA protocol prescribes the use of a radioactive, 70 cm long, line source surrounded by four attenuating metal sleeves so that the annihilations cannot 'escape' from the source. The system sensitivity of the PET component of present human whole-body MR-PET devices ranges from 7–23 cps kBq^{-1}. This variation depends primarily on differences in axial FOV and crystal length. For comparison, the system sensitivity of the first LSO-scanner introduced by Siemens, the ACCEL PET, was 6.4 cps kBq^{-1}.[8]

5.7.4 Accuracy

The NEMA test of accuracy and count losses exploits the data recorded for the NEC rate. It examines the error of the PET measurement at a certain radioactivity concentration in relation to the result obtained at a very low radioactivity concentration, *i.e.* when random coincidences are negligible. In this way, the upper limit of radioactivity concentration can be determined when the correction of dead time losses is no longer properly corrected. This test is especially important for PET studies with short-lived positron emitters, which are applied with high amounts of injected radioactivity so that enough counts can be collected avoiding images with too much noise.

5.7.5 Image Quality

The quality of a PET image may be characterised using different aspects. An important aspect in clinical examinations is image contrast, which is a useful metric with which to compare regions with high and low uptake of radioactivity. This became particularly relevant with the increasing application of FDG-PET in oncological diagnostics. Here physicians want to define diagnostically important radioactivity uptake in metastases and must be able to distinguish the uptake in a suspicious region against the low and noisy background of FDG in the surrounding tissue. Aside from biological factors, the image contrast may depend on instrumental factors, such as the detector system, image reconstruction method and data corrections (see also Section 7.2). To allow for an objective comparison of the instrumental factors, the NEMA protocol defines a body-shaped phantom with spheres of different diameters to simulate a clinical imaging situation (Figure 5.7). It is recommended to fill four small spheres with a 4 : 1 or 8 : 1 radioactivity contrast relative to the background and to fill the two big spheres with just water to simulate cold regions simulating no radioactivity uptake.

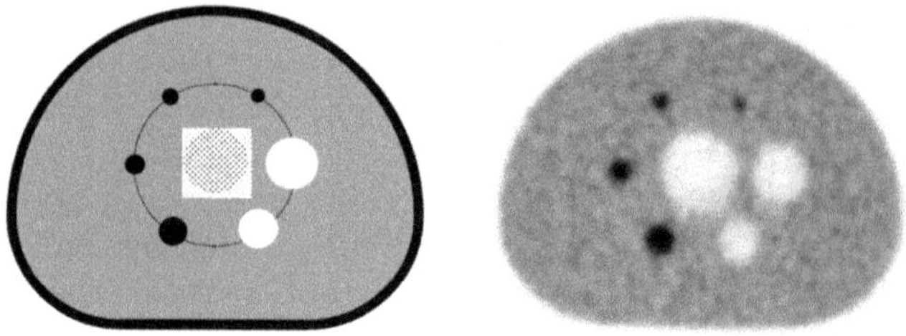

Figure 5.7 Left: sketch of NEMA-phantom used for examining image quality (adapted from NEMA Nu 2-2012 with permission[6]). The black spheres with diameters of 10, 13, 17 and 22 mm are to be filled with a radioactive water solution. The white spheres with diameters of 28 and 37 mm are filled with water. The big chamber is to be filled with 1/4 or 1/8 of activity concentration of the black spheres. The central region is designed to simulate lung tissue. Right: image resulting from an NEMA-guided image quality test. Adapted from ref. 10 with permission from John Wiley and Sons, Copyright © 2016 The Authors. Published by American Association of Physicists in Medicine and John Wiley & Sons Ltd.

5.8 Partial Volume Effect and its Correction

Due to the limited spatial resolution of PET, with current values being about 4 mm in the centre of whole-body PET detectors, an image of an organ with high radioactivity concentration inside and no/low radioactivity outside has blurred edges instead of sharp edges at the border of the organ. This leads to image voxels outside the organ containing spurious 'radioactivity concentration', whereas the radioactivity concentration measured inside at the border of the organ is underestimated. This effect is known as the partial volume effect (PVE). Image regions with different contrast 'spillover', that is, a portion of the activity of one region is assigned to another (in fact this happens even if regions have the same radioactive concentration). With the steady progress of PET towards better image resolution, the PVE has decreased but still has to be considered in the case of small or thin structures, such as small metastases and cortical grey matter.

To correct for the PVE, a number of procedures have been proposed which initially focused on brain imaging,[12,13] but were later extended to tumour imaging.[14] A common base of all methods is the tracing of the true organ border by MRI or CT.

There are other methods which aim to reduce the PVE by using different software techniques to make improvements in the image resolution. These include, image deconvolution (with the disadvantage of noise amplification), iterative reconstruction involving the point spread function across the detector FOV, and iterative reconstruction using anatomical priors obtained from MRI or CT (see also Section 7.1.2.4).

The PVE is dependent not only on image resolution but also on the energy of the emitted positron. A high positron energy leads to a longer positron range. As mentioned in Section 5.2, the positron range is a fundamental limit to the spatial resolution of PET. Thus, images based on high-energy positrons emitters, such as ^{82}Rb (mean positron energy 1.4 MeV), present a larger PVE than images of a radiopharmaceutical labelled with the low-energy emitter ^{11}C (mean positron energy 0.39 MeV). Here again, deconvolution techniques[15] or iterative reconstruction[16] have been proposed to improve resolution and reduce PVE.

5.9 Acquisition Modes

Most PET images are acquired as static images. Using this mode, the duration of image acquisition depends on the amount of injected activity, the half-life of the positron emitter, the uptake in the organ to be studied and the required image quality. The acquisition of a static image is usually started when the uptake has reached an approximately stable level. Thus, in oncological FDG-PET a wait of at least 30 min before starting the scan is to be expected. Sometimes the temporal behaviour of uptake and the release of a radiopharmaceutical is of interest, *e.g.* for the purpose of kinetic modelling (see Chapter 8). In this case, a dynamic acquisition is chosen with a sequence of image frames (often with different durations) lasting from seconds (during the phase of radiotracer uptake) to several minutes (when temporal changes of the radiotracer uptake are small). Sometimes it may be advisable to avoid the decision between static and dynamic imaging, *e.g.* if a new radiopharmaceutical is tested or patient motion is expected, necessitating motion correction (see Chapter 11). In this case, a list mode acquisition can be chosen. The list mode data contain the time points, the energy, and the positions of all accepted coincidences resulting in very large sets of data. Subsequently, one can decide as to the most appropriate way to frame the data—either into a static frame or a dynamic sequence of choice.

References

1. S. R. Cherry, M. Dahlbom and E. J. Hoffman, *J. Comput. Assist. Tomogr*, 1991, **15**, 655.
2. W. W. Moses, *IEEE Trans. Nucl. Sci.*, 2003, **50**, 1325.
3. S. Surti and J. S. Karp, *Phys. Med.*, 2016, **32**, 12.
4. P. Bartzakos and C. J. Thompson, *Phys. Med. Biol.*, 1991, **36**, 735.
5. S. R. Cherry, J. A. Sorenson and M. E. Phelps, *Physics in Nuclear Medicine*, Saunders Elsevier, Philadelphia, 4th edn, 2012, ch. 18, p. 317f.
6. NEMA NU 2–2012, *Performance Measurements of Positron Emission Tomographs*. National Electrical Manufacturers Association, Rosslyn, VA, USA, 2013.
7. S. C. Strother, M. E. Casey and E. J. Hoffman, *IEEE Trans. Nucl. Sci.*, 1990, **37**, 783.
8. H. Herzog, L. Tellmann, C. Hocke, U. Pietrzyk, M. E. Casey and T. Kuwert, *IEEE Trans. Nucl. Sci.*, 2004, **51**, 2662.

9. G. Delso, S. Fürst, B. Jakoby, R. Ladebeck, C. Ganter, S. G. Nekolla, M. Schwaiger and S. I. Ziegler, *J. Nucl. Med.*, 2011, **52**, 1914.
10. A. M. Grant, T. W. Deller, M. M. Khalighi, S. H. Maramraju, G. Delso and C. S. Levin, *Med. Phys.*, 2016, **43**, 2334.
11. D. L. Bailey, T. Jones and T. J. Spinks, *Eur. J. Nucl. Med.*, 1991, **18**, 374.
12. O. G. Rousset, Y. Ma and A. C. Evans, *J. Nucl. Med.*, 1998, **39**, 904.
13. C. C. Meltzer, P. E. Kinahan, P. J. Greer, T. E. Nichols, C. Comtat, M. N. Cantwell, M. P. Lin and J. C. Price, *J. Nucl. Med.*, 1999, **40**, 2053.
14. M. Soret, S. L. Bacharach and I. Buvat, *J. Nucl. Med.*, 2007, **48**, 932.
15. R. Laforest and X. Liu, *J. Nucl. Med.*, 2009, **50**(suppl. 2), 1482.
16. O. Bertolli, A. Eleftheriou, M. Cecchetti, N. Camarlinghi, N. Belcari and C. Tsoumpas, *Phys Med.*, 2016, **32**, 323.

CHAPTER 6

Positron Emission Tomography Instrumentation

C. W. LERCHE*[a], U. PIETRZYK[a,b] AND M. LENZ[a,b]

[a]Forschungszentrum Jülich, Institute of Neuroscience and Medicine – 4, Wilhelm-Johnen-Straße 1, 52428 Jülich, Germany; [b]University of Wuppertal, Faculty of Mathematics and Natural Sciences, Gauss-Str. 20, 42109, Wuppertal, Germany
*E-mail: c.lerche@fz-juelich.de

6.1 Basics of Signal Detection in PET

As previously highlighted in Chapter 5, imaging with positron emission tomography (PET) involves the detection of 511 keV photons, which originate from the annihilation of a positron and an electron. The energy of annihilation photons is relatively high compared to that of a γ-photon during imaging with a gamma camera or SPECT. Consequently, although both systems use radionuclides, the detection system of a modern PET scanner is different to that of gamma cameras and has not radically changed since the first developments in the mid-1970s. Detectors for PET consist of: (1) a scintillator crystal to stop the incoming 511 keV photons; and (2) a photodetector (or photosensor) to detect the scintillation light created during the interaction of the photons with the crystal.

New Developments in NMR No. 19
Hybrid MR-PET Imaging: Systems, Methods and Applications
Edited by N. Jon Shah
© The Royal Society of Chemistry 2019
Published by the Royal Society of Chemistry, www.rsc.org

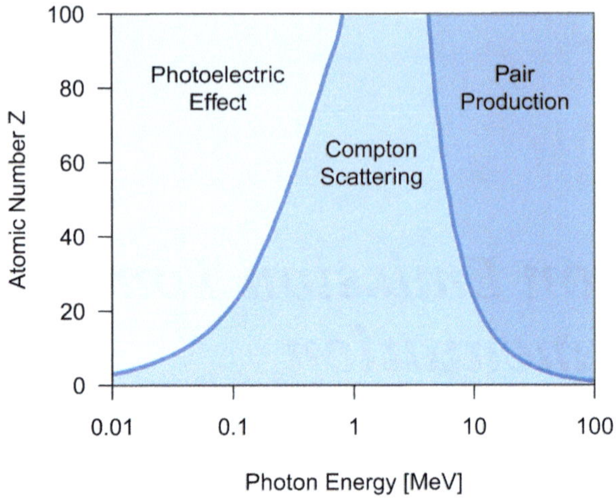

Figure 6.1 The strength of photoelectric effect, Compton scattering, and pair production depends on the photon energy and the atomic number of the absorbing material. The respective areas show parameter ranges where an interaction is dominant.

The selection of an appropriate scintillator for use in a PET system depends on a number of criteria. To stop a 511 keV photon, the scintillator needs to have a high density and high effective atomic number. This feature is often called the 'stopping power' and refers to the ability to completely stop a photon of high energy within a single crystal. However, as there are a large number of materials that have this feature, the effective atomic number has been introduced to ease comparison: the effective atomic number being the weighted average atomic number of all the components of a material. The overall stopping power is also related to the physical size of a crystal, which is directly related to the ability of the detector to provide good spatial resolution. Both factors, size and stopping power, can be optimised by selecting a crystal material and determining the fraction of events where a 511 keV photon can be completely stopped within the crystal. For this, also the fraction between interactions through photoelectric effect and all interactions (photo-fraction) is relevant. Figure 6.1 gives an overview of dominant elementary photon interactions dependent on the absorber's atomic number Z and the photon's energy.

6.2 Components

6.2.1 Scintillators

The 'stopping power' is only one of a number of properties used to differentiate between different types of scintillator crystals. Table 6.1 summarises such properties for several scintillators used in modern PET-Systems. These properties are: density, effective atomic number, decay time, photon yield, index of refraction, hygroscopy, and peak emission wavelength. While the

Table 6.1 Properties of common scintillators for PET and SPECT imaging. Adapted from S. Cherry *et al.*[1,2]

Property/scintillator	NaI(Tl)	BGO	LSO (Ce)	LYSO	GSO (Ce)	$LaBr_3$(Ce)
Density/g cm^{-3}	3.67	7.13	7.40	7.40	6.71	5.3
Effective atomic number	50	73	66	66	59	46
Decay time/ns	230	300	40	40	60	35
Photon yield/per keV	38	8	26	30	12–15	61
Index of refraction	1.85	2.15	1.82	1.82	1.85	1.9
Hygroscopy	Yes	No	No	No	No	Yes
Peak emission/nm	415	480	420	375	430	358

importance of most properties is obvious, the effects of decay time and photon yield are briefly discussed in more detail. The process of scintillation can be understood as the emission of a sequence of optical photons. The creation of the photons is described by a rise time, with the number of scintillation photons per gamma photon energy unit being defined as light yield. Finally, the decay time determines the time when the emission rate has decreased to $1/e$ of the maximum. The collection time of scintillation photons is necessarily longer in larger crystals, and the shape of the crystals can influence the performance of a scintillation detector. Furthermore, the number of photons created should be sufficiently large to trigger the photodetector and the readout electronics to form an electric pulse strong enough to obtain good estimates of the energy and the arrival time of the detected annihilation photon.

In order to yield the best performance, both crystal and photodetector are tightly connected to each other to ensure that there is no loss of light during the transition of the light from the crystal to the photodetector. To find optimal transition, an effective optical coupling is necessary between the scintillation crystal and the photodetector, which normally have high refractive indices.

For the application of scintillation detectors in PET, very good timing capabilities are mandatory. Fast detectors allow the coincidence time window to be kept as short as possible to bring the benefit of fewer random coincidences. Furthermore, good timing capabilities allow the subsequent data analysis system to utilise the time-of-flight information, giving the advantage of images with reduced noise and enhanced contrast, which in turn is beneficial for the patient.

6.2.2 Photodetectors

The important task of the photodetector is to collect the greatest possible amount of available scintillation light entering from the scintillation crystal and to convert it into a measurable electric signal. First PET devices were equipped with photomultiplier tubes (PMT), which make use of the photoelectric effect in a photocathode (see Section 5.3). Typical gains are in the order of 10^6 to 10^7, making PMTs interesting photodetectors for standalone PET or combined PET/CT.

However, as the classical PMT cannot be operated in strong magnetic fields and environments with rapidly varying radio frequencies without serious distortions, its use in hybrid MR-PET scanners is not possible. A further disadvantage is the requirement of a high operating voltage, which is in the range of 1–2 kV and which has to be highly stabilised, since the multiplication factor strongly depends on dynode voltages. By choosing dedicated materials for the photocathode and entry window, the wavelength can be set at maximum sensitivity according to the emission spectrum of the scintillator.

The wavelength-dependent quantum efficiency, *i.e.* the ratio of emitted electrons and incident photons, typically amounts to 20–25% at maximum sensitivity wavelength for a common bi-alkali photocathode, although efficiencies of up to 35% and nearly 45% at maximum sensitivity wavelength have been reached with the super bi-alkali photocathode and the ultra-photocathode, respectively.[3]

Silicon-based, modern photodetectors, such as APD (avalanche photodiode) or SiPM (silicon photomultiplier) can be operated in the magnetic fields present in MR scanners and will be described in the following.

While silicon photodiodes can be used to directly measure photons of high energy, they can also be combined with a scintillator crystal, as the energy of scintillation light is high enough to release electron–hole pairs in the depletion zone of the diode. The induced charge is proportional to the number of light photons hitting the photodetector. Such photodetectors typically have a quantum efficiency of between 60% and 80%, which is considerably higher than that of a photomultiplier tube. However, in contrast to PMTs, photodiodes do not have inherent signal amplification, and hence the amplification factor is 1 compared to 10^6–10^7. Consequently, such photodetectors would require specific readout electronics with very high gain paired with very low noise. Figure 6.2 shows the emission spectrum of LSO/LYSO and typical quantum efficiencies for a PMT and an APD.

The problem of a low amplification factor can be partly overcome by using avalanche photodiodes (APD), which are operated at a reverse bias voltage close to the breakdown voltage. Compared to a photodiode, APDs have an additional, highly doped p- or n-layer next to the depletion region. This layer enhances the electric field strength in the so-called multiplication zone. Therefore, primary electron–hole pairs are accelerated to sufficiently high energies to create further electron–hole-pairs by impact ionisation. Bias voltage-dependent charge multiplication factors in the range of 100–500 can be achieved in this operation mode, which is also known as linear mode.[5] However, to prevent the gain fluctuating with the temperature and lowering the energy resolution, operation at very stable

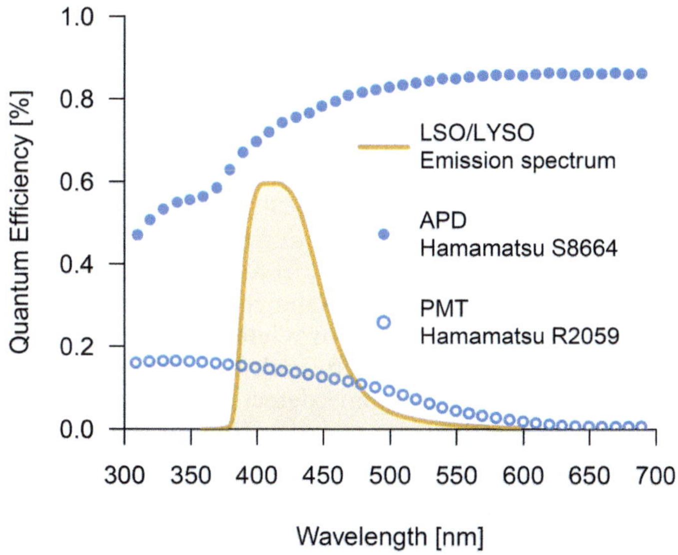

Figure 6.2 Emission spectrum of LSO/LYSO and the quantum efficiency of a PMT and APD. Care has to be taken that the photodetector's sensitive spectral range matches the emission spectrum of the scintillator. Data adapted from Chen *et al.*[4]

temperatures is necessary, along with low noise readout electronics with voltage gains of 10^4–10^5.

In addition to the linear mode, APDs can also be operated in the 'Geiger-mode' at reverse bias voltage, which is higher than the breakdown voltage. APDs optimised for this operation mode are called single photon avalanche diodes (SPAD). As the name already indicates, a self-sustaining avalanche can be triggered by a single photon. The avalanche is self-sustaining and thus, after having registered the avalanche current as a standardised photon count signal, a quenching circuit is needed to stop the avalanche. The output signal of the SPAD is independent of the energy of the incoming radiation, thus giving it the name 'Geiger-mode APD', which is analogous to a Geiger counter. Only when the SPAD's reverse bias voltage has been lowered below the breakdown voltage and has subsequently been restored to the initial reverse bias, photons can be counted again. In the simplest case, a series resistor is used for quenching (passive quenching): the avalanche current causes a voltage drop above the resistor, which lowers the voltage over the SPAD so that the avalanche is extinguished and, consequently, the current through the quench resistor and the corresponding voltage drop vanishes. Thus, the reverse bias voltage can recover and the depletion zone of the SPAD is recharged.

Following this sequence, the SPAD recharges to bias voltage. This process can take several 100 ns, during which the SPAD cannot detect any additional photons. Shorter dead times can be achieved with active quenching circuits, lowering the range to between 1 ns and 10 ns, but these require space on the substrate and lower the photosensitive area. Similar to other detectors, SPADs have internal noise sources. Thermally generated electron hole pairs can also trigger an avalanche, even if the SPAD is operated in total darkness. Therefore, this type of noise is referred to as dark counts, and the dark count rate (DCR) increases with temperature. Another source of noise is afterpulsing: during an avalanche, carriers can get trapped in metastable energy states, from which they are released several nanoseconds after the original avalanche is quenched. They subsequently cause an avalanche, which is correlated in time. Although cooling is beneficial for lowering the DCR, afterpulsing increases when lowering the temperature. Afterpulses are only detectable when the lifetime of a trapped carrier is longer than the recharge time of the SPAD. Since this lifetime becomes shorter with increasing temperature, the probability for afterpulses decreases accordingly.

SPADs allow amplifications of between 104 and 107, but a single SPAD cell does not allow the discrimination of multiple photons, which arrive during the recharge phase. To circumvent this problem, so-called multi-pixel photon counters (MPPC) or silicon photomultipliers (SiPM) have been developed. These consist of a parallel circuit of thousands of very small micro-cells, where a micro-cell is a unit of a Geiger-mode APD and its related quenching resistor. Each micro-cell contributes to the output signal. The equivalent circuit of a SiPM is shown in Figure 6.3.

The analysis of the output signal shape can be used to determine the number of counted photons, especially since the time integral of the SiPM signal is proportional to the produced charge.

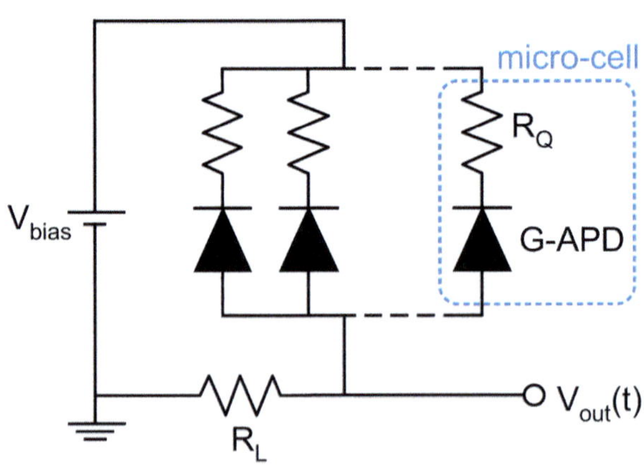

Figure 6.3 Simplified equivalent circuit of a SiPM: The micro-cells, which consist of a G-APD and a quenching resistor R_Q, are connected in parallel and supplied by a bias voltage V_{bias}. The load resistor R_L is used to ground the micro-cells and to measure the time-dependent output signal V_{out}. The output signal is the sum of the individual micro-cell signals.

Besides dark counts and afterpulsing, crosstalk plays an important role in SiPM, but not in single SPADs. Ideally, a photon only contributes to the signal of a single cell after it has been absorbed in the active volume of that specific cell. However, a photon can also cause a signal in a neighbouring cell, which is called crosstalk. In the case of optical crosstalk, the carriers generated in the high field region can recombine and emit secondary photons. These secondary photons can trigger avalanches in the adjacent cells. Electrical crosstalk describes the lateral diffusion of carriers into the neighbouring cell. The latter can be addressed by junction isolation, while optical crosstalk can be suppressed by extension of the cell distance and optical trenches. However, this approach also reduces the fill factor of the SiPM. The fill factor gives the ratio of the photosensitive surface of all micro-cells and the total surface of the SiPM. Multiplying the fill factor with the quantum efficiency (ratio of generated electron hole pairs and incident photons) and the probability that an incident photon causes a cell discharge gives the photon detection efficiency (PDE). The fill factor is lowered by all structures where no avalanche can be triggered, *i.e.* quench resistors, guard rings, connections, *etc.* SiPMs can be saturated, since each micro-cell can only count one photon per acquisition cycle (until the cell is recharged).[6]

While traditional SiPMs sum the signals of single SPADs to an analogue output signal, digital SiPMs use CMOS logic circuits to count independent avalanches from individual SPADs. This allows digitisation in an early stage of the signal-processing pipeline and supersedes additional technologies for signal amplification, shaping and conversion. The output signal consists of the number of cell discharges per pixel and a timestamp, which is generated by a time-to-digital-converter (TDC). Devices from Philips Digital Photon Counting, currently the only manufacturer of dSiPMs, have the special features of a customisable trigger and an event validation scheme, which can be applied with an on-board FPGA. This FPGA also synchronises all dSiPMs on the board, and furthermore, can be used to optimise the dark count rate by recording and storing an inhibit map. This inhibit is then loaded into the device and disables all micro-cells with excessively high dark count rates.[7]

Due to excellent timing resolutions of below 100 ps, high quantum efficiencies and compact design, SiPMs are currently renowned detector components for MR-PET.

6.3 PET System Architecture

6.3.1 Scintillation Detector

The scintillation detector is the most essential building block of most PET scanners. Its function is to stop the annihilation radiation emitted by the administered radiotracer and to convert it into an electric pulse. This photo-conversion is a multistep process. First, the γ-ray is converted inside the scintillator into a large number of optical photons. Second, the optical photons are guided to the photodetectors *via* light-guides and by reflecting material surrounding the crystal and the photodetectors convert the individual

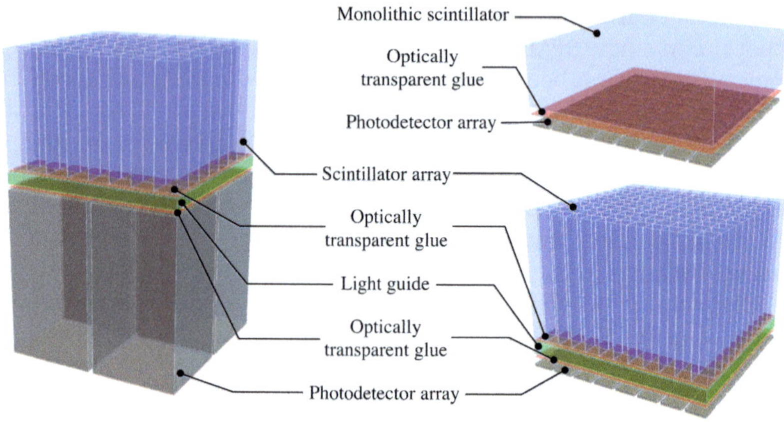

Figure 6.4 Schematic drawings of three possible realisations of scintillation detectors for PET. To the left, the block detector is shown.[8] The block detector, which is used the most for human PET scanners, is built from four single channel PMTs and a scintillator array. To the right, scintillation detectors typically used in preclinical PET imaging are shown. The upper right image shows a configuration with a large monolithic crystal where the scintillation crystal itself can act as a light guide and an additional light guide may not be required. The lower right image shows an implementation with a photodetector array and a small pitch scintillator array for preclinical PET.

optical photons into electrical charges. Finally, the charges are digitised and analysed. Figure 6.4 shows the most common realisations of scintillation detectors for PET.

Either large monolithic scintillation crystals or arrays of many small, optically separated scintillator pixels can be used for building scintillation detectors. For optimal optical light collection, the scintillators are usually covered with reflecting coatings on all faces, with the exception of the face that is coupled optically to the photodetectors. In order to calculate the position of a scintillation event from the measured light distribution, either a position sensitive photodetector or an array of photodetectors must be used. With the development of compact SiPMs, direct optical coupling of photodetector pixels to individual crystal pixels from the scintillator array is now possible. In this case, the position of the γ-ray detection is the position of the triggered photodetector pixel. However, it is still necessary to digitise the electrical pulses from the individual photodetector pixel, requiring a very large number of electronic readout channels. Therefore, in most cases, the light sharing principle is used in combination with the computation of the weighted mean, which is also known as Anger logic, due to its inventor.[1] Anger's method is illustrated in Figure 6.5. For this method, a light guide is inserted between the scintillator and the photodetectors in order to spread out the optical scintillation photons emitted during the detection of a single γ-ray and inside a single crystal over at

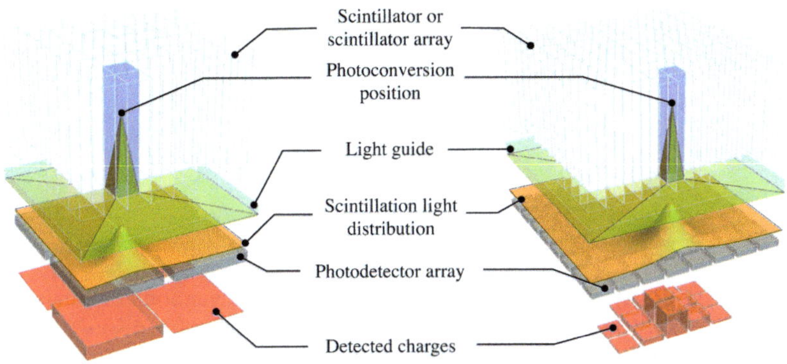

Figure 6.5 Illustration of the light sharing principle. To the left, light sharing for a block detector for a human PET scanner is shown. The scintillation light is emitted isotropically at the photo-conversion position and the photons are guided towards the photodetectors by the reflecting surface coatings of the scintillator pixel. The light guide spreads the optical photons over all four photodetectors, where they are then converted to charges and used to compute the weighted mean of the scintillation light distribution according to eqn (6.1). To the right, light sharing in a scintillation detector for preclinical PET is illustrated. Scintillation detectors for preclinical PET are characterised by having higher granularity of the photodetector array and a smaller pitch of the crystal array in order to achieve better spatial resolution.[1]

least four photodetector pixels. The smaller the distance from the scintillating crystal to the photodetector, the more light will be detected by the underlying and neighbouring photodetector pixels, giving rise to characteristic signal patterns for each scintillation pixel. The sum of all the detected charges $Q_{i,j}$ from one scintillation event is proportional to the energy E_γ of the γ-ray and can therefore be used for its estimation, according to eqn (6.1), left. Since the exact positions x_i and y_i of the photodetector pixels are known, the position of photo conversion of the γ-ray can be estimated by computing the weighted means $\overline{X}_\gamma$ and $\overline{Y}_\gamma$ from the detected charges $Q_{i,j}$ and x_i and y_i according to eqn (6.1) middle and right

$$E_\gamma = \sum_i^{N_x}\sum_j^{N_y} Q_{i,j}, \quad \overline{X}_\gamma = \frac{1}{E_\gamma}\sum_i^{N_x}\sum_j^{N_y} x_i Q_{i,j}, \quad \overline{Y}_\gamma = \frac{1}{E_\gamma}\sum_i^{N_x}\sum_j^{N_y} y_i Q_{i,j} \tag{6.1}$$

N_x and N_y in this equation denote the number of photodetector pixels in x and y direction, respectively.

Since the area under the curve is proportional to the energy of the converted γ-ray and also to the amplitude of $I_{i,j}(t)$, the charges $Q_{i,j}$ itself are obtained by the integration of the photodetector currents $I_{i,j}(t)$ over the time from the start to the end of the scintillation pulses. Therefore, the detected γ-rays can be discriminated according to their energy by using a leading-edge

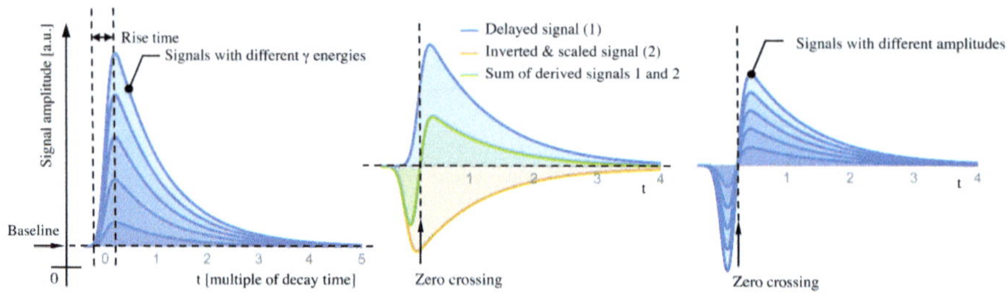

Figure 6.6 Left: Current pulses $I(t)$ from the photo-conversions of γ-rays with different energies. Middle: functional principle of the CFD. An inverted signal and a delayed signal are derived from the original current $I(t)$. The sum of these derived signals gives an alternate signal with exactly one zero crossing, which can be detected. Right: The zero-crossing time is independent of the signal amplitude and therefore a reliable estimate for the γ-ray arrival time.[9]

trigger. As can be seen from Figure 6.6, the constant leading edge trigger level, together with the rise time of $I_{i,j}(t)$, leads to a fluctuation of the trigger time. This effect is called 'walk' and is normally avoided by using constant fraction discriminators (CFD) for the generation of the integration start and estimation of the arrival time of the γ-ray (Figure 6.6 middle and right). In general, rise times are dominated by the photodetector and the scintillation process dominates decay times. For correct energy discrimination with the leading-edge trigger, it is necessary to remove potentially existent baseline pedestals. The integration time only depends on the rise and decay times of the scintillator and the photodetector and is, therefore, the same for all scintillation events.

A coincidence detection unit processes all single detected γ-rays, whose energy, arrival time and photo-conversion position have been determined. The most important tasks of this unit are the identification of coincident scintillation events caused by the detection of both γ-rays from the positron annihilation and the estimation of random coincidence event rates. A coincidence event is registered when any pair of scintillation detectors detects two γ-rays with energies within a predefined interval around 511 keV and within a predefined coincidence time window. In the case that the LOR is outside the PET field of view (*e.g.* adjacent detectors), or more than two γ-rays have been detected (multiple coincident event), the coincidence event is discarded. A random coincidence event occurs when two γ-rays are detected with energies within a predefined energy interval of around 511 keV and within a predefined coincidence window, but which originate from different positron annihilations (see Figure 5.4). The random coincidence event rate depends quadratically on the activity inside the PET FOV and is, in general, different for different detector pairs. The rate can be estimated either by the delayed coincidence method or by the pre-scaled singles method (Figure 6.7). Other tasks of the coincidence unit can be scintillation detector dead time estimation and energy correction and filtering of the detected events.

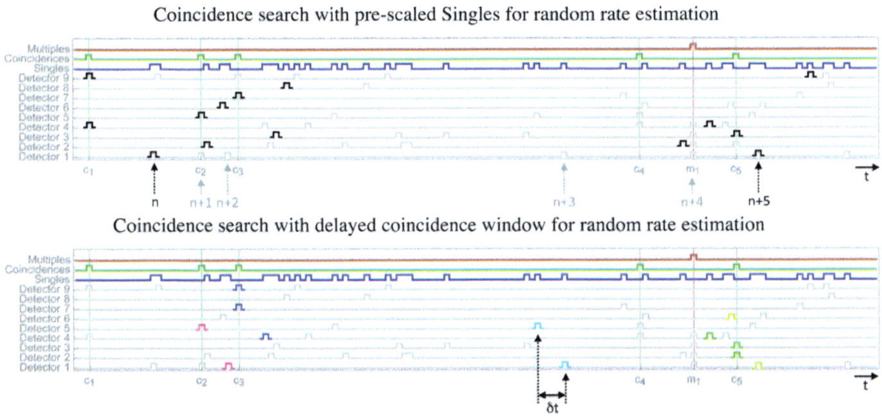

Figure 6.7 Illustration of coincidence search with random rate estimation for a unit with nine scintillation detectors and corresponding detector activity diagrams. Every rectangular pulse represents a single detected γ-ray event (single) within the predefined energy interval. Detected coincident events are called c_i, multiple coincident events are called m_k. When two pulses occur at the same time, a coincidence is detected (green line). The singles rate (blue line) can be used to estimate the activity in the FOV for dead time correction. Upper figure: estimation of random coincidence rate with the pre-scaled singles method. For each detector, the single detection rate is measured. The random coincidence rate for an arbitrary detector pair is then proportional to the product of both single rates and the duration of the coincidence window. In order to avoid saturation of the processing hardware, only a fraction of the singles is evaluated for the random rate estimation; in the figure above, only every fifth single from each detector is used. Lower figure: estimation of the random coincidence rate with the delayed window method. In this method, signals from one of each detector pair are delayed by a time δt larger than the length of the coincidence window, and coincidence events with this delay are registered. This gives a direct estimate for the random coincidence rate for the corresponding detector pair.

6.3.2 PET System Design

The need to stop and detect both annihilation γ-rays determines the general design and the most relevant performance parameter of PET scanners. During the positron annihilation process, both γ-rays are emitted in nearly exactly opposed directions, and their flight directions form a straight line between both scintillation pixels where the γ-rays were detected. This line is called the line-of-response (LOR) and determines the flight direction of the γ-rays, which cannot be determined by other means. The detection of this process is known as coincidence detection. The annihilation position along this LOR can only be estimated approximately by measuring the difference of arrival times (see Section 5.4). This technique is called time-of-flight (TOF) PET. Modern PET systems achieve coincidence resolution times (CRT) of ≈300 ps, which is, however, insufficient for the exact determination of the

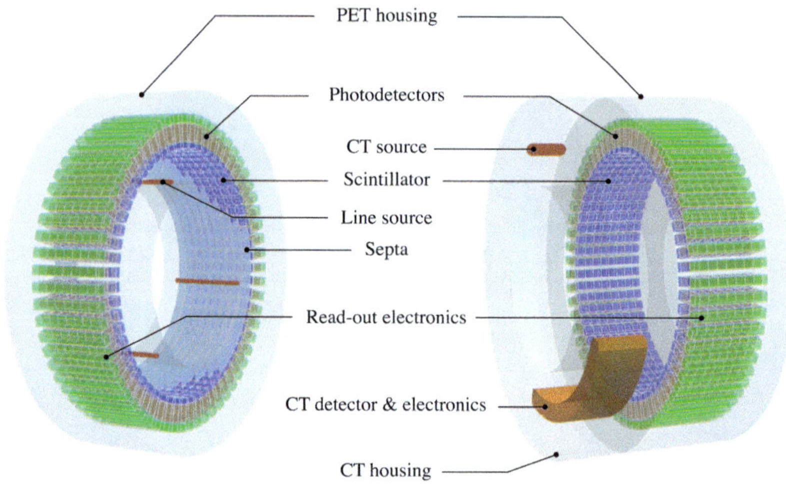

Figure 6.8 Left: PET scanner implementation with rotating transmission sources. Right: Combined PET/CT scanner. In both cases, the transmission data and emission data will be acquired sequentially. In the PET/CT scanner (left), the PET ring and the CT ring are arranged in a close in-line formation. Additional γ-ray shields, similar to the septa, may be placed between the CT component and the PET component to minimise detection events scattered between the modalities.

annihilation position. Thus, PET system designs always consist of annular arrangements of many scintillation detectors building an entire ring, as shown in Figure 6.8.

For accurate PET image reconstruction, the attenuation of the γ-rays along the different LORs must be known (see Sections 7.2.1 and 11.2), thus all PET systems have a system for measuring these attenuations. Originally, the attenuations for the different LORs were determined with transmission sources. These consist of a long-lived radioactive isotope emitting γ-rays with same or at least similar energy as the annihilation radiation. This is arranged inside a rod-like capsule to realise a line source. These line sources rotate around the patient. Usually, multiple line sources are integrated into the PET scanner in order to lower the transmission scan-time. For the correct determination of the attenuations, oblique γ-rays from the transmission sources are supressed by introducing septa between different rings of scintillation detectors. These septa are made from a highly γ-ray absorbing material like tungsten. Since the beginning of the 21st century, PET scanners have been combined with computer tomography (CT) scanners to combine the benefits of the structural information provided by CT imaging with the metabolic information available from the PET image. As attenuation of X-rays in biological tissues is very similar to the attenuation of γ-rays, CT images are used in combined PET/CT scanners for determining the attenuation. The combination of PET with CT also made the use of septa redundant. The PET scanner configurations in Figure 6.8 only allow a limited axial region of the patient to be imaged. Examinations of regions that exceed the axial extension of the PET FOV are realised by changing the axial position of the patient bed with respect to the PET FOV.

6.3.3 PET Systems

PET systems intended for different applications will differ significantly in their design and imaging performance. This is a consequence of various aspects, including the high costs associated with key components, *i.e.* scintillator and photodetectors, fundamental physical effects and the requirements of the specific application, besides others. For instance, the scintillation detector rings for human PET scanners must have a large diameter and have a spatial resolution of 5 mm. In contrast, PET scanners for mice require a spatial resolution close to 1 mm but do not need a large diameter. The most important performance parameter of any PET scanner is the detection sensitivity and the scanner's spatial resolution. The spatial resolution of a PET scanner is generally not isotropic, and due to the depth of interaction (DoI) effect (see Section 5.5), it is also not homogeneous over the whole PET FOV (Figure 6.9). Pre-clinical PET scanners are normally built with smaller scintillation pixels and therefore have a better spatial resolution than PET scanners for humans. However, the DoI effect (also called radial elongation)

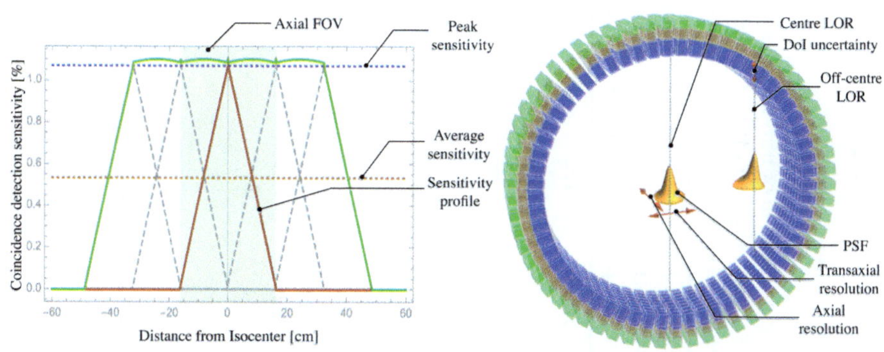

Figure 6.9 Left: Typical coincidence detection sensitivity profile in the axial direction for a human PET scanner. The peak sensitivity is reached at the centre of the FOV. The detection sensitivity falls approximately linearly along the axial direction of the FOV and reaches zero at the edges, giving rise to a triangular shape (red line). Therefore, the average detection sensitivity for a single bed position is approximately half of the peak sensitivity. In the case of examinations requiring several bed positions, the distance between two bed positions is generally set to half of the axial FOV, since in this case, an approximately flat sensitivity profile is reached (green line). Right: Illustration of transaxial resolution, axial resolution and radial elongation due to the DoI effect. For positron annihilations at off-centre positions, the generally unknown DoI leads to an uncertainty in the transaxial direction; also giving rise to a wider point spread function (PSF) in the transaxial direction.

Table 6.2 Performance parameters for a selection of human whole-body PET scanners, dedicated human head scanners and preclinical scanners.[10] The spatial resolution is given in the transaxial direction (first value) and the axial direction (second value). The coincidence timing resolution (CTR) and the noise equivalent count rate (NECR) have been explained in Chapter 5.

Type	Axial FOV/cm	Number crystal pixels	Sensitivity (%)	Spatial resolution at centre	Peak NECR/kcps	CTR/ps	Remark
Siemens ECAT HRRT	25.2	119 808	4.3	2.4 × 2.4	230 @ 37 kBq mL^{-1}	<2000	Human head scanner with depth of interaction detection.
Siemens Biograph mCT Flow	21.8	32 448	0.95	4.4 × 4.5	175 @ 28 kBq mL^{-1}	540	Human scanner.
Philips Ingenuity	18	28 336	0.74	4.7 × 4.7	120 @ 19 kBq mL^{-1}	591	Human scanner.
Philips Vereos	16.3	23 040	1	4.0 × 4.0	400 @ 30 kBq mL^{-1}	307	Human scanner with direct optical coupling.
GE Discovery 710	15.7	13 824	0.75	4.9 × 5.6	130 @ 29.5 kBq mL^{-1}	544	Human scanner.
Siemens Inveon	12.7	25 600	9.3	1.8 × 1.8	1670 @ 130 kBq mL^{-1}	<3432	Preclinical system.
Mediso Nanoscan PET 122S	10	36 504	8	1.25 × 1.25	850 @ 130 kBq mL^{-1}	1200	Preclinical system.
Bruker Albira SI	8	24	9	1.2 × 1.2	560 @ 53 kBq mL^{-1}	NA	Preclinical system with monolithic scintillators.

is more pronounced for preclinical PET scanners, owing to the smaller detector ring diameter. Several methods for detecting the depth of interaction with specialised scintillation detectors have been developed.[11] Owing to the large detector ring diameter and the resulting reduced solid angle coverage, human whole-body PET scanners generally exhibit the smallest coincidence detection sensitivity. Common to all PET scanner is that the coincidence detection sensitivity achieves its maximum at the centre of the FOV (isocentre) and falls to zero at both edges of the axial FOV, as shown in Figure 6.9. A list of the typical performance parameters for several different PET systems are given in Table 6.2.

References

1. S. R. Cherry, J. A. Sorenson and M. E. Phelps, *Phys. Nucl. Med. e-Book*, Elsevier Health Sciences, 2012, p. 195.
2. C. W. E. van Eijk, *Radiat. Prot. Dosim.*, 2008, **1**(3), 13.
3. K. Nakamura, Y. Hamana, Y. Ishigami and T. Matsui, *Nucl. Instrum. Methods Phys. Res. A*, 2010, **623**(1), 276.
4. J. Chen, L. Zhang and R. Zhu, *IEEE Nucl. Sci. Symp. Conf. Rec.*, 2005, **1**, 254.
5. D. Renker and E. Lorenz, *J. Instrum.*, 2009, **4**(1), 04004.
6. F. Zappa, S. Tisa, A. Tosi and S. Cova, *Sens. Actuators, A*, 2007, **140**(1), 103.
7. T. Frach, G. Prescher, C. Degenhardt, R. de Gruyter, A. Schmitz and R. Ballizany, *IEEE Nucl. Sci. Symp. Conf. Rec.*, 2009, **1**, 1959.
8. M. E. Casey and R. Nutt, *IEEE Trans. Nucl. Sci.*, 1986, **33**(1), 460.
9. K. Iniewski, *Electronics for Radiation Detection*, CRC Press, 2010, p. 315.
10. P. J. Slomka, T. Pan, D. S. Bermann and G. Germano, *Prog. Cardiovasc. Dis.*, 2015, **57**(6), 566.
11. M. Ito, S. J. Hong and J. S. Lee, *Biomed. Eng. Lett.*, 2011, **1**(2), 70.

CHAPTER 7

PET Quantification

J. J. SCHEINS*, E. ROTA KOPS, L. CALDEIRA AND B. MA

Forschungszentrum Jülich, Institute of Neuroscience and Medicine – 4, Wilhelm-Johnen-Straße 1, 52428 Jülich, Germany
*E-mail: j.scheins@fz-juelich.de

7.1 Tomographic Image Reconstruction

In PET, the three-dimensional (3D) tracer activity concentration of a subject is obtained by acquiring projections for different angular views with a PET scanner and image reconstruction algorithms are then used to solve the inversion problem. All spatial projections are acquired as a large aggregation of simultaneously observed, so-called, lines-of-response (LOR), which can be sorted according to projection angles (Figure. 7.1). Each LOR j holds a statistical sample of activity $\hat{y}_j = y_j/\Delta T$ along its specific pathway through the subject. This is given by the detected number of coincidence events y_j within the acquisition time interval ΔT. Broadly speaking, there are two basic conceptual approaches available for the reconstruction, analytical and iterative methods. A brief overview is given in the following, while more details can be found in the wider literature.[1–3]

7.1.1 Analytical Methods

Analytical reconstruction methods are essentially founded on the Fourier Slice Theorem (FST)—also known as central slice theorem or projection slice theorem—that allows analytical expressions to be derived to solve the

New Developments in NMR No. 19
Hybrid MR-PET Imaging: Systems, Methods and Applications
Edited by N. Jon Shah
© The Royal Society of Chemistry 2019
Published by the Royal Society of Chemistry, www.rsc.org

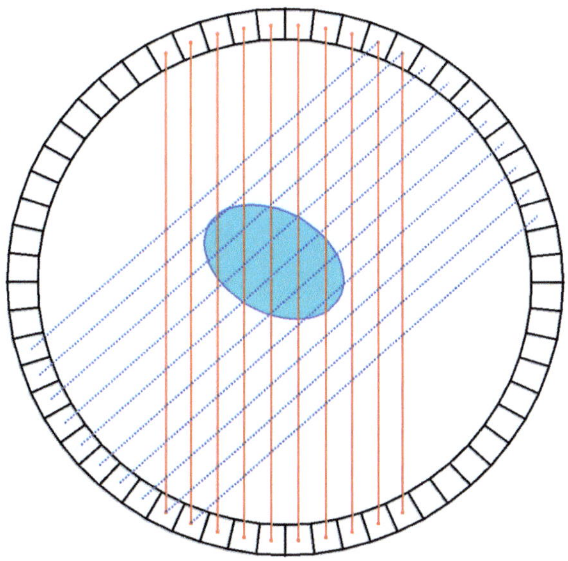

Figure 7.1 Sketch of single PET detector ring operating in coincidence mode with pairs of opposing detectors. Each detector combination represents a so-called line-of-response (LOR), while all parallel rays for a specific projection angle provide unique projection data (group of blue and red lines).

inversion problem. For example, the continuous 2D function $f(x, y)$ can be directly obtained from its 2D Radon Transform (RT) (eqn (7.1))

$$P(\theta, x_\theta) = \int_{-\infty}^{+\infty} f(x_\theta \cos\theta - y_\theta \sin\theta, x_\theta \sin\theta + y_\theta \cos\theta) \mathrm{d}y_\theta \tag{7.1}$$

where the RT can be understood as a complete set of parallel projections which address all LORs (Figure 7.2), *i.e.* all line integrals for all projection angles $\theta = [0, \pi]$ according to the equation of a line under rotation (eqn (7.2))

$$x \cos\theta + y \sin\theta = x_\theta \tag{7.2}$$

Hence, $P(\theta, x_\theta)$ is a function of the projection angle θ and the radial coordinate x_θ (Figure 7.2 and 7.3).

The FST denotes that the inverse 2D Fourier Transform (FT) of the 1D FT of $P(\theta, x_\theta)$ provides $f(x, y)$ again. In this context, the RT represents an idealised model of the PET data acquisition process. However, real scanner data require a transition from continuous, ideal input data to discretised data with finite resolution as well as noise. The noise is unavoidable due to the stochastic character of the photon detection process. The projection space sampling is limited due to the finite detector resolution. That means that the limited

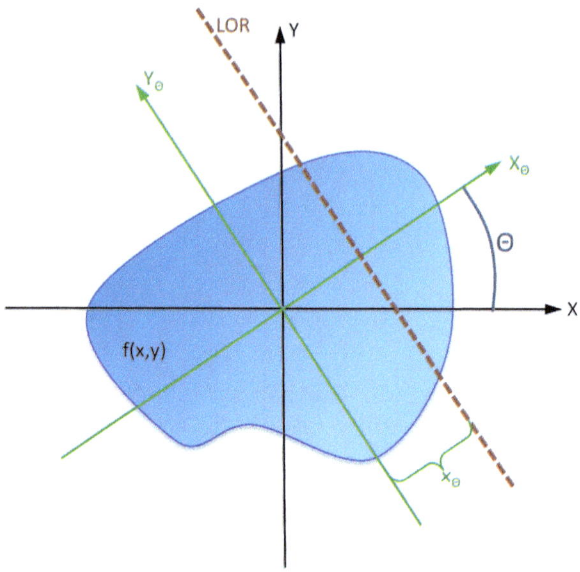

Figure 7.2 Any LOR can be directly addressed by introducing a rotated coordinate system (X_θ, Y_θ). The rotation angle θ is defined by the LOR such that the LOR becomes perpendicular to the X_θ axis and the radial shift x_θ is constant along the LOR. LOR projection values are obtained by integrating the function $f(x, y)$ according to eqn (7.1) and (7.2), respectively.

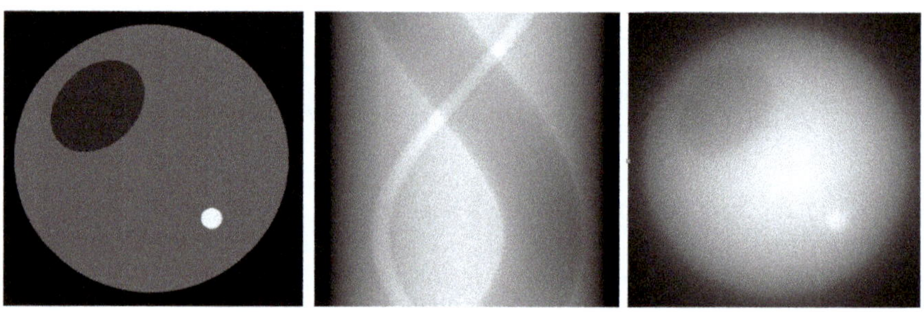

Figure 7.3 Mathematical phantom image $f(x, y)$ (left) and the corresponding projections $P(\theta, x_\theta)$ under the terms of eqn (7.1) visualised as so-called sinogram (middle) with projection angle θ changing along the vertical direction. The pure backprojection from projection to image space domain provides a blurred image (right) due to the intrinsically resulting convolution by $1/x_\theta$.

accuracy of the projection data propagates into poor solutions if no effective regularisation is applied to control the resolution-noise trade-off. Generally, the inversion problem is ill-posed and thus analytical methods usually tend to evoke rather large variations of the solution for small changes or fluctuations of the input projection data. Therefore, the inverse 2D FT provides relatively poor results, because small uncertainties are strongly amplified in the solution image as a consequence of error propagation. More robust analytical reconstruction requires some refinement of the exact solution. This leads directly to the use of the so-called filtered back-projection algorithm (FBP) as the standard analytical method in PET. FBP uses various derivatives of convolution kernels (filters in the frequency domain) for the purpose of regularisation. In this way, the specific choice of the convolution kernel allows one to control the resolution-noise properties of the final images. The implementation of FBP is straightforward and can be realised computationally. However, it is still known to amplify noise as visible streak artefacts in the image domain. Generally, all analytical methods inherently ignore the Poisson-nature of the projection data. In addition, the complex photon detection process (see also Chapter 6) causes systematic errors requiring obligatory pre-corrections (see Section 7.2). Although the (over)-simplified data model allows very fast image reconstruction, a compromised image quality becomes unavoidable.

7.1.2 Iterative Methods

When compared to analytical approaches, iterative reconstruction methods are known to provide superior image quality at the cost of high computational challenges, especially for 3D PET acquisition data. Better image quality can be obtained by accurately modelling the characteristic detection process and by considering the Poisson-nature of emission data using an appropriate likelihood function. Meanwhile, iterative methods have evolved to become the standard approach for all modern clinical and pre-clinical PET systems. This is due to the tremendous progress in computing power over the last two decades in terms of numerical throughput.

7.1.2.1 General Recipe

The 3D image volume of tracer activity $\hat{\nu}$ is usually discretised and represented by a finite number I of values $\hat{\nu}_i$ assigned to specific basis functions of the image space, *e.g.* Cartesian voxels or polar voxels (Figure 7.4). A linear model then allows one to calculate expectation values $\mathbf{Y} = \{Y_j\}$ for all LORs depending on the values $\nu_i = \hat{\nu}_i \cdot \Delta T$ as an estimate of the image by evaluating linear equations of the form

$$Y_j = \sum_{i=1}^{I} g_{ij} \cdot \nu_i \quad \text{with } j = [1..J] \text{ and } i = [1..I], i.e.\ \boldsymbol{Y} = \mathbf{G} \cdot \mathbf{v} \qquad (7.3)$$

which is denoted as the forward projection of the image $\boldsymbol{\nu} = \{\nu_i\}$. In eqn (7.3), the matrix elements g_{ij} define the detection probabilities (DP) of recording coincidence events with the detector combination addressed by LOR j, which originates from the voxel i. In this way, eqn (7.3) determines the expectation value as a function of the estimated activity distribution for each LOR. Essentially, all matrix elements build the so-called system response matrix (SRM), which models the system-specific properties for coincidence gamma photon detection. Now, for a given SRM, the iterative reconstruction allows an image solution to be found whose forward projection $\boldsymbol{Y}$ optimally corresponds to the measured data $y = \{y_i\}$. This requires two conceptual components. First, an objective function or cost function needs to be defined which provides a measure of the goodness of any image $\boldsymbol{v}$ with respect to the measured data. Second, an optimisation algorithm finds a solution, which minimises the cost function.

7.1.2.2 System Response Matrix

Obviously, the accuracy of the system model (represented by the SRM) has a crucial impact on the resulting image quality.[4] Analytical methods play a major role in approximating the DP weights of the SRM. Here, a variety of methods exist to address the trade-off between the accuracy of the calculated DPs and the computational effort, *e.g.* single ray tracing,[5] multiple ray tracing,[6] distance-driven methods[7] or calculations of volumes-of-intersection.[8] As an example, Figure 7.4 illustrates the numerical burden used to calculate length-of-intersection (LOI) weights within a conventional Cartesian voxel grid (left) or volumes-of-intersection (VOI), within a polar voxel configuration (right) as for example applied in ref. 9. The former method is easy to implement and is fast. The latter method applies so-called tubes-of-response (TORs) (instead of LORs) that considers the pixel sizes of the PET detectors more accurately, but at the cost of increased calculation effort.[9]

The purely geometrical matrix $\boldsymbol{G}$ can be regarded as the central component for the imaging process, which can have further factorisable contributions as follows

$$\mathrm{G} = \mathrm{N \cdot P \cdot G \cdot I} \tag{7.4}$$

to consider additional physical effects within the SRM formalism.[10] N denotes a simple diagonal matrix to account for detector normalisation factors (see Section 7.2.3) and attenuation correction factors (see Section 7.2.1) separately for each LOR. P holds convolution kernels in the projection space domain to model detector cross-talk and detector blurring. Additionally, I holds blurring kernels in the image space domain. Usually, the number of LORs, J, is much larger than the number of voxels, I. Thus, the solution for the image according to eqn (7.4) is sufficiently constrained. However, this also implies, that the computational evaluation of P is quite costly compared to that of I.

Therefore, pure image space convolution has been proposed to consider the finite system resolution by exploiting the so-called point spread functions (PSF) in image space. Such PSF modelling allows improvements in image quality but can also produce the well-known Gibbs artefacts due to the physical limitations of this technique.[11]

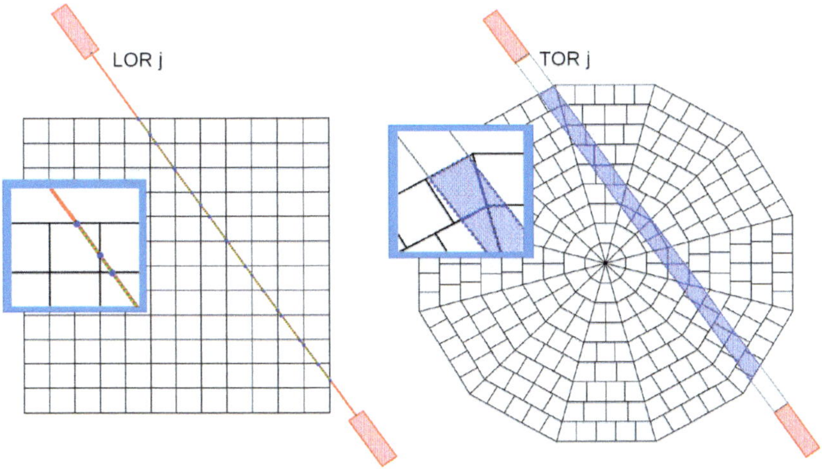

Figure 7.4 Cartesian voxels (left) and a polar voxel assembly (right) as two examples of applied basis functions in the image space domain; length-of-intersection weights assuming LORs (left) and area-of-intersection weights (2D) resp. volumes-of-intersection weights (3D) using tubes-of-response (right) are indicated. For simplicity only the 2D case is drawn, whereas the 3D case simply requires a copy of the grid in the axial direction to generate a stack of subsequent transaxial planes.

Detector simulations offer another possibility for SRM calculations by providing a detailed model of the detection process.[12,13] However, as a drawback, a long simulation time is unavoidable without supercomputing and the use of affordable CPU time results in significant statistical uncertainties in the SRM elements.

The major challenge for the SRM organisation and evaluation is the total matrix size, *i.e.* typical values for clinical systems with $J \approx 0(10^8)$ and $J \approx 0(10^7)$ result in $0(10^{15})$ system matrix elements. Considering the high sparsity, *i.e.* the fraction of non-zero elements is very low, the SRM can be implemented as a sparse matrix; nevertheless, a typical matrix size would be in the order of $0(10^{11})$ elements and would require several TB of memory. In this context, two basic strategies can be applied and also combined to address this problem.

Many implementations exploit symmetries of the system to significantly reduce the matrix size by eliminating redundant information. Symmetries evoke congruent geometrical cases, *i.e.* matrix weights are identical for symmetric LORs with symmetrically addressed voxels. Thus, weights only need to be stored once but they can be used for several LORs simultaneously. SRM compression, due to the exploitation of symmetries, allows a pre-calculated system matrix to be stored.[9] Furthermore, symmetry-driven SRM compression immediately supports the single instruction multiple data (SIMD) technique, which optimises numerical throughput for evaluation.[14]

The effort can be reduced by merging several 3D physical LORs to single 'representative' LORs. This reduces the projection space, J, to downsize the SRM. Here, several techniques of axial and transaxial data compression exist. In addition, or alternatively, simple geometrical models of DP weights cause higher SRM sparsity and lower size, thus reducing the overall calculation effort during matrix evaluation. For instance, the LOI weighting scheme usually provides significantly less non-zero elements compared to the VOI weighting scheme. This is because a LOR naturally addresses less voxels than a corresponding TOR counterpart. However, all introduced simplifications of the system model usually result in some compromise in the achievable image quality.

7.1.2.3 Example—MLEM

One of the most prominent paradigms in PET image reconstruction is represented by the maximum-likelihood-expectation-maximisation algorithm (MLEM)[15] and some advantageous variants thereof, *e.g.* ordered-subset-expectation-maximisation (OSEM)[16] or ordinary Poisson OSEM (OP-OSEM).[17] The foundation of the MLEM is the Poisson statistics, which become typical for low probability counting measurements. Thus, the Poisson likelihood function L is applied as a cost function, which can be formulated as a product over all contributing LORs as follows (eqn (7.5))

$$\mathcal{L}\left(y \middle| Y(\nu)\right) = \prod_{J} \frac{\left(Y_j\right)^{y_j}}{y_j!} \mathrm{e}^{-Y_j} \tag{7.5}$$

which gives the likelihood to observe the measured data y, while having the expectation values $Y(\nu)$, which are given by the image $\boldsymbol{\nu}$ according to eqn (7.3). The solution can be found by iteratively using eqn (7.6)

$$\nu_i^{(k+1)} = \nu_i^{(k)} \cdot \frac{1}{\sum_j \hat{g}_{ij}} \sum_j \hat{g}_{ij} \frac{y_i}{Y_j^{(k)}} \tag{7.6}$$

where (k) and $(k + 1)$ indicate the values for the kth resp. $(k + 1)^{\text{th}}$ iteration. As an example, Figure 7.5 illustrates the performance of the MLEM algorithm for 1 million (top row) and 10 million (bottom row) counts of a simulated 2D phantom. Evidently, the signal-to-noise ratio in the reconstructed images depends on the statistical uncertainties in the projection data as well as the number of applied iterations, *i.e.* for increasing number of iterations the resolution improves; however, the level of noise rises as an opposing trend. In terms of the mean square error (MSE) of reconstructed images with respect to the true image, MSE values usually reach a minimum after a specific number of iterations (Figure 7.5(b) and (e)) and then start to rise where noise becomes the dominating component in the solutions. This is an effect of the limited statistics of the input projection data, as shown in Figure 7.6. As expected from the MLEM algorithm, the log-likelihood function under the terms of

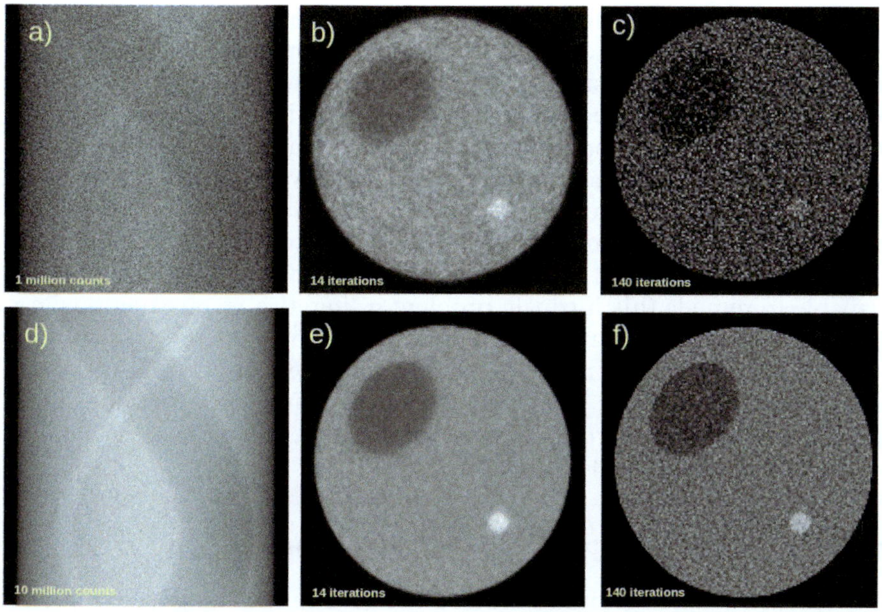

Figure 7.5 Simulated 2D phantom data with 1 million counts (top row) and 10 million counts (bottom row) visualised as sinograms with 300 projection angles and 300 radial bins (left column) and corresponding images from MLEM reconstruction for different iterations; in (b) and (e) images with minimal mean square errors (MSE) values are shown.

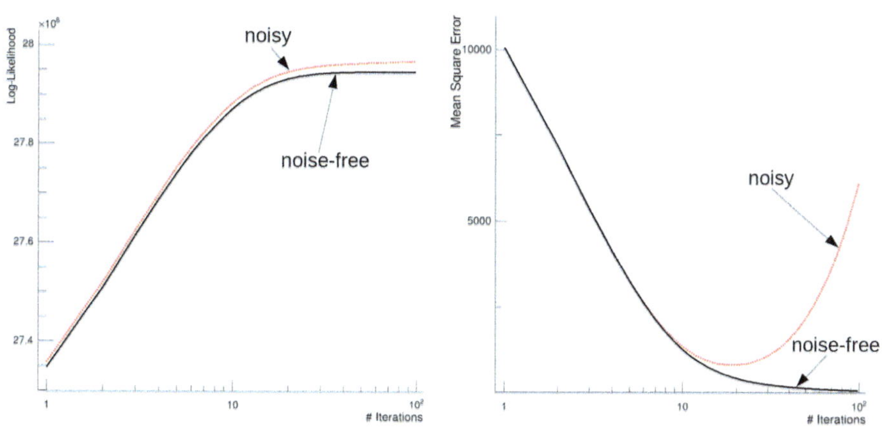

Figure 7.6 Log-likelihood values (left) and MSE values of reconstructed images with respect to the true images as function of the number of iterations for the simulated 2D phantom (see Figure 6.4) for the noise-free case (black curves) and noisy data with 10 million counts (red curves).

eqn (7.5) increases monotonically for any case, *i.e.* noisy or noise-free input data. The logarithm of the likelihood is also a monotonic surrogate of the cost function, which is easier to calculate because the product's overall terms become a sum of simplified expressions, while usually disregarding constant contributions. Furthermore, the MSE values decrease monotonically in the case of noise-free data (black curve, right plot). However, in the case of noisy data, *i.e.* ill-conditioned inversion problem, MSE values are observed to grow after reaching a minimum. This behaviour addresses the degree of regularisation, which is required to obtain stable solutions. Various strategies are available to remedy this intrinsic problem of iterative reconstructions, *e.g.* the introduction of stopping rules to limit the number of iterations, applying a post-reconstruction filter like 3D Gaussian filter with adjusted kernel sizes, or introducing a Bayesian prior term into the cost function which penalises unwanted solutions according to prior information about the activity distribution (see Section 7.1.2.4).

7.1.2.4 *Maximum* A Posteriori *(MAP) Reconstruction*

MAP reconstruction algorithms are an alternative to obtaining reconstructed PET images with a reduced level of noise due to additional regularisation.[2] This approach offers a flexible and principled way to control the noise. Instead of finding the solution according to eqn (7.5), an additional term is introduced (eqn (7.7))

$$\mathcal{L}^{\mathrm{MAP}}\left(y|\nu\right)=\mathcal{L}(y\,|\,\boldsymbol{Y}(\boldsymbol{\nu}))\cdot \mathrm{e}^{-\beta R(\nu)} \tag{7.7}$$

where β is a regularisation parameter (Bayes parameter) and $R(\boldsymbol{\nu})$ *is* the Gibbs energy function, also known as a prior or regulariser. In accordance to eqn (7.6) the iteration formula is extended as follows (eqn (7.8))

$$\nu_i^{(k+1)}=\nu_i^{(k)}\cdot\frac{1}{\sum_j \hat{g}_{ij}+\beta\dfrac{\partial R(\nu)}{\partial \nu_i}}\sum_j \hat{g}_{ij}\frac{y_i}{Y_j^{(k)}} \tag{7.8}$$

Effectively, $R(\boldsymbol{\nu})$ provides a penalty on disfavoured solutions, *i.e.* solutions which disregard the desirable properties as formulated by the prior, *e.g.* the smoothness of images. Due to the introduced regularisation, MAP reconstruction usually reaches a point where iteration updates become very small, thus indicating approximate convergence of the solution.

The Bayes parameter is a free parameter and allows one to control the influence of the applied prior. Thus, a data dependent (empirical) adjustment is usually required. This implies a shortcoming of MAP reconstruction for routine usage. If the Bayes parameter becomes too large, convergence is no longer guaranteed, and images can suffer due

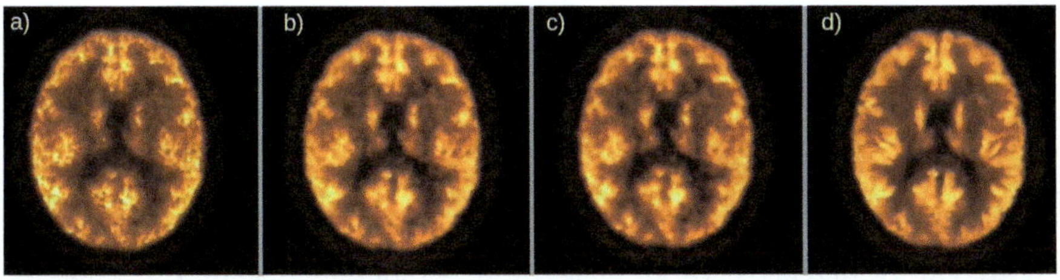

Figure 7.7 Reconstructed images for a human brain dataset: (a) MLEM using stop early rule (25 iterations); (b) MAP without anatomy quadratic prior with $\beta = 1$ (90 iterations); (c) MAP with boundaries quadratic prior with $\beta = 0.9$ (90 iterations); and (d) MAP with Bowsher quadratic prior with $\beta = 0.1$ (90 iterations).

to blurring. If β is chosen to be too small, the influence becomes marginal, *i.e.* images remain noisy while the MAP solution crossed over to the MLEM solution.

The regularisation tends to cause blurring of object edges, *i.e.* noise reduction is closely linked to losses of resolution. However, high resolution is also of clinical importance.[2] One could select the Median Root prior or an edge-preserving Gibbs prior to minimise this blurring.[18] However, the problem with the Gibbs prior is that one needs to decide the threshold between a small and a large difference between voxels. A small difference is considered noise and should not be allowed, whereas a large difference is considered a tissue border and it is thus allowed. Another method to overcome edge blurring is to define anatomical boundaries where smoothing is forbidden. The boundary method requires good agreement between anatomical boundaries (usually from MRI or CT) and the functional boundaries in PET (Figure 7.7). This is usually considered true for brain imaging. For MR-PET human brain imaging, in particular, more complex multi-parametric anato-functional priors have been proposed.[19]

7.2 Data Corrections

Acquired PET data need to be corrected for different inherent physical effects, which are either related to the processes gamma radiation undergoes while moving through matter or related to its biased and imperfect detection.

7.2.1 Attenuation of Radiation

Quantitative PET imaging requires an attenuation map to correct for attenuation of radiation in matter, which is one dominant source of qualitative and quantitative bias (see Section 6.3.2 and Section 11.2). Figure 7.8 illustrates the effect of attenuation correction (AC) by showing images with and without such a correction applied.

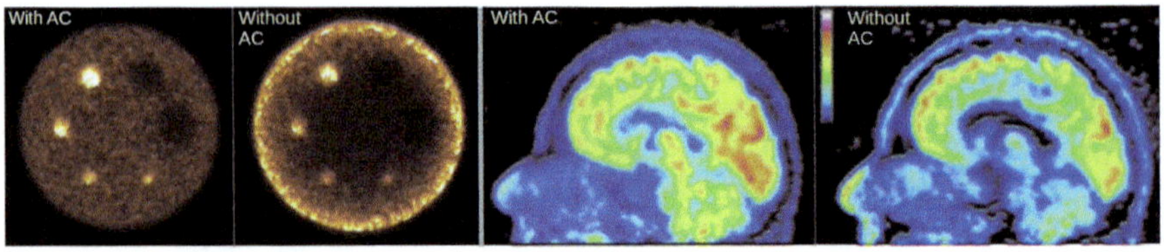

Figure 7.8 Difference between images with/without attenuation correction for a cylinder phantom (left) and FDG human brain scan data (right).

When photons travel through matter they undergo stochastic scattering and absorption depending on the electron density of the material (see Section 6.1). If photons are travelling along a straight line in tissue, only a fraction of them arrive at distance L without scattering or absorption. The effective attenuation is a path integral as follows

$$a_j = e^{-\int_{L_j} \mu(l)\mathrm{d}l} \tag{7.9}$$

where a_j is the attenuation fraction, and $\mu(l)$ is the spatial-dependent linear attenuation coefficient of the tissue in units of cm^{-1} along the LOR j. The attenuation coefficient depends on the energy of the photons and the electron density in the tissue. The attenuation is essentially proportional to the electron density due to the Compton effect and photo effect, both of which dominate any electron–photon interaction. Thus, a higher tissue density results in a larger attenuation fraction. For instance, the 511 keV attenuation coefficients of important tissues are: lung 0.018–0.027 cm^{-1}, adipose tissue 0.086–0.093 cm^{-1}, soft/brain tissue 0.094–0.100 cm^{-1}, spongious bone 0.110 cm^{-1} and cortical bone 0.130–0.172 cm^{-1}. According to eqn (7.9), the attenuation correction factor (ACF) $A_j = 1/a_j$ compensates the attenuation along the pathway L_j as a multiplicative correction with individual values A_j for all LORs. In order to correct for attenuation, the 3D map of the attenuation coefficients μ (also called μ-map) of the investigated subject is required.

On standalone PET scanners, such attenuation maps are often directly measured. Blank scan data (B_j) and transmission scan data (T_j) are typically acquired with rotating positron emitter (line) sources. Then, the ratio $A_j = T_j/B_j$ represents an unbiased (but noisy) estimator of the true ACF value, which can be considered in the diagonal matrix N of eqn (7.4). Variance-reduced estimators can be obtained by tomographic reconstruction of the μ map using (B_j) and (T_j) directly.[20]

A different situation is given for PET/CT scanners. The available CT images represent transmission-based attenuation maps, but for significantly lower photon energy than 511 keV. Therefore, they have other physical units (Hounsfield Units, HU) and values (air: −1000 HU, water: 0 HU).

Consequently, an extrapolation from CT attenuation to PET attenuation coefficients μ is mandatory.[21]

For practical reasons, in MR-PET scanners, the needed μ map of the subject can usually only be derived from acquired MR images. However, no direct correlation between MR image intensity and electron density exists, *i.e.* no immediate conversion from MR images to the attenuation map is possible. Several methods have been developed up to now, which can be found in Section 11.2.

7.2.2 Compton Scattering

Compton scattering is a dominant source of quantitative bias. Emitted photons substantially interact with electrons when they pass through any matter (*e.g.* tissue, scanner bed) on their path towards the detectors. For photons with 511 keV energy, the predominant interaction is Compton scattering. This is illustrated in Figure 5.4. Without any Compton scattering the originating positron emission can be correctly located along the detected LOR, because annihilation photon pairs are emitted in nearly opposite directions. However, if one or both photons undergo Compton scattering, the detected LOR no longer corresponds to the location of the emission. Such misidentifications qualitatively decrease the image contrast and lead to a quantitative bias in the reconstructed activity image.

Scattered photons have a reduced energy E' according to eqn (7.10)

$$E' = \frac{E}{1 + \frac{E}{m_0 c^2}\left(1 - \cos\theta\right)} \tag{7.10}$$

where θ is the scatter angle, E the initial energy and m_0c^2 is the resting energy of the electron. Thus, to some extent, scatter events can be identified and rejected depending on the energy resolution of the photon detectors of the scanner. The number of events affected by Compton scattering essentially depends on the amount of matter and related electron density along the pathway of the photons, *i.e.* size, geometry and specific density of the measured subject, but also depends on any other passive material of the scanner, such as the cover, bed, MR coil or shielding material between the subject and the radiation detectors. For 3D PET data acquisition, the scatter fraction (ratio of number of scatter events to the total number of events) can reach up to 40% for human brain measurements and 50% for whole-body scans.[22] Consequently, an accurate scatter correction is necessary to obtain quantitative results.

The probability of Compton scattering of a photon at a single electron is given by the Klein–Nishina formula. This formula allows modelling of the process of scattering, which is dependent on the electron density distribution (basically represented by the attenuation map, see Section 11.2), and on the activity distribution (emission image). However, here, one inherent problem becomes obvious regarding the initial emission image. Usually, a reconstructed image, which has been reconstructed without the application of any scatter correction, is used for the estimation of the scatter component. Although in this way, the first estimate of the Compton scattering component is biased, the bias can be minimised by iteratively re-calculating the scatter correction from an already scatter-corrected image with additional follow-up image reconstruction.

Monte Carlo methods (MC) have been identified as the most accurate method with which to estimate the fraction of scattered events for each individual LOR. The general idea of MC is to simulate the stochastic processes of physical interactions between photons and matter in detail, thus also taking multiple scattering and appropriate detection efficiencies accurately into account. After full simulation, a simulated dataset (separately including unscattered and scattered coincidences) is obtained and can be easily scaled to the measured data, thus providing the fraction of Compton events for each LOR. However, the major disadvantage of MC methods is that they are very time consuming, making them impractical for clinical applications.

As an alternative, the Single Scatter Simulation (SSS) approach is the most widely used method for scatter correction in clinical PET systems.[23–25] SSS is based on the analytical model that only one of the two annihilation photons undergoes a single Compton scattering process. The magnitude and spatial distribution are analytically approximated by the Klein–Nishina formula regarding a restricted number of sampling points, which are exclusively considered to estimate probabilities of scattering. As further simplification, usually, only a reduced number of detector elements are considered to minimise the calculation effort. An interpolation procedure is then performed to expand the scatter contribution estimates to the full projection space domain. In this approach, simple detection efficiency factors are used to model the detector physics and account for factors such as the incidence angle and the penetration depth of considered rays. Finally, the estimated scatter distribution is scaled to match to the measured data. The SSS method is very fast but has intrinsic approximations, *i.e.* it neglects the effect of multiple scattering and inhomogeneous detection probabilities. Another drawback is the necessity of performing a so-called tail fitting of the obtained scatter distribution; tail fitting addresses all LORs which are not touching the subject, and thus do not contain any true coincidences. Consequently, all such LORs exclusively contain scattered coincidences, which can be used to directly adjust the scatter fraction from the SSS according to the measured LOR data. The contribution from random coincidences (see Section 7.2.4) can be subtracted to obtain estimates of the pure scatter event contribution. However, such a tail fitting technique is error prone and tends to introduce systematic uncertainties, especially for dynamic imaging with rather poor statistics. Therefore, more robust fitting methods have been proposed for the scatter estimation. The adjustment of the scatter background can be performed during the iterative reconstruction by fixing the scale factor for all LORs simultaneously such that the likelihood function (see Section 7.1.2.3) is further minimised.

7.2.3 Detector Normalisation

Due to the typical block design of PET detectors (see Section 6.3), the sampled LORs usually have inhomogeneous sensitivities for several reasons, *e.g.* geometrical dependencies of photon absorption probabilities, tolerances of scintillator light yield and electronic gain of each detector element. Information about these inhomogeneities is required to obtain artefact-free images from the reconstruction algorithm. This system-specific correction is called normalisation and essentially comprises individual normalisation factors N_j, *i.e.* one specific value for each LOR j. In the case of iterative image reconstruction, the normalisation factors can easily be incorporated in the system response matrix (see Section 7.1.2.2).

As a standard strategy, such normalisation factors can be derived from simple phantom measurements with known activity distribution and which illuminate all relevant scanner LORs. A simple phantom allows the accurate calculation of the expectation value of true coincidences (eventually corrected for random coincidences as well as unavoidable photon attenuation and Compton scattering occurring in the phantom). Special normalisation phantoms, typically containing ^{68}Ge for convenience of handling, *e.g.* either rotating thin plane sources or orbiting line sources, allow a homogeneous illumination of the entire field-of-view (identical expectation values for all LORs) while attenuation and Compton scattering can be neglected. In this way, a normalisation factor for each LOR can be derived by calculating the ratio between measured coincidences and expected true coincidences. Such a simple approach is denoted as 'direct normalisation' and usually suffers from statistical uncertainties because even for extremely long periods of data acquisition (>1 day), the number of counts per LOR remains rather poor. Such statistical uncertainties of the normalisation factors directly propagate into an increased level of noise in any reconstructed image. Additional methods of so-called component-based variance reduction[26] allow the statistical uncertainties to be substantially reduced. Here, normalisation factors are modelled as unique products of the individual crystal efficiencies involved and a strongly limited number of geometric factors that characterise variations of LOR efficiencies for different geometrical cases of photon incident angles. For example, due to intrinsic scanner symmetries, specific sets of symmetric LORs have congruent geometrical properties, which can be consistently characterised by the same geometrical efficiency values. Effectively, the normalisation factors are no longer independent, *i.e.* each normalisation factor is factorised into a few specific components. Thus, the number of unknowns can be reduced from the number of LORs to the number of components. In this way, the statistical uncertainties of the extracted components can be coherently reduced, thus reducing the uncertainties of the calculated normalisation factors according to the applied component model.[26]

In contrast to all geometrical dependencies, which naturally remain time-invariant, the individual crystal efficiencies are usually undergoing slowly varying drifts of detector gain and timing due to the change in ambient conditions of the acquisition system, *e.g.* temperature, high voltage, ageing. Therefore, routine normalisation measurements are acquired to update the normalisation factors. Normally, the update intervals depend on the overall detector system stability but are conducted on an obligatory basis every few months.

7.2.4 Random Coincidences

In the case of simultaneous emission, the coincidence detection of two single photons arising from the same annihilation process is realised due to their temporal correlation (see Figure 5.4 and 6.7). That means, two detected single photons are accepted as true coincidence if the time difference of their occurrence is shorter than a characteristic time period τ, also denoted as coincidence timing window. Consequently, uncorrelated single photons can be misidentified as true coincidence when accidentally occurring sufficiently close together in time (see Figure 6.7). Such unavoidable misclassification of detected single photon events are called random coincidences. Obviously, the rate of detected random coincidence for a specific detector combination R_{AB} increases with the rate of single events (S_A, S_B) for each detector, as well as the timing window τ under the terms of

$$R_{AB} = 2\tau S_A S_B \tag{7.11}$$

From eqn (7.11), a good timing resolution of the PET detectors to minimise the contribution of random coincidences becomes evident. Usually, it is necessary to appropriately estimate and subtract the contribution of random coincidences from the measured coincidences (prompts) to yield quantitative images.

To identify random coincidences, two standard straight-forward methods are used, namely directly utilising measured single rates, S_A, of all detectors according to eqn (7.11) or exclusively detecting/classifying random coincidences within a so-called delayed coincidence window. For the latter method, the timestamps from one detector are delayed by a time significantly greater than the coincidence timing window, τ. In this particular way, no true coincidences, only random coincidences, are classified for each LOR and the found number of random coincidences represents a good estimate of random coincidences in the prompt signal. An advantage of the delayed window technique is that the detected delayed events have dead-time properties identical to the detected prompt events. However, the statistical uncertainties for the random sample from the delayed window are much larger than if single rates according to eqn (7.11) are directly used, because S_A, S_B are orders of magnitude larger quantities than a sample of R_{AB}. Thus, for the delayed window approach an additional variance-deduction technique is proposed to overcome the problem of higher statistical uncertainties by calculating pseudo-single rates.[27]

7.2.5 Detector Deadtime

Any PET acquisition system requires a minimum processing time for any individual single event or coincidence event. During the data processing time of one triggered event, other subsequently occurring events in the same detector unit cannot be processed and recorded. Consequently, the global sensitivity of the system is reduced. This processing time is the so-called detector deadtime (DT). DT is related to many aspects, including the overall detector architecture, the characteristic signal integration and response time of SiPM or PMT, the electronic circuits, *etc.* This phenomenon becomes especially important at high detector count rates and leads to non-linear changes in the measured number of coincidences, depending on the overall activity in the field-of-view. To obtain unbiased quantitative results, the acquired data need to be corrected for efficiency losses due to DT.

DT can be classified into two categories: paralysing DT and non-paralysing DT.[28–30] In the case of paralysing DT, each incident single event causes DT for a period of time, ξ, regardless of whether or not the acquisition hardware is already processing an event. For example, if another event occurs within the DT cycle of a preceding event, the new event will be ignored, however this second event 'restarts' the current DT cycle (Figure 7.9) accordingly. That means the DT cycle is stretched further. The occurrence of incident events

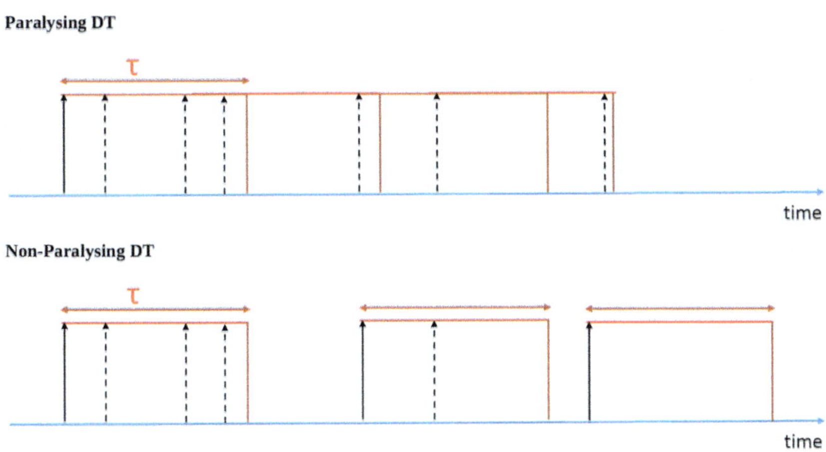

Figure 7.9 Differences of the measured counts between paralysing DT (top) and non-paralysing DT (bottom) for the same series of incident events. For the seven incident events, the paralysing mode only allows one event to be detected, whereas in the non-paralysing mode three events can be recorded.

obeys Poisson statistics; thus, the paralysing DT provides an effective single rate as follows (eqn (7.12a))

$$S_{\mathrm{A,eff}} = S_{\mathrm{A}}\mathrm{e}^{-S_{\mathrm{A}}\xi} \tag{7.12a}$$

with ξ as a system-specific time constant of the paralysing DT. For example, the interaction of the photons with the detecting crystals is a paralysing component. In contrast, non-paralysing DT implies that if an event occurs in the DT cycle of a preceding event, it will be simply ignored without further affecting the current detection process. The relationship between the measured counting rates and the incident counting rates can be expressed as follows (eqn (7.12b))

$$S'_{\mathrm{A,eff}} = S_{\mathrm{A}}/(1-\xi') \tag{7.12b}$$

with ξ' as the time constant to characterise the non-paralysing contribution to DT. For example, the DT caused by the coincidence-processing unit behaves essentially in a non-paralysing way, because arriving events are simply ignored while a coincidence is processed.

Effectively, DT correction is usually performed by modelling the dead-time losses as a combination of paralysing and non-paralysing components and obtaining parameters for the model. These can be derived from measurements of a decaying source to characterise the detection efficiencies for changing, calculable activity levels. The simplest way for dead correction is to build a look-up table, which provides correction factors for each detector unit as a function of the single count rate S_{A}.

7.2.6 Radioactive Decay

Usually, PET data are acquired over time periods that are significantly longer than the half-life of the applied radiotracer (see Section 5.2). Hence, the reduction in activity levels over the complete acquisition time needs to be corrected to obtain unbiased time activity curves (TAC) with respect to the starting time of the measurement. The decay correction factor $D(t)$ describes the reduction of activity as function of time due to the radioactive decay with isotope half-life $T_{1/2}$ that is eqn (7.13)

$$D(t) = \mathrm{e}^{-\ln 2 \frac{t}{T_{1/2}}} \tag{7.13}$$

Because the data are usually reconstructed as a series of time frames, the appropriate mathematical integration provides the effective decay correction factor for a frame starting at t_0, with a duration of Δt, as follows (eqn (7.14))

$$D_{\mathrm{eff}}(t,\Delta t) = D(t)\cdot\frac{(1-\mathrm{e}^{-\alpha})}{\alpha} \text{ with } \alpha = \ln 2\cdot\frac{\Delta t}{T_{1/2}} \tag{7.14}$$

which is denoted as image-frame decay correction.

7.2.7 Detector Calibration

Once, all corrections as described above are taken accurately into account, images can be reconstructed which have no artefacts and the obtained intensity values describe the spatial distribution of the radiotracer concentration in the subject. Usually, in many clinical PET applications, this level of quantification is sufficient because the 3D images are interpreted visually without addressing absolute voxel values. This also includes methods that consider a reference tissue compartment to characterise the input function for kinetic modelling. In this way, any global calibration factor is effectively immediately cancelled out. In contrast, for more advanced kinetic modelling (see Chapter 14) PET image values must be correlated to the tracer concentration in blood samples taken during the dynamic PET measurement. Then, it is necessary to have an accurate cross calibration between the reconstructed activity of the PET images and externally measured blood samples. This can be realised by measuring a phantom with uniform activity concentration in the PET scanner and reconstructing an image while applying all data correction in the same way as for the protocol of the study. The calibration factor can then be extracted easily by comparing the numerical values in a region-of-interest of the reconstructed PET image with the activity of an aliquot taken from the phantom, which can be measured with the same device as used for measuring the blood samples. Note, that such a calibration still does not necessarily provide values in absolute units of radioactivity concentration [Bq cm^{-3}] of image voxel intensities unless the true absolute concentration in the aliquot is accurately determined by the external device.

7.3 Patient Motion

Usually, with increasing acquisition time the probability of patient motion increases, thus motion compensation methods are desirable to maintain an unbiased quantification in dynamic studies. However, in the presence of motion, the appropriate handling of the above data corrections becomes more cumbersome. That means, the physical correlations between subject and detectors change according to the motion track, thus data corrections are no longer invariant over the full duration of the measurement. For example, mismatches arise between the moving subject and the attenuation map, which is usually acquired in only one specific reference position. Also, the detector normalisation needs special attention because any LOR of the projection space domain, which addresses the subject in a fixed frame of reference, is sampled by different physical LORs with different sensitivities (Figure 7.10). Thus, LOR-based motion correction methods need to calculate effective normalisation factors depending on the relative motion between the subject and scanner gantry.[31] In cases where the motion parameters can be estimated or directly measured, various correction methods are directly available to eliminate motion-blurring artefacts. More details relating to commonly applied motion correction methods in PET can be found in Chapter 12.

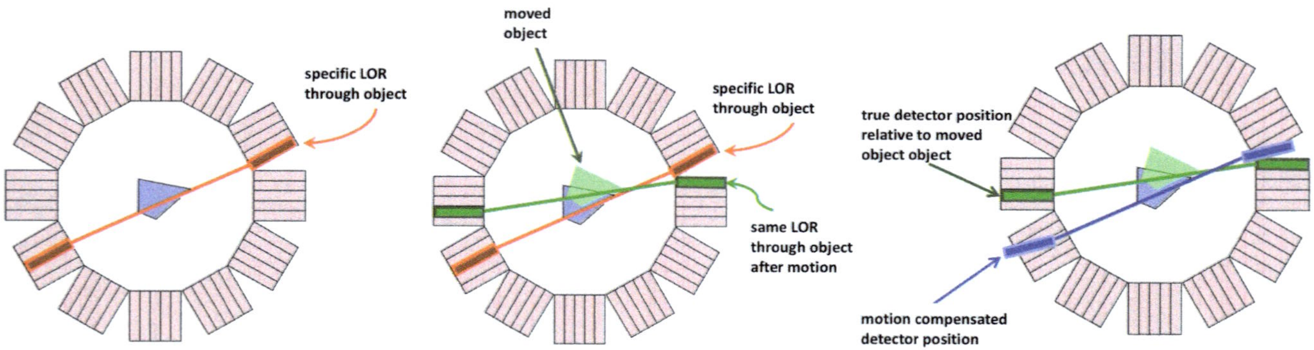

Figure 7.10 Illustration of the effect subject motion relative to the PET detectors. In the initial position of the object a specific LOR through the object (red line) is addressed by a specific detector combination (left). After subject motion the same LOR (in relation to the subject) is now moved (green line) with respect to the detectors (middle), thus another detector combination is addressed. Motion compensation requires transformation of the LOR (green) back to the initial position (right) with accordingly modified detector positions (blue).

References

1. D. L. Bailey, D. W. Townsend, P. E. Valk and M. N. Maisey, *Positron Emission Tomography, Basic Sciences*, Springer, London, 2005.
2. M. N. Wernick and J. Aarsfold, *Emission Tomography, The Fundamentals of PET and SPECT*, Elsevier, 2004.
3. S. R. Cherry, J. A. Sorenson and M. E. Phelps, *Physics in Nuclear Medicine*, Elsevier Saunders, 2012.
4. A. Iriarte, R. Marabini, S. Matej, C. O. S. Sorzano and R. M. Lewitt, *Comput. Med. Imaging Graph.*, 2016, **48**, 30.
5. R. L. Siddon, *Med. Phys.*, 1985, **12**, 252.
6. S. Moehrs, M. Defrise, N. Belcari, A. DelGuerra, A. Bartoli, S. Fabbri and G. Zanetti, *Phys. Med. Biol.*, 2008, **53**, 6925.
7. B. De Man and S. Basu, *Phys. Med. Biol.*, 2004, **49**, 2463.
8. J. Scheins, F. Boschen and H. Herzog, *IEEE Trans. Med. Imaging*, 2006, **25**, 1363.
9. J. Scheins, H. Herzog and N. J. Shah, *IEEE Trans. Med. Imaging*, 2011, **30**, 879.
10. J. Qi, R. M. Leahy, S. R. Cherry, A. Chatziioannou and T. H. Farquhar, *Phys. Med. Biol.*, 1998, **43**, 1001.
11. S. Stute and C. Comtat, *Phys. Med. Biol.*, 2013, **58**, 3849.
12. S. Jan, G. Santin, D. Strul, S. Staelens, K. Assié, D. Autret, S. Avner, R. Barbier, M. Bardiès, P. M. Bloomfield, D. Brasse, V. Breton, P. Bruyndonckx, I. Buvat, A. F. Chatziioannou, Y. Choi, Y. H. Chung, C. Comtat, D. Donnarieix, L. Ferrer, S. J. Glick, C. J. Groiselle, D. Guez, P.-F. Honore, S. Kerhoas-Cavata, A. S. Kirov, V. Kohli, M. Koole, M. Krieguer, D. J. van der Laan, F. Lamare, G. Largeron, C. Lartizien, D. Lazaro, M. C. Maas, L. Maigne, F. Mayet, F. Melot, C. Merheb, E. Pennacchio, J. Perez, U. Pietrzyk, F. R. Rannou, M. Rey, D. R. Schaart, C. R. Schmidtlein, L. Simon, T. Y. Song, J.-M. Vieira, D. Visvikis, R. Van de Walle, E. Wieërs and C. Morel, *Phys. Med. Biol.*, 2004, **49**, 4543.
13. J. Cabello and M. Rafecas, *Phys. Med. Biol.*, 2012, **57**, 1759.
14. J. Scheins, K. Vahedipour, U. Pietrzyk and N. J. Shah, *Phys. Med. Biol.*, 2015, **60**, 9349.
15. K. Lange and R. Carson, *J. Comput. Assist. Tomogr.*, 1984, **8**, 306.
16. H. Hudson and R. S. Larkin, *IEEE Trans. Med. Imaging*, 1994, **13**(4), 601.
17. C. Michel, M. Sibomana, A. Boi, X. Bernard, M. Lonneux, M. Defrise, C. Comtat, P. E. Kinahan and D. W. Townsend, *Proc. IEEE Nucl. Sci. Symp. Med. Imag. Conf.*, 1998, **2**, 1323.
18. L. Caldeira, J. Scheins, P. Almeida and H. Herzog, *Nucl. Instrum. Methods Phys. Res., Sect. A*, 2013, **702**, 141.
19. A. Mehranian, M. A. Belzunce, F. Niccolini, M. Politis, C. Prieto, F. Turkheimer, A. Hammers and A. J. Reader, *Phys. Med. Biol.*, 2017, **62**, 5975.
20. H. Rothfuss, V. Panin, A. Moor, J. Young, I. Hong, C. Michel, J. Hamill and M. Casey, *Phys. Med. Biol.*, 2014, **59**, 5483.
21. J. P. T. Carney, D. W. Townsend, V. Rappoport and B. Bendriem, *Med. Phys.*, 2006, **33**, 976.

22. H. Zaidi and K. F. Koral, *Eur. J. Nucl. Med. Mol. Imaging*, 2004, **31**, 761.
23. C. C. Watson, D. Newport and M. E. Casey, *Radiol. Nucl. Med.*, 1996, **4**, 255.
24. C. C. Watson, D. Newport, M. E. Casey, R. A. deKemp, R. S. Beanlands and M. Schmand, *IEEE Trans. Nucl. Sci.*, 1997, **44**, 90.
25. J. M. Ollinger, *Phys. Med. Biol.*, 1996, **41**, 153.
26. R. D. Badawi and P. K. Marsden, *Phys. Med. Biol.*, 1999, **44**, 571.
27. L. G. Byars, M. Sibomana, Z. Burbar, J. Jones, V. Panin, W. C. Barker, J.-S. Liow, R. E. Carson and C. Michel, *Proc. IEEE Nucl. Sci. Symp. Med. Imag. Conf.*, 2005, **5**, 2622.
28. C. Moisan, J. G. Rogers and J. L. Douglas, *IEEE Trans. Nucl. Sci.*, 1997, **44**, 1219.
29. L. Eriksson, K. Wienhard and M. Dahlbom, *IEEE Trans. Nucl. Sci.*, 1994, **41**, 1566.
30. G. Germano and E. J. Hoffman, *J. Comput. Assist. Tomogr.*, 1988, **12**, 836.
31. V. W. Zhou, A. Z. Kyme, S. R. Meikle and R. Fulton, *Phys. Med. Biol.*, 2009, **54**, 5321.

CHAPTER 8

Kinetic Modelling and Extraction of Metabolic Parameters

H. HERZOG*[a], H. IIDA[b,c] AND L. CALDEIRA[a]

[a]Forschungszentrum Jülich, Institute of Neuroscience and Medicine – 4, Wilhelm-Johnen-Straße 1, 52428 Jülich, Germany; [b]Department of Investigative Radiology, National Cerebral and Cardiovascular Centre, Osaka, Japan; [c]Department of Medical Physics, Turku University Hospital, Department of Biomedicine, University of Turku, Turku, Finland

*E-mail: h.herzog@fz-juelich.de

8.1 Introduction

While magnetic resonance imaging (MRI) is primarily considered as a tool with which to image anatomical details with high resolution, positron emission tomography (PET) can be characterised as being the tracer method used to observe physiological and metabolic functions without any disturbance thereof by the measurement procedure. Often just a qualitative image of the uptake of the radiotracer labelled with a positron emitter gives sufficient information about a disease. This is demonstrated in Figure 8.1 where the lack of uptake of [^{18}F]-fluoro-deoxyglucose (FDG) in the occipito-temporal cortex gives a clear hint towards Alzheimer's disease. However, if one wishes to ascertain whether the radiotracer uptake of the

New Developments in NMR No. 19
Hybrid MR-PET Imaging: Systems, Methods and Applications
Edited by N. Jon Shah
© The Royal Society of Chemistry 2019
Published by the Royal Society of Chemistry, www.rsc.org

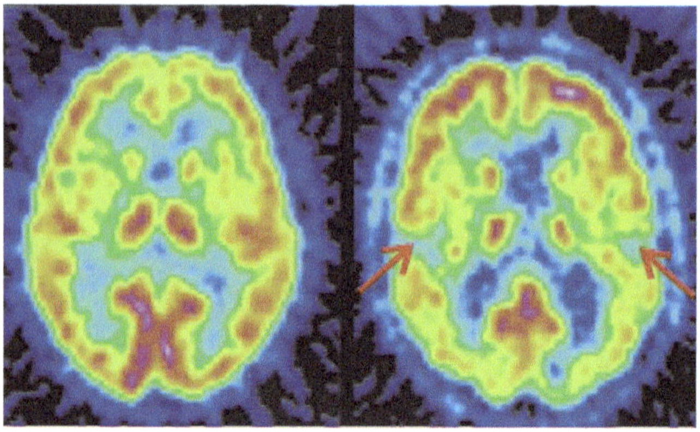

Figure 8.1 Qualitative images comparing the cerebral uptake of [^{18}F]-fluoro-deoxy-glucose (FDG) in a normal volunteer and a patient with Alzheimer's disease. The red arrows mark the loss of neurons in the occipitotemporal cortex, which is an indicator for this kind of dementia.

whole organ is too low, a simple qualitative evaluation of the PET image does not deliver the right answer. This is because, in addition to the possibility of low uptake being caused by a disease of the examined organ, low uptake may also be caused by too little injected activity or by a great body volume, so that a high percentage of the radiotracer is taken up by other body parts. In this situation the so-called standard uptake value (SUV) is helpful, which is defined as[1]

$$\mathrm{SUV} = \frac{\text{radioactivity concentration in the organ of interest}\left(\mathrm{kBq\ mL^{-1}}\right)}{\text{injected radioactivity (MBq)/body mass (kg)}} \tag{8.1}$$

Assuming 1 mL is equivalent to 1 g; the SUV is without dimension. By using the SUV, both the amount of injected activity and the body volume are considered. In clinical practice, the SUV is the parameter used most often for a semi-quantitative evaluation of radiotracer accumulation. For example, a decreased SUV of FDG in a metastasis before and after treatment indicates the success of the intervention.

Although the use of SUV is regarded as useful and is widespread, it is not sufficient to obtain a deeper insight into body functions in quantitative terms, *i.e.* it does not show how much cerebral blood flow increases during stimulation or the percentage of neuroreceptors blocked by a psychopharmaceutical. Table 8.1 presents a list of PET-radiopharmaceuticals and the corresponding functions.

To assess such functions based on quantitative PET data, different approaches have been developed. Although these approaches are very rarely applied in clinical practice, they have enabled clinical and basic research *in-vivo* to analyse many aspects of human (patho)-physiology and biochemistry.

Table 8.1 Using appropriate kinetic models, different physiological and metabolic functions can be derived from images of corresponding PET-radiopharmaceuticals.

PET-radiopharmaceutical	Function
$[^{18}F]$-fluoro-deoxyglucose	Glucose consumption
$[^{15}O]$-water	Perfusion
$[^{15}O]$-butanol	Perfusion
$[^{15}O]$-oxygen	Oxygen consumption
$[^{15}O]$-carbon monoxide	Blood volume
$[^{11}C]$-raclopride	Binding potential to and density of dopamine receptors
$[^{11}C]$-flumazenil	Binding potential to and density of benzodiazepine receptors
$[^{18}F]$-altanserin	Binding potential to and density of serotonin receptors
$[^{11}C]$-carfentanil	Binding potential to and density of opioid receptors

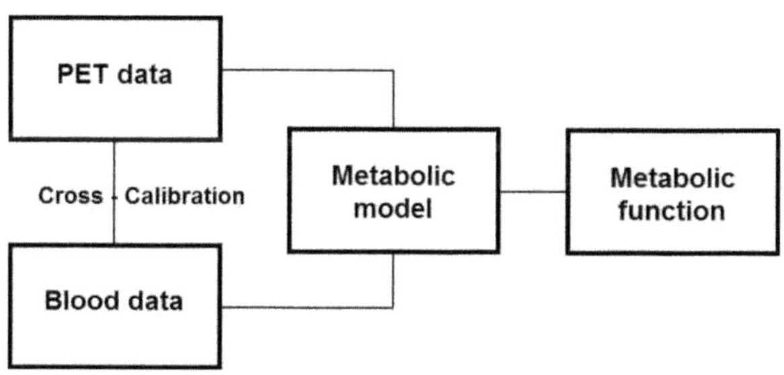

Figure 8.2 Flowchart for the calculation of metabolic functions. Quantitative dynamic or static PET data and time-activity data of the radioactivity concentration in the blood, which are cross-calibrated to each other, are required as input into a mathematical metabolic model.

This chapter introduces the different approaches leading from the radioactivity concentration quantified by PET to body functions.

8.2 The Three Components Required for Kinetic Modelling

The general concept applied to evaluate a metabolic function in a quantitative way with PET is based on three components (Figure 8.2): (1) the PET data that describe the spatial and temporal distribution of the radiopharmaceutical within the body and/or an organ; (2) the simultaneously measured data about the temporal course of the radioactivity within the blood or plasma; and (3) the metabolic model that yields a mathematical relationship between PET, blood data and the function of interest.

Point 1: The reconstructed PET images mostly deliver time-activity data of a region- or volume-of-interest (ROI or VOI) or of the entire image volume (Figure 8.3A). Some modelling procedures require just a single so-called autoradiographic image, resulting from a static PET acquisition or from summing up a temporal sequence of images (Figure 8.3B).

Point 2: Ideally, the temporal course of the radioactivity within the blood or plasma is obtained from the artery supplying the organ to be examined. Therefore, this time-activity curve is also called the input function. As it is generally not convenient or possible to obtain this function directly from the artery supplying the organ, blood samples from the radial or brachial artery are preferred. When measuring blood flow or blood volume, it is sufficient to know the radioactivity concentration of (whole) blood. Most applications, however, require the concentration of the radiotracer in the plasma, since the plasma concentration is decisive for the gradient of the radiotracer between the intra and extravascular space. After centrifuging the blood samples, plasma probes become available and a multi-probe well counter, which has been cross-calibrated with respect to the PET scanner, is used to measure their radioactivity. The cross-calibration ensures that the activity concentrations of both the blood and PET images have the same unit, kBq cc^{-1}. Often arterial blood sampling is regarded as too burdensome or not possible, and

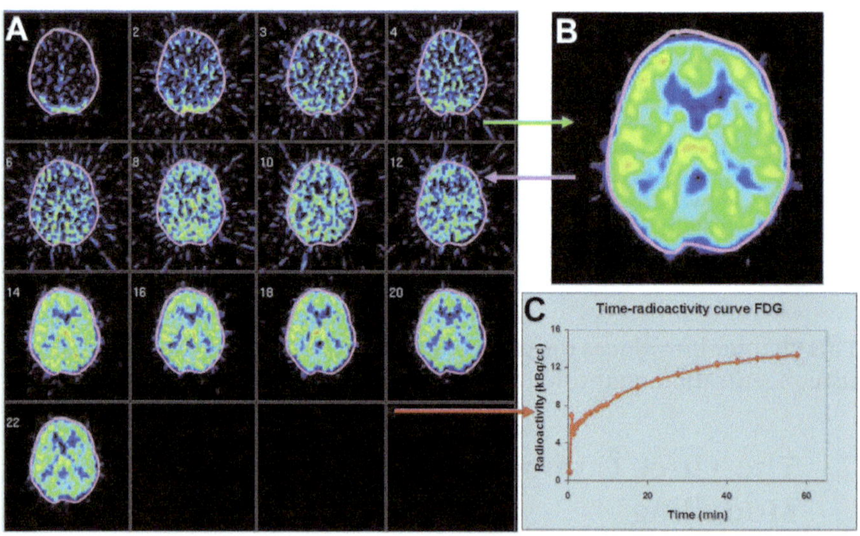

Figure 8.3 Quantitative PET data of [^{18}F]-fluoro-deoxyglucose (FDG) presented as a dynamic sequence of images ranging from shortly after injection until 55 min, when a nearly constant level of tracer of uptake is reached (A). Averaged image of cerebral FDG uptake by summing images of the dynamic sequence from 30–55 min (B), (green arrow). In order to obtain a time-activity of the cerebral FDG dynamics, a region of interest (ROI) is defined on the averaged image and then transferred to the dynamic images (pink arrow). The time-activity curve is obtained by integrating the radioactivity concentration of each image within the ROI and by plotting the integrated values over time (C).

in such a case venous blood is withdrawn from, *e.g.* an antecubital vein. However, directly after the injection of the radiopharmaceutical, there is a considerable discrepancy between arterial and venous radioactivity, so that the latter leads to an inaccurate input function—especially in the time shortly following the injection. This consequently leads to errors in the model analysis. In order to overcome this, the forearm, from which the blood samples are withdrawn, is heated by warm water or air *etc.* to arterialise the venous blood. Although this approach is not ideal, it may be acceptable if the heating leads to a venous oxygen content of more than 80%. If new radiopharmaceuticals are to be introduced and analysed, arterial blood sampling should be applied first. Later, the errors may be checked by using arterialised venous blood sampling.

In neuroreceptor studies, in particular, knowing the time-activity curve of the plasma may not represent the correct input function and is not sufficient for accurate kinetic modelling, since the radiotracer is most often metabolised within the liver. In this case, the plasma contains both the originally injected radiotracer and also its metabolised end product—both labelled with the same positron emitter. To distinguish between these two components, plasma probes are analysed by thin-layer chromatography, for example, so that the time-activity curve of the plasma containing the non-metabolised radiotracer (also called free ligand) can be determined as an input function (Figure 8.4).

Point 3: A general starting point with which to combine the three components shown in Figure 8.2 is a balance. A balance considers the sum of fluxes of radioactivity into and out of an organ together with the change of radioactivity over the time within the organ. For this purpose, the biological system to be analysed is divided into compartments. A compartment is a real

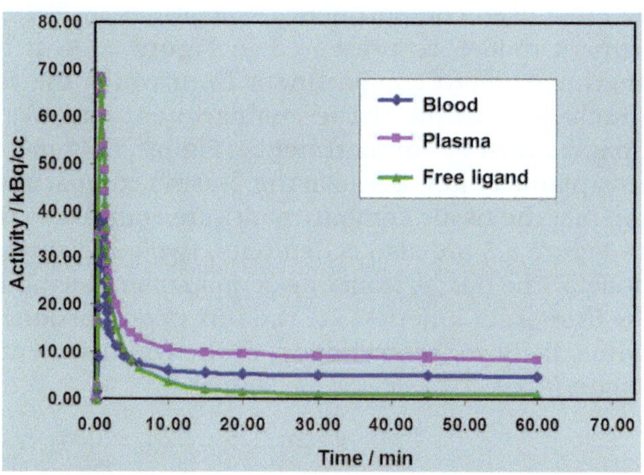

Figure 8.4 Time-activity curves of a PET radiotracer for neuroreceptor studies measured in blood, in plasma and as free ligand, *i.e.* non-metabolised radiotracer.

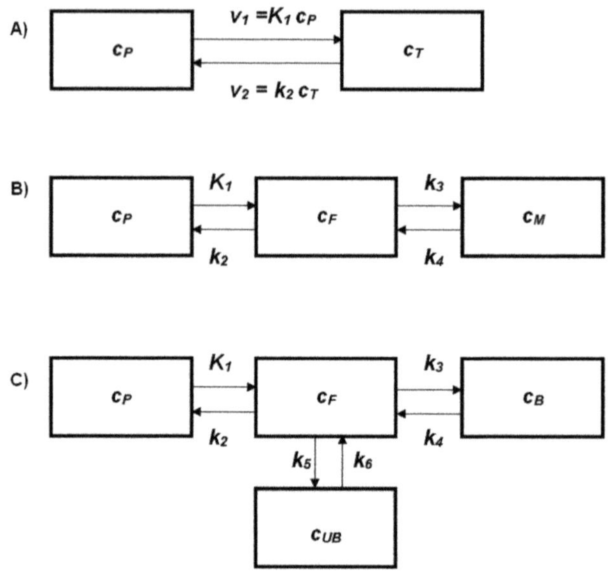

Figure 8.5 The most used compartment models in PET. (A) One-tissue compartment model applied for flow radiotracers or for the simplified description of models of reversibly bound neuroreceptor ligands. (B) Two-tissue compartment model applied for FDG and neuroreceptor ligands. (C) Three-tissue compartment model applied for neuroreceptor ligands if a compartment for the non-specifically bound radiotracer is to be considered.

or virtual volume in which substances are assumed to be instantaneously and homogeneously distributed.

One of the most used compartment descriptions in PET is the simple two-compartment model, as presented in Figure 8.5A. It consists of a plasma (sometimes blood) compartment from which the radiopharmaceutical is transferred to the tissue compartment and *vice versa*. However, there may be further compartments (Figure 8.5B and C), the roles of which are explained later. Because the plasma compartment is always present, often just the tissue compartments are counted so that the models shown in Figure 8.5 are also called one-, two-, and three-tissue compartment models. The flux, v, from one compartment to another is mostly controlled by first-order kinetics, *i.e.* the flux is proportional to the substance or radioactivity concentration, c, of the compartment from which v originates (eqn (8.2))

$$v = kc \tag{8.2}$$

The proportionality factor k has the unit min^{-1} and is called rate constant. In the case of first-order kinetics, k is a true constant. Using this relation, the one-tissue compartment model is expressed as in Figure 8.5A. Commonly, the transfer between the compartments is only indicated by the rate constants. Figure 8.5 also contains the conventional order of indices used in the model setup in PET.

The corresponding mathematical equation, with the balance of the fluxes as a function of time t, reads (eqn (8.3))

$$\frac{dc_{\mathrm{T}}(t)}{dt} = K_1 c_{\mathrm{P}}(t) - k_2 c_{\mathrm{T}}(t) \tag{8.3}$$

$c_{\mathrm{T}}(t)$ is the radioactivity concentration over time within the tissue as observed by PET. Mostly, capital K_1 is used instead of k_1, because it describes a constant of a clearance rate with the unit mL g^{-1} min^{-1} rather than a simple rate constant with the unit min^{-1}. At this point, it is emphasised that the rate constants do not represent just a mathematical entity. As will be seen in the following, they are the basic entities used to obtain the quantified parameters of the observed biological function.

8.3 Kinetic Analysis of a One-tissue Compartment Model

The one-tissue compartment model, as defined in Figure 8.5A and eqn (8.3), is an often applied, sometimes simplifying, approach for radiopharmaceuticals whose uptake into the tissue is promptly followed by the outwash back into blood vessels, *i.e.* they show reversible kinetics. Whereas the measured time functions $c_{\mathrm{P}}(t)$ and $c_{\mathrm{T}}(t)$ are known, the rate constants k_i, which are the characterising parameters of the compartment model, are unknown and to be determined. For this purpose, eqn (8.3) must be solved. This can be achieved analytically or by numerical integration. The analytical solution of eqn (8.3), which may be obtained with the help of the Laplace transform, reads

$$c_{\mathrm{T}}(t) = K_1 \int_0^t c_{\mathrm{P}}(\tau)\, e^{-k_2(t-\tau)} d\tau \tag{8.4}$$

This equation represents the convolution integral of the input function $c_{\mathrm{P}}(t)$ with the exponential term $e^{-k_2 t}$. The parameters k_i are numerically determined by fitting the left side of eqn (8.4), *i.e.* the theoretical $c_{\mathrm{T}}(t)$ to $c_{\mathrm{T}}^{\mathrm{PET}}(t)$ measured by PET. For this purpose, initial values for K_1 and k_2 are assumed, and using the measured $c_{\mathrm{P}}(t)$, a theoretical $c_{\mathrm{T}}^{\mathrm{theor}}(t)$ is calculated, which most probably does not fit $c_{\mathrm{T}}^{\mathrm{PET}}(t)$. The comparison between $c_{\mathrm{T}}^{\mathrm{theor}}(t)$ and $c_{\mathrm{T}}^{\mathrm{PET}}(t)$ is commonly done by calculating the root square mean of the differences between the two functions

$$\mathrm{RMS} = \sqrt{\sum \left(c_{\mathrm{T}}^{\mathrm{theor}} - c_{\mathrm{T}}^{\mathrm{PET}}\right)^2} \tag{8.5}$$

This RMS is to be minimised so that $c_{\mathrm{T}}^{\mathrm{theor}}(t)$ fits $c_{\mathrm{T}}^{\mathrm{PET}}(t)$ at its best. Since not all data have the same statistical quality, *e.g.* in the case of short frames with increased noise, the squared differences of eqn (8.5) are often weighted by

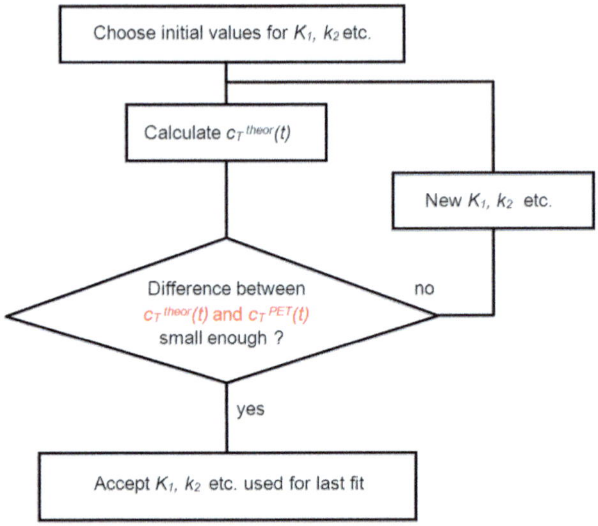

Figure 8.6 Block diagram indicating the algorithm for iterative curve fitting.

appropriate factors. If RMS is above a given threshold, the parameters k_i are varied by using a specific strategy and RMS is recalculated based the updated $c_{\mathrm{T}}^{\mathrm{theor}}(t)$ and so on.

This iterative approach is called non-linear least square fitting and is visualised in Figure 8.6. There are different strategies to vary the k_i in such a way so that RMS is minimised as quickly as possible. The two approaches mostly applied to the kinetic analysis of PET data are the Simplex[2,3] and the Marquardt–Levenberg[4,5] algorithms. Finally, when $c_{\mathrm{T}}^{\mathrm{theor}}(t)$ fits $c_{\mathrm{T}}^{\mathrm{meas.}}(t)$, the resulting k_i are regarded as those model parameters that describe the metabolic function in an optimum way. Figure 8.7 demonstrates the beginning and end of a curve fitting procedure using the program PMOD (Zürich; Switzerland) for biomedical image quantification.[6]

K_1 characterises the influx of the radiotracer into the tissue. The greater K_1, the faster the influx. Generally,

$$K_1 = EF \tag{8.6}$$

is the product of extraction fraction (E) and blood flow (F) (eqn (8.6)). The extraction fraction represents the difference between arterial and venous concentration divided by the arterial concentration, $(c_{\mathrm{A}} - c_{\mathrm{V}})/c_{\mathrm{A}}$. The closer E approaches to 1, *i.e.* c_{V} becomes very small, the more the parameter K_1 is governed by blood flow.

If E equals 1, then K_1 represents the physiological function F, so that eqn (8.4) can be written as eqn (8.7)

$$c_{\mathrm{T}}(t) = F\int_0^t c_{\mathrm{P}}(\tau)\, e^{-k_2(t-\tau)} d\tau \tag{8.7}$$

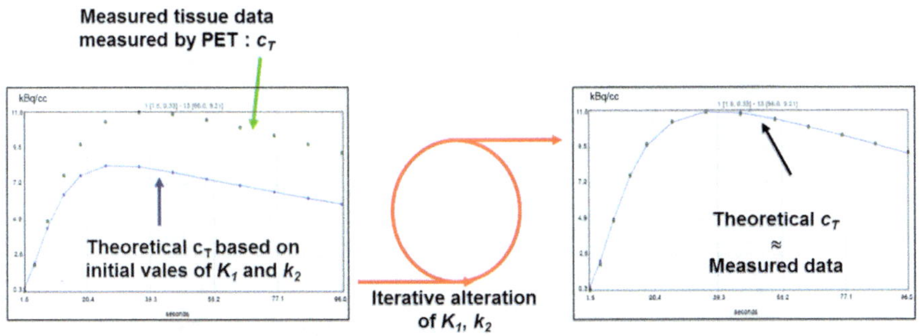

Figure 8.7 Example of iterative curve fitting for a one-tissue compartment model using the program PMOD (Zürich; Switzerland) for biomedical image quantification. The left plot shows the start of the fitting procedure with a great discrepancy between the measured kinetic PET data and the theoretical model solution, for which initial values for K_1 and k_2 were assumed. The right plot documents the end of the fitting procedure with a satisfying match between measured PET data and theoretical curve.

Figure 8.8 shows the cerebral distribution of the cerebral blood flow (CBF) radiotracer [^{15}O]-butanol and time-activity curves measured in a cortical ROI. Although the images summed over a period of 2 min are very similar, the time-activity curve on the right-hand side indicates both faster uptake and outwash of the radiotracer compared to the left curve. The curve fitting based on eqn (8.7) resulted in cortical CBF values of 61 mL min^{-1} 100 g^{-1} (left) and 174 mL min^{-1} 100 g^{-1} (right). This difference was caused by an experimental change of blood CO_2 from 30–60 mm Hg.[7]

There are kinetic processes, *e.g.* when studying reversible neurotransmitter ligands such as [^{11}C]-raclopride, where the radiotracer uptake shows an equilibrium at time T_1 of a PET study so that (eqn (8.8) and (8.9))

$$\frac{dc_{\mathrm{T}}(t)}{dt} = K_1 c_{\mathrm{P}}(t) - k_2 c_{\mathrm{T}}(t) = 0 \quad \text{for}\, t = T_1 \tag{8.8}$$

and

$$\frac{K_1}{k_2} = \frac{c_{\mathrm{T}}(T_1)}{c_{\mathrm{P}}(T_1)} = \mathrm{DV} \tag{8.9}$$

This ratio is a frequently used metabolic parameter and is called the distribution volume (DV). In neurotransmitter receptor analysis, DV is regarded as an indicator of the available receptors. Instead of calculating DV as the ratio $c_{\mathrm{T}}(T_1)$ over $c_{\mathrm{P}}(T_1)$, DV may be determined *via* the fitting procedure, which yields the K_1 and k_2. Even in cases when a satisfactory determination of the single k_i is not possible, the DV, as their ratio, shows a stable outcome.

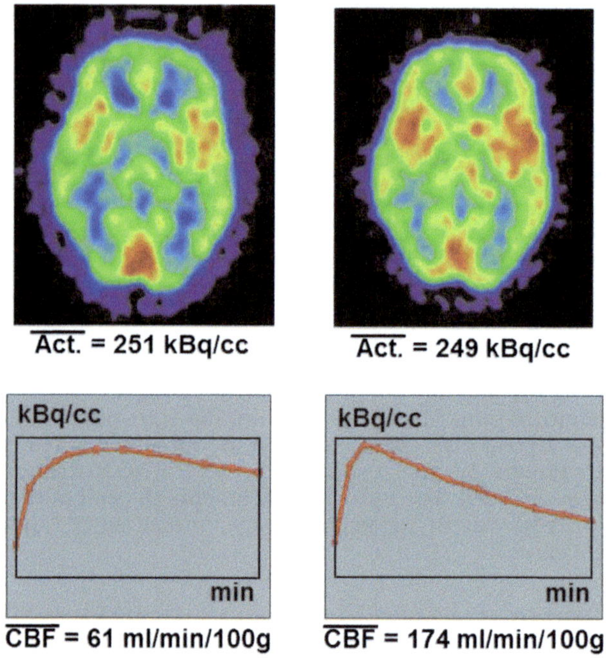

Figure 8.8 Images of the cerebral blood flow (CBF) radiotracer [^{15}O]-butanol and time-activity curves measured in a cortical region of interest (ROI). Whereas the images are very similar, the time-activity curve on the right-hand side indicates both faster uptake and outwash of the radiotracer compared to the left curve. Correspondingly, the curve fitting based on eqn (8.7) resulted in cortical CBF values of 61 mL min^{-1} 100 g^{-1} (left) and 174 mL min^{-1} 100 g^{-1} (right). This difference was caused by an experimental change of blood CO_2 from 30–60 mm Hg.[7]

8.4 The Kinetic Analysis of a Two- or Three-tissue Compartment Model

The application of a two-tissue compartment is appropriate if the radiotracer experiences a biochemical reaction after crossing from blood to tissue. In this case, the first tissue compartment is assumed to contain the free radiotracer $c_F(t)$ and the second tissue compartment represents the metabolised or bound tracer ($c_M(t)$ or $c_B(t)$) (Figure 8.5B).

The following differential equations (eqn (8.10) and (8.11)) allow for a mathematical analysis of a two-tissue compartment model

$$\frac{dc_F}{dt} = K_1 c_P - (k_2 + k_3) c_F + k_4 c_M \tag{8.10}$$

$$\frac{dc_M}{dt} = k_3 c_F - k_4 c_M \tag{8.11}$$

The analytical solution for $c_T(t)$, which is $c_F(t)$ plus $c_M(t)$ as measured by PET is given by eqn (8.12a)

$$c_T^{PET}(t) = \frac{K_1}{a_2 - a_1} \int_0^t c_P(t-\tau)\left[(k_3 + k_4 - a_1)e^{-a_1\tau} + (a_2 - k_3 - k_4)e^{-a_2\tau}\right]d\tau \qquad (8.12a)$$

with eqn (8.12b)

$$a_2, a_1 = \frac{(k_2 + k_3 + k_4) \pm \sqrt{(k_2 + k_3 + k_4)^2 - 4k_2k_4}}{2} \qquad (8.12b)$$

The rate constants K_1 to k_4 can be obtained by a non-linear curve fitting approach as described in Section 8.3 and Figure 8.6.

The analysis of the cerebral glucose metabolism, with the help of FDG, utilises the two-tissue compartment model.[8,9] In this case, $c_P(t)$ and $c_F(t)$ contain the unchanged FDG, whereas $c_M(t)$ represents the FDG-6-phosphate which has undergone the hexokinase reaction. Knowing the rate constants K_1 to k_3 and neglecting k_4 (k_4 is neglected here because it is much smaller than k_3), the local cerebral metabolic rate of glucose consumption (LCMRGlc) results from this operational equation (eqn (8.13))

$$\text{LCMRGlc} = \frac{c_P^{Glc}\left[c_T(T) - K_1 e^{-(k_2+k_3)T} \int_0^T c_P(t)e^{(k_2+k_3)t}dt\right]}{\text{LC}\left[\int_0^T c_P(t)dt - e^{-(k_2+k_3)T} \int_0^T c_P(t)e^{(k_2+k_3)t}dt\right]} \qquad (8.13)$$

c_P^{Glc} represents the plasma glucose level and LC is a so-called lumped constant, which incorporates the biochemical difference between glucose and FDG. $c_T(T)$ is the FDG concentration measured in a tissue of interest at time T. Whereas c_P^{Glc} is the individual value of the subject investigated, standard literature values are used for LC and also for K_1 to k_3, if eqn (8.13) is to be applied to all voxels of the FDG image recorded at time T. As explained below, individual constants K_1 to k_3 cannot be reliably determined by fitting the time-activity data of single voxels.

The two-tissue compartment model is also applied in neuroreceptor analyses where the radiolabelled ligand is bound to the neuroreceptor ($c_B(t)$) (Figure 8.9). As described in Section 8.5, the binding potential (BP) or the receptor affinity as parameters of interest, can be derived from the rate constants calculated by curve fitting of the model solution as described by eqn (8.12a) and (8.12b). Often, the neurotransmitter molecules bind not only to the specific receptor but also to other binding sites in an unspecific way. The latter binding becomes obvious when the specific binding is blocked by the injection of a specific antagonist prior to the PET measurement. To describe this biomedical behaviour, a three-tissue compartment (Figure 8.5C) is the appropriate model. Here, with the help of the fitting approach, the rate constants K_1 to k_6 are the parameters to be extracted from the kinetic PET study.

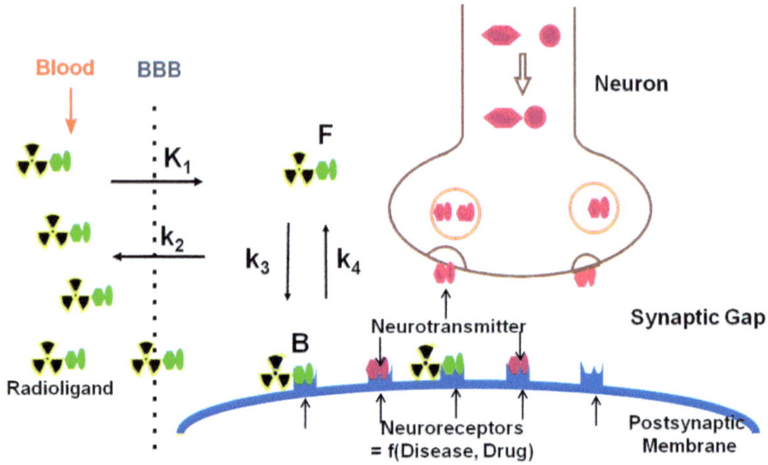

Figure 8.9 Scheme of the uptake and binding of a radioligand with an affinity to neuroreceptors. The radioligand enters the extracellular space (F) at the synaptic gap from the blood *via* the blood brain barrier (BBB) and part of it binds to neuroreceptors at the postsynaptic membrane. Other neuroreceptors are occupied by internal neurotransmitter molecules, which are released from the presynaptic neuron. Figure 8.5B and eqn (8.10) and (8.11) present the corresponding compartment model and mathematical description. If appropriate, F may be subdivided in compartments with unbound and non-specifically (not to neuroreceptors) bound radioligand (see Figure 8.5C).

The fitting of four or even six rate constants suffers, however, from a general difficulty: the identifiability of the individual k_i. Often a perfect fit is observed, but the validity of the k_i must be questioned. There are several reasons for this uncertainty. The quality function RMS (k_i) may have local or relative minima in addition to the absolute minimum. Furthermore, several similar 'absolute' minima may exist, as indicated in Figure 8.10. Although a unique solution exists for the theoretical $c(t)$, this is not obvious for RMS(k_i), which is influenced by errors and noise of both the measured PET and blood data. In particular, in models with two- and three-tissue compartments, similar values of RMS are obtained for totally different sets of k_i depending on the initial values with which the fitting procedure is started. This situation is indicated in Figure 8.10. In order to obtain the 'true' rate constants, appropriate initial values should be selected and the most probable range of the k_i should be taken from literature or animal studies.

8.5 Model-based Analysis of Neuroreceptor Metabolism

Although Figure 8.5C represents the appropriate model for neuroreceptor metabolism, this model is often simplified to obtain more reliable rate constants from the fitting procedure. When a fast exchange between the two

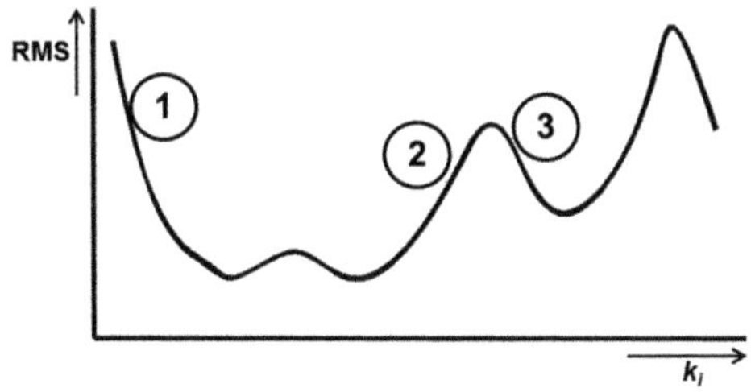

Parameter	K_1	k_2	$k3$	$k4$	$k5$	$k6$	
Given	**0.107**	**0.201**	**0.201**	**0.060**	**0.029**	**0.026**	RMS
Fit 1	0.110	0.171	0.000	0.652	0.191	0.058	10.351
Fit 2	0.152	0.554	0.231	0.048	0.181	0.045	10.281
Fit 3	0.112	0.186	0.003	0.058	0.203	0.058	10.279

Figure 8.10 Top: A plot of the root mean square (RMS) values according to eqn (8.5) over the rate constants k_i demonstrates that the fitting procedure may end in different minima, depending on the initial values (indicated by the numbers 1, 2 or 3) of k_i. Bottom: The curve fitting procedure of a simulated three-tissue compartment model produces very different sets of k_i in spite of nearly same RMS.

compartments of free and non-specifically bound (*i.e.* not bound to neuroreceptors) radiotracers occurs, they can be combined in to one compartment, resulting in a two-tissue compartment model.

Furthermore, sometimes all tissue compartments are lumped into one single tissue compartment. Although the fits for such a simplified model will probably be not ideal, it is possible to achieve stable K_1 and k_2 parameters, with their ratio equalling DV, which is an often-used indicator for the availability of neuroreceptors. DV may be decreased due to a disease or because drugs are occupying (blocking) the neuroreceptors.

When applying and fitting the two-tissue compartment model of neuroreceptor metabolism, the binding potential BP of a radioligand can be calculated from eqn (8.14)

$$\text{BP} = \frac{k_3}{k_4} \tag{8.14}$$

BP represents the interplay between the two rate constants and refers to the binding of the radioligand to and its release from the neuroreceptor.

If the concentration of the available neuroreceptors (B_{max}), which may be altered by a disease, is to be quantified, additional information is necessary. This can be achieved from two separate PET studies together with corresponding curve fitting procedures. In one study, a radioligand with high specific activity is administered and in a second study, it is administered with low specific activity. The specific activity is the ratio between the mass of the radiolabelled ligand and the mass of the non-radiolabelled (cold) ligand. This approach has been suggested by Mintun *et al.*, whose paper is recommended for further information.[10] The double study delivers B_{max} and K_{D}. K_{D} is a measure of the affinity between the radioligand and the neuroreceptor and becomes smaller with higher affinity (eqn (8.15))

$$\text{Affinity} = 1/K_D \tag{8.15}$$

Furthermore, BP can be expressed as eqn (8.16)

$$\text{BP} = B_{max}/K_D \tag{8.16}$$

K_D is very small for those radioligands which bind strongly to the neuroreceptors. In this case, BP becomes large. Assuming that K_D is not influenced by a drug or disease, BP is correlated with B_{max} and a measure for the maximum available neuroreceptors.

There is another way to assess BP *via* the measurement of distribution volumes. For this purpose, in addition to the tissue with specific binding by neuroreceptors, a second so-called reference tissue is considered, which is devoid of neuroreceptors and expresses only unspecific binding. Often the cerebellum is regarded as a reference tissue. For both of these tissues, DV is determined resulting in DV_{Spec} and DV_{Ref}.

The distribution volume ratio (DVR) results from (eqn (8.17))

$$\text{DVR} = \text{DV}_{Spec}/\text{DV}_{Ref}, \tag{8.17}$$

BP can be calculated from DVR (eqn (8.18))

$$\text{BP} = \text{DVR} - 1 \tag{8.18}$$

The present section does not aim at a comprehensive description of the many published suggestions to assess the outcome parameters of the neuroreceptor metabolism. Excellent reviews on the strategies to study the neuroreceptor metabolism have been written by Laruelle *et al.*[11] and Slifstein *et al.*[12] An important paper entitled "Consensus Nomenclature for *in-vivo* Imaging of Reversibly Binding Radioligands" has been published by Innis *et al.*[13] Furthermore, the webpages of the Turku PET Centre, which contain a great deal of additional information, are recommended.[14]

8.6 Parameter Extraction by Linearisation

The noise contained in the measured PET decreases as the size of the ROI increases. In contrast, the noise contained in the measured PET increases with a small ROI. This is especially the case if an ROI comprises of just a few voxels or even one voxel. The latter situation is true if the radioactivity image is to be converted voxel-by-voxel into a parametric image of the metabolic function to be studied. In this situation, the non-linear fitting procedure explained above should not be applied. Instead, a linear approach must be chosen. For this purpose, the differential equation describing the model is rearranged with the result of a linear algebraic equation. Then, the outcome parameters can be determined for each voxel by a linear regression, which supplies a unique solution.

The following first example aims to calculate the parametric images of CBF, where a one-tissue compartment model described by a differential equation as shown in eqn (8.3) is applied. In this case, K_1 is substituted by F. The linearisation is achieved first by integrating the differential model equation (eqn (8.19)) and second by appropriate rearranging (eqn (8.20))[15]

$$c_{\mathrm{T}}(t) = F\int_0^t c_{\mathrm{P}}(\tau)d\tau - k_2\int_0^t c_{\mathrm{T}}(\tau)d\tau \tag{8.19}$$

$$\frac{c_{\mathrm{T}}(t)}{\int_0^t c_{\mathrm{P}}(\tau)d\tau} = F - k_2\frac{\int_0^t c_{\mathrm{T}}(\tau)d\tau}{\int_0^t c_{\mathrm{P}}(\tau)d\tau} \tag{8.20}$$

This equation defines the relationship for a straight line (eqn (8.21))

$$y(t) = F - k_2\ x(t) \tag{8.21}$$

Then, by calculating y and x for each time point t and by arranging $y(t)$ over $x(t)$, a plot such as that shown in Figure 8.11 is obtained. If a linear regression line fits the straight part of this plot, F results as the intercept of this line. To obtain a parametric image, the linear regression procedure is applied voxel-by-voxel.

As mentioned above, the DV is an important outcome parameter in neuroreceptor studies. If the kinetic model of a neuroreceptor ligand can be described by, or simplified into a one-tissue compartment model, eqn (8.3) can be integrated and then rearranged into another linear equation

$$\frac{\int_0^t c_{\mathrm{T}}(\tau)d\tau}{c_{\mathrm{T}}(t)} = -\frac{1}{k_2} + \frac{K_1}{k_2}\frac{\int_0^t c_{\mathrm{P}}(\tau)d\tau}{c_{\mathrm{T}}(t)}. \tag{8.22}$$

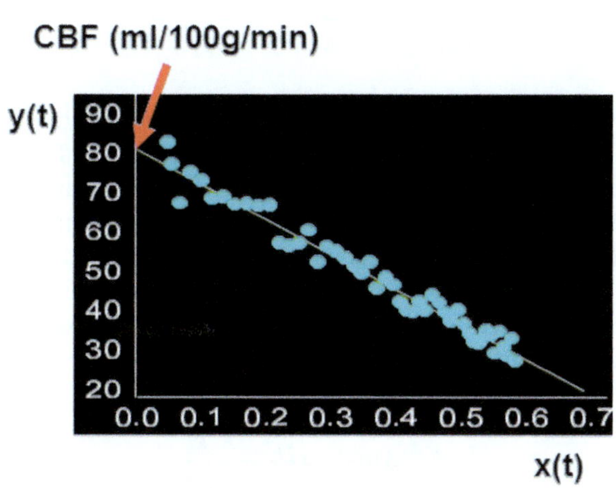

Figure 8.11 Example of determining CBF by linear regression with application of eqn (8.21).

With K_1/k_2 = DV, the distribution volume of the neuroreceptor ligand can be obtained voxel-by-voxel as the slope of eqn (8.22). Eqn (8.22) is similar to that of the Logan-plot (whereas the Logan plot considers the plasma volume fraction V_p, it is neglected here, so that the intercept is just $-1/k_2$) and is frequently used in the so-called graphical analysis of reversible neurotransmitter-receptor radioligands, *i.e.* radioligands released from the neuroreceptor during the time of the PET study.[16] Furthermore, the derivation of the Logan plot also includes the two-tissue compartment model for reversible radiotracers.

BP can be assessed on a voxel-by-voxel basis by applying a suggestion of Logan *et al.* to calculate the DVR (eqn (8.17) and (8.18)) *via* a linearisation procedure.[17] This approach does not require blood sampling but uses a reference tissue such as the cerebellum as the input function. The corresponding formula is then (eqn (8.23))

$$\frac{\int_0^t c_{\mathrm{T}}(\tau)d\tau}{c_{\mathrm{T}}(t)} = \mathrm{Int} + \mathrm{DVR}\frac{\int_0^t c_{\mathrm{Ref}}(\tau)d\tau}{c_{\mathrm{T}}(t)}. \tag{8.23}$$

At this point, an often used reference tissue approach introduced by Lammertsma and Hume may be indicated: the simplified reference tissue model.[18] Although originally developed for the evaluation of ROI data, it can deliver parametric images when applying calculations voxel-by-voxel with basic functions rather than linearisation as suggested by Gunn *et al.*[19]

Another famous linearisation approach is that of the Patlak-plot, which is applied for two-tissue compartment models with k_4 equal to zero, *e.g.* in the case of irreversible binding (eqn (8.24))[20]

$$\frac{c_{\mathrm{T}}(t)}{c_{\mathrm{P}}(t)} = V_0 + \frac{K_1 k_3}{k_2 + k_3}\frac{\int_0^t c_{\mathrm{P}}(\tau)d\tau}{c_{\mathrm{P}}(t)} \tag{8.24}$$

The slope $(K_1k_3)/(k_2 + k_3)$ represents the influx rate of the radiotracer into the irreversible compartment. In the case of FDG, the phosphorylated FDG is nearly completely trapped in such a compartment, so that the rate constant k_4 can be neglected. The application of eqn (8.13) for obtaining a parametric image of LCMRGlc requires the use of general, non-individual rate constants. The Patlak-plot opens the way to an image of LCMRGlc calculated voxel-by-voxel considering individual rate constants, when the slope determined by eqn (8.24) is utilised in the following formula proposed by Phelps *et al.* (eqn (8.25))[8]

$$\mathrm{LCMRGlc} = \frac{c_{\mathrm{P}}^{\mathrm{Glc}}}{\mathrm{LC}}\frac{K_1 k_3}{k_2 + k_3} \tag{8.25}$$

8.7 MR-PET Protocol for Neuroreceptor Studies

Figure 8.12 describes a possible protocol of a simultaneous MR-PET study. The PET investigation uses a reversible radioligand and a bolus-infusion[21] scheme for administering the radioligand. In this way, an equilibrium between tissue and input function can be achieved so that eqn (8.9) can be applied, *i.e.* the DV_{Spec} in the specific target ROI and DV_{Ref} in the reference ROI must be measured.

With eqn (8.26)

$$DV_{Spec} = c_{Spec}(t)/c_P(t) \text{ and } DV_{Ref} = c_{Ref}(t)/c_P(t) \tag{8.26}$$

one gets eqn (8.27)

$$DVR = DV_{Spec}/DV_{Ref} = c_{Spec}(t)/c_{Ref}(t) \tag{8.27}$$

so that an input function is not required, and the only necessary data are the activity concentrations $c_{Spec}(t)$ and $c_{Ref}(t)$ measured by PET.

One aim of such a study may be to consider a change of DV_{Spec} caused by a challenge or stimulation during the second part of the acquisition so that one can compare DVR both with and without the challenge. Using eqn (8.18), changes in BP follow directly from changes in DVR.

While PET gives information about changes in a receptor binding due to a challenge, MRI can supply additional and complementary data. For example, as indicated in Figure 8.12, extra information about CBF with arterial spin labelling (ASL) or about the concentration of neurotransmitters with magnetic resonance spectroscopy (MRS) may be obtained. There are

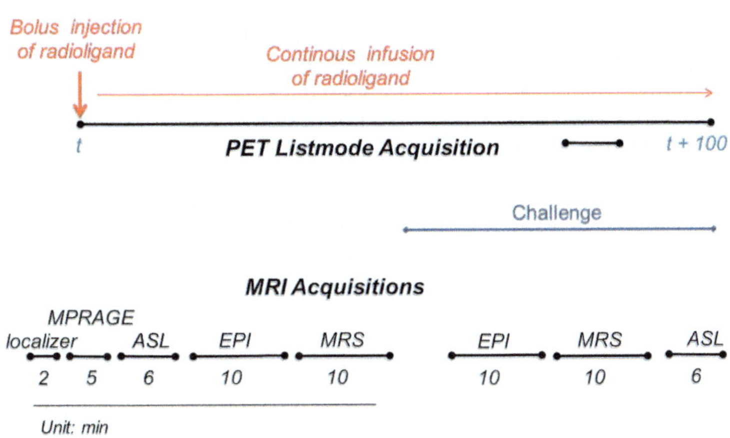

Figure 8.12 Possible protocol of a simultaneous MR-PET study to assess different aspects of neuroreceptor metabolism.

uncertainties[22] to the extent to which uptake of a radioligand is flow-dependent.[22] Therefore, MRI-based measurements of CBF can be used to ensure that changes of DVR are caused by altered neuroreceptor metabolism and not by altered CBF.

8.8 Summary

Kinetic modelling is a very useful tool for research and specific clinical applications. The use of kinetic models is not trivial, because it requires detailed knowledge about the behaviour of the radiopharmaceutical, which may, interestingly enough, be obtained with the help of the kinetic analysis itself. Some models are sensitive to noise, especially when applied to single voxels instead of larger ROIs, making simplifications such as the reduction in the number of compartments, the use of reference tissue or linearised approaches necessary. Nevertheless, there has been a lot of research on the topic, which guides us through the most used and useful methods. These methods are summarised in Table 8.2. All methods require quantified PET

Table 8.2 Overview of different compartment models and analysis (plot) tools with the necessary input data and the achievable outcome parameters. IF = input function. The other abbreviations are explained in the text.

Section	Model/analysis tool	Necessary data	Outcome
8.3	One-tissue compartment (example [^{15}O]-water or [^{11}C]-raclopride simplified from two-tissue compartment).	• Quantified PET data • IF ($c_P(t)$)	K_1 and k_2 F, if E is known or assumed 1. DV = K_1/k_2
8.4 and 8.5	Two-tissue compartment (example FDG or simplification of neurotransmitter three-tissue compartment, when free and unspecific compartments are lumped into one).	• Quantified PET data • IF ($c_P(t)$) • FDG: plasma glucose level of subject and lumped constant (standard value)	K_1 to k_4 FDG: general LCMR$_{Glc}$ (if $k_4 = 0$) Neurotransmitters: BP = k_3/k_4 B_{max} and K_D (high and low affinity study)
8.5	Two-tissue compartment for neurotransmitter.	• Quantified PET data • Reference tissue as IF	DVR = DV$_{Spec}$/DV$_{Ref}$ BP = DVR − 1
8.6	Logan plot (reversible tracers).	• Quantified PET data • IF ($c_P(t)$)	DV (slope)
8.6	Logan plot with reference tissue.	• Quantified PET data • Reference tissue as IF	DVR (slope) BP = DVR − 1
8.6	Patlak plot (irreversible tracers).	• Quantified PET data • IF ($c_P(t)$)	FDG: LCMR$_{Glc}$ (if $k_4 = 0$)
8.7	Analysis of reversible neurotransmitter with bolus-infusion-protocol.	• Quantified PET data • IF ($c_P(t)$) or • Reference tissue as IF	DV$_{Spec}$ at equilibrium DV$_{Ref}$ at equilibrium DVR = DV$_{Spec}$/DV$_{Ref}$

images and many of them need the time course of the PET radiopharmaceutical in the blood (input function, IF). Finally, bolus-infusion methods are suggested for neuroreceptor studies with MR-PET.

References

1. K. R. Zasadny and R. L. Wahl, *Radiology*, 1993, **189**, 847.
2. G. Dantzig, *Programming in a Linear Structure*, U.S. Air Force Comptroller, USAF, Washington, D.C., 1948.
3. J. A. Nelder and R. Mead, *Comput. J.*, 1965, **7**, 308.
4. D. W. Marquardt, *J. Soc. Ind. Appl. Math.*, 1963, **11**, 431.
5. K. Levenberg, *Q. Appl. Math.*, 1944, **2**, 164.
6. C. Burger and A. Buc, *J. Nucl. Med.*, 1998, **38**, 1818.
7. L. J. Kemna, S. Posse, H. Herzog, V. G. Kiselev, S. Wiese, D. Gembris and L. Tellmann, *NeuroImage*, 1999, **9**(6, Part 2), S281.
8. M. E. Phelps, S. C. Huang, E. J. Hoffman, C. Selin, L. Sokoloff and D. E. Kuhl, *Ann. Neurol.*, 1979, **6**, 371.
9. M. Reivich, D. Kuhl, A. Wolf, J. Greenberg, M. Phelps, T. Ido, V. Casella, J. Fowler, E. Hoffman, A. Alavi, P. Som and L. Sokoloff, *Circ. Res.*, 1979, **44**, 127.
10. M. A. Mintun, M. E. Raichle, M. R. Kilbourn, G. F. Wooten and M. J. Welch, A quantitative model for the in vivo assessment of drug binding sites with positron emission tomography, *Ann. Neurol.*, 1984, **15**, 217.
11. M. Laruelle, M. Slifstein and Y. Huang, *Methods*, 2002, **27**, 287.
12. M. Slifstein and M. Laruelle, *Nucl. Med. Biol.*, 2001, **28**, 595.
13. R. B. Innis, *et al.*, *J. Cereb. Blood Flow Metab.*, 2007, **27**, 1533.
14. http://www.turkupetcentre.net/petanalysis/tags.html#Modeling.
15. T. Yokoi, H. Iida, H. Itoh and I. Kanno, A new graphic plot analysis for cerebral blood flow and partition coefficient with iodine-123-iodoamphetamine and dynamic SPECT validation studies using oxygen-15-water and PET, *J. Nucl. Med.*, 1993, **34**, 498.
16. J. Logan, J. S. Fowler, N. D. Volkow, A. P. Wolf, S. L. Dewey, D. J. Schlyer, R. R. MacGregor, R. Hitzemann, B. Bendriem, S. J. Gatley and D. R. Christman, *J. Cereb. Blood Flow Metab.*, 1990, **10**, 740.
17. J. Logan, J. S. Fowler, N. D. Volkow, G. J. Wang, Y. S. Ding and D. L. Alexoff, *J. Cereb. Blood Flow Metab.*, 1996, **16**, 834.
18. A. A. Lammertsma and S. P. Hume, *NeuroImage*, 1996, **4**, 153.
19. R. N. Gunn, A. A. Lammertsma, S. P. Hume and V. J. Cunningham, *Neuroimage*, 1997, **6**, 279.
20. C. S. Patlak, R. G. Blasberg and J. D. Fenstermacher, *J. Cereb. Blood Flow Metab.*, March 1983, **3**, 1.
21. R. E. Carson, M. A. Channing, R. G. Blasberg, B. B. Dunn, R. M. Cohen, K. C. Rice and P. Herscovitch, *J. Cereb. Blood Flow Metab.*, 1993, **13**, 24.
22. N. M. Alpert, R. D. Badgaiyan, E. Livni and A. J. Fischman, *Neuroimage*, 2003, **19**, 1049.

Part B

Hybrid MR-PET Imaging: technical overview

Section I

Hardware

CHAPTER 9

Introduction and Historical Overview

H. HERZOG*[a] AND N. J. SHAH[a,b,c]

[a]Forschungszentrum Jülich, Institute of Neuroscience and Medicine – 4, Wilhelm-Johnen-Straße 1, 52428 Jülich, Germany; [b]JARA - BRAIN - Translational Medicine, Aachen, Germany; [c]Department of Neurology, RWTH Aachen University, Aachen, Germany
*E-mail: h.herzog@fz-juelich.de

9.1 Introduction

Comprehensive structural and functional insights into the body can be obtained by using different tomographic methods available in clinical diagnostics and medical research. Whereas computed tomography (CT) and magnetic resonance imaging (MRI) primarily produce images of anatomical details with high-resolution, single photon emission computed tomography (SPECT) and positron emission tomography (PET) are able to observe a great variety of functional, *i.e.* physiological and biochemical, processes. For many years, the complementary diagnostic information available from the different procedures could only be obtained by using dedicated imaging devices separately. This situation changed with the combination of PET and CT into one single instrument capable of delivering corresponding functional and structural information in one examination.[1,2] In PET/CT the two modalities

New Developments in NMR No. 19
Hybrid MR-PET Imaging: Systems, Methods and Applications
Edited by N. Jon Shah
© The Royal Society of Chemistry 2019
Published by the Royal Society of Chemistry, www.rsc.org

are positioned one after the other and have a common patient bed. Although first ideas around PET/CT aimed at simultaneous imaging, the ideas lacked feasibility and only scanners capable of sequential measurements were produced. With the development of spiral CT, with more and more slices, the acquisition time of the CT measurement could be shortened to just a couple of seconds so that the subsequent PET study became the time-limiting factor. If the patient does not move, the detailed anatomical information afforded by the CT supports the evaluation of the PET images—beyond the diagnostic value of the CT itself. Often the images obtained by PET and CT deliver complementary data, which improves both the sensitivity and specificity of the diagnostic outcome. In particular, oncology has benefited from the successful combination of CT with PET, especially when using [^{18}F]-fluorodeoxyglucose (FDG)[3–6] as a radiopharmaceutical, and, as a consequence, all major manufactures now only produce PET/CT scanners instead of PET-only scanners. In addition, SPECT/CT, originally proposed by Hasegawa *et al.*[7] and currently offered by the majority of manufacturers, is becoming more and more popular. More details comparing PET/CT *vs.* SPECT/CT have been reported by Delbeke *et al.*,[8] Bockisch *et al.*[9] and Hicks *et al.*[10]

In the 1970s, MRI was introduced as another imaging tool and was primarily used to show structural details of the body. Nowadays, it is available in clinical departments and practices. Since its beginnings, MRI has demonstrated a steady increase in resolution and in versatility so that anatomical details can be revealed in many different ways. Furthermore, MR methods, such as functional MRI (fMRI), diffusion-weighted imaging (DWI) and chemical shift imaging (CSI) provide functional and metabolic information beyond anatomy. When compared to CT, MRI provides much better contrast in soft tissues such as liver and brain. Moreover, unlike the imaging procedures mentioned above, MRI does not involve ionising radiation making it suitable for repeated scanning or, in particular, in the paediatric population.

Having the advantages of MRI in mind and being aware of the success of PET/CT, the idea of combining MRI with PET in one device is obvious. Previously, PET and MRI had already been used in combination, especially where images of the two modalities were registered and fused by appropriate software tools.[11,12] Such a combination had been particularly successful in brain imaging where just a rigid-body registration is necessary. In whole-body imaging, however, the combination of MRI and PET in one instrument with one patient bed, as with PET/CT, is most advantageous. Utilising the specific advantages of MRI and PET, an integrated PET/MR-scanner or MR-PET scanner—as it is commonly called—allows a multi-parametric investigation in a 'one-stop-shop' way.

9.2 Historical Overview

When combining PET and CT, the technology of the two modalities remained basically unchanged. Only the PET component was modified by no longer offering a transmission facility, because the data for attenuation correction could be derived from the CT data. PET- and MR-scanners, however, cannot be simply combined in one single device. It quickly became apparent during the first conceptual studies of MR-PET that the two modalities are not directly compatible. The principal problem is related to the scintillation detectors required for PET. Photomultipliers (PMTs) are the commonly used devices for converting the scintillation light emitted into an electronic signal (see Section 5.3 and Section 6.2.2). Photomultipliers, however, are sensitive, even to very small magnetic fields (such as that of the Earth) and must be shielded appropriately. This makes combination impossible due to the strong magnetic field found in MRI. Even if it were possible to protect the photomultipliers using ferromagnetic shielding, large forces would act on the shielding and completely destroy the homogeneity of the magnetic field, thus making MR imaging impossible. Another problem relates to the PET signal processing electronics, which could affect the MR images adversely. Thus, special measures must be taken to shield the PET electronics (see also Section 6.2.2).

In 1997, the group of S. Cherry and P. Marsden at UCLA reported the first simultaneously acquired PET and MR images of a phantom acquired in a 0.2 T MRI system.[13] Interferences between PET and MRI were avoided by connecting the scintillation crystals and the PMTs by 4 m long optical fibres so that the PET detector was kept away from the (comparatively weak) magnetic field. As the scintillation crystals were the only PET components in the MR field of view (FOV), interferences from the PET detector and its electronics onto MRI were minimised. Figure 9.1 shows the equivalent experiment of the same group using a 1.5 T MRI.[14]

Before the manufacturers of clinical imaging devices started their own developments towards MR-PET, several experimental groups worked on preclinical MR-PET adequate for small animal imaging. PMTs are still commonly used in PET detectors since they achieve a very high signal gain of 10^6 with low noise and are not sensitive to temperature changes. Thus, the strategy of keeping the PMTs outside the magnetic field and coupling them to scintillation crystals by long optical fibres was further pursued. Raylman *et al.*[15] and Mackewn *et al.*[16,17] reported good quality of MRI and PET in simultaneous measurements in mice or rats when they placed the crystal frontend of self-constructed PET detector rings, with an inner diameter of about 8 cm, in clinical 3 T MR scanners. The distance between crystals and PMTs was 2.5 m and 3.5 m, respectively. Lucas *et al.* positioned the detector modules of a Siemens small-animal PET scanner microPET-Focus F120 in the centre of a split MRI and connected them with 1.2 m long fibres to shielded PMTs outside the magnet.[18] However, the use of long optical connections leads to the loss of light along the fibres, resulting in a decreased energy resolution and consequently reduced PET performance. This finding motivated the investigation of alternative approaches beyond the use of PMTs.

The performance of photosensitive solid-state components, such as Avalanche photodiodes (APDs), cannot be influenced, even by a strong magnetic field (see Section 6.2.2). After basic studies on APDs by Pichler *et al.*[19] at the

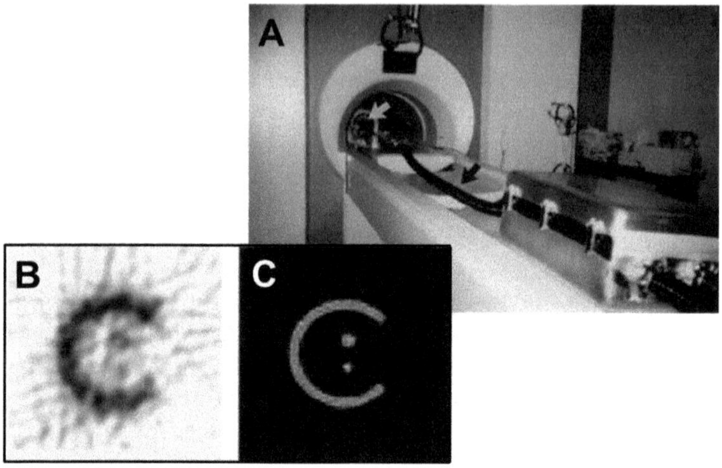

Figure 9.1 The very first simultaneous MR-PET acquisition performed by the group of Simon Cherry and Paul Marsden. (a) The crystals of the PET detector are located inside a 1.5 T MR scanner (white arrow) and connected *via* optical fibres (black arrow) to a metal box containing the detector electronics with its photomultipliers. The PET image (b) and the MR image (c) of a C-shaped phantom filled with a solution of $NiSO_4$ and ^{68}Ga. Adapted from ref. 14 with permission from John Wiley and Sons, Copyright © 1999 Wiley-Liss, Inc.

Technical University of Munich, Germany, the first APD-based PET detector was developed by Cherry's group at UC Davis by Pichler,[20] Catana and colleagues.[21] The 8 × 8 arrays of LSO-crystals of the PET detector were coupled *via* 15 cm long optical fibres to position sensitive APDs. This detector was placed in a Bruker 7 T Biospec animal MRI system. PET images of a mouse injected with [^{18}F]-fluoride could be acquired *ex vivo* simultaneously with MR images, without showing remarkable artefacts. A direct coupling between LSO crystals and APDs was further realised by Judenhofer working as part of Pichler's group at the University of Tübingen.[22] This MR-compatible PET was placed in a small animal 7 T MR scanner.

Schlyer *et al.* at Brookhaven National Laboratories also applied APDs as PET readout electronics and constructed a small portable MR-compatible PET detector ring called 'Rat Conscious Animal PET' (RatCAP) that can be operated in a 9.4 T MRI.[23] This PET detector can be fixed around the neck and head of a rat in a way that the animal carries the PET ring and can remain conscious.

All the above-mentioned systems result from non-industrial developments. In 2011, the first commercial small animal PET/MR (offered by Mediso Ltd, Hungary) became available. The Mediso nanoScan® PET/MR is a compact, sequential PET/MR system with LYSO crystals coupled to position sensitive PMTs.[24] The special permanent magnet 1 T MR subsystem minimises external magnetic field effects so that the distance between the centre of the PET FOV and the MR magnet bore opening is only around 6 cm. The first tests

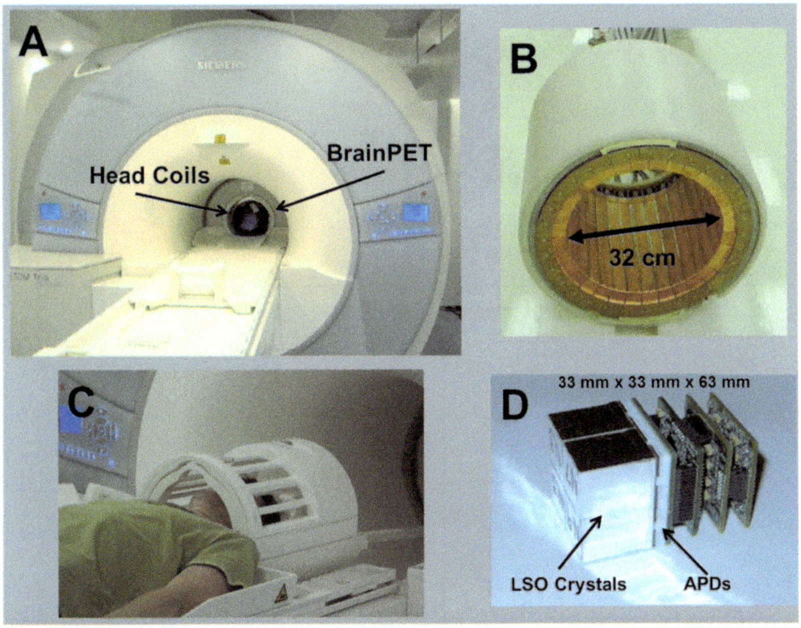

Figure 9.2 (a) The 3 T MR-BrainPET as installed in the Forschungszentrum, Jülich. (b) The BrainPET detector constructed as an insert to 3 T MRI and consisting of 32 copper-shielded cassettes. (c) A subject positioned with his head inside the headquarters but outside the MR bore. (d) The compact MR-compatible detector block with its 12 × 12 array of LSO crystals and adjacent APDs.

showed that this PET/MR imaging system was able to fulfil the original PET and MR specifications.

Recently, other manufacturers have been offering small PET/MR systems for both simultaneous and sequential acquisition. More details about these systems, which utilise silicon photomultipliers (SiPMTs) as solid-state readout electronics, are given below.

A first MR-PET scanner for human brain imaging was presented by Siemens (Munich, Germany) in 2006.[25] The cylindrical BrainPET, with two RF head coils placed within its inner diameter, just fits into the 60 cm bore of a standard Siemens 3 T MAGNETOM Trio MR scanner (Figure 9.2). The basic design of the PET detector is similar to that of Pichler's group.[22] The BrainPET consists of 32 copper-shielded cassettes, each containing six detector modules in a row, covering an axial FOV of 19.2 cm. The individual LSO crystals of a detector module block are coupled to a 3 × 3 array of APDs. Each of the 32 cassettes is linked by a 10 m long cable to the filter plate of the MRI cabin and the data acquisition electronics. The central image resolution of the BrainPET is approximately 3 mm. Between 2006 and 2010, a limited number of four prototypes of this 3 T MR-BrainPET were installed in two German laboratories at Tübingen and Jülich and also in Boston and Atlanta. Reports of tests and experiences relating to this hybrid scanner revealed no

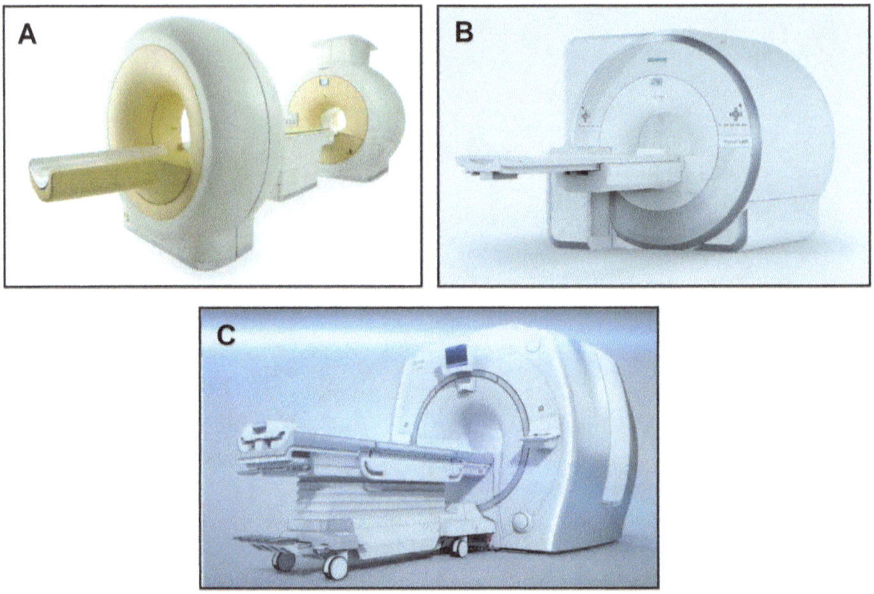

Figure 9.3 The whole-body MR-PET offered by Siemens, Philips and General Electric in the order of their introduction: (a) Philips Ingenuity TF PET/MR;[35] (b) Siemens Biograph mMR;[36] and (c) General Electric SIGNA PET/MR.[37]

remarkable drawbacks in either the MR or PET performance and no interferences between the two modalities.[26–28] Presently, the Jülich and Boston 3 T MR-BrainPET scanners are still in operation.

In 2010, Philips (Eindhoven, Netherlands) issued a sequential MR-PET system and Siemens a simultaneous one, both appropriate for whole-body studies. The Ingenuity TF PET/MR, offered by Philips, combines two well-established components, the Philips Achieva 3 T MR and the time-of-flight PET part of Philips' PET/CT systems. The two scanners are linked by a common turntable bed so that the patient need not move between the scans (Figure 9.3(a)). Although the centres of the FOVs of PET and MRI are 4.2 m apart, additional shielding was installed to ensure proper operation of the PET, which is equipped with PMTs.[29] Presently, 11 Ingenuity TF PET/MR are installed worldwide. More details and aspects comparing this sequential MR-PET with a simultaneous or integrated MR-PET are described by Herzog and van den Hoff.[30]

Based on the experiences obtained with the prototypes of the BrainPET/MRI, Siemens designed a simultaneous whole-body MR-PET called Biograph mMR (molecular MR) (Figure 9.3(b)). The PET detector is also APD-based and is integrated with a 3 T MR scanner with essentially the same specifications as the Siemens Verio MRI. The PET component is installed between the gradient and radiofrequency transmit-coils. In this way, it is fully integrated in the MRI scanner and not visible from outside the MR gantry, which has a 60 cm bore. The PET ring has a transaxial FOV of 59.4 cm and an axial one of 25.8 cm.[31]

The 56 copper shielded cassettes of the PET ring are similar to those of the above mentioned BrainPET. The 8 × 8 matrix of individual 4 × 4 × 20 mm^3 LSO crystals of a detector module is coupled to a 3 × 3 array of APDs. The PET image resolution is 4.8 mm (FWHM) at a radial distance of 1 cm from the FOV's centre. As already seen with the 3 T MR-BrainPET, the mMR shows no drawbacks between MR and PET. About 70 Biograph mMR are installed worldwide.

Although APDs were the decisive components for constructing an MR-compatible PET detector, it quickly became clear that they were not really optimal. Compared to PMTs, with their gain of 10^6, the gain of APDs is just 10^2. Furthermore, the rise time of APDs, with more than 1 ns, is longer than that of PMTs, so that time-of-flight (TOF) PET, which has been successfully introduced in PET/CT scanners, cannot be realised in an APD-based PET detector. In this situation, silicon photomultipliers (SiPMs)—also called Geiger-mode APDs—offer an appropriate alternative with a gain of 10^6 and a rise time of 300 ps (see Section 6.2.2). Furthermore, the supply voltage of SiPMs is approximately 40 V compared with approximately 400 V necessary for APDs. With the increasing interest in MR-PET, the research towards SiPM applied for PET, was intensified.[32–34] In 2014, GE introduced an integrated, simultaneous MR-PET for whole-body imaging, the SIGNA PET/MR, with SiPM-based PET technology (Figure 9.3(c)). The PET detector has a timing resolution of less than 400 ps FWHM so that it has TOF capability. The scintillator crystals of sizes 4 × 5.3 × 25 mm^3 are made of LYSO and read out by 3 × 6 SiPMs. The analogue PET signals, with their spatial, energy, and timing information, are digitised and the digital data are transmitted over optical fibres. The axial FOV of the PET is 25 cm. The PET resolution ($X/Y/Z$) is 4.2/4.2/5.7 mm (FWHM) at r = 10 mm. The PET detector is integrated in a modified GE Discovery 750 w 3 T MRI scanner and mounted between the RF shield and the gradient coil.[38] Presently, about 25 SIGNA PET/MR systems are operated worldwide.

In recent years, a number of preclinical MR-PET systems, also using SiPM-based readout electronics, were developed by research groups[39] and different manufacturers such as Bruker (Rheinstetten, Germany), MR-Solutions (Guildford, Great Britain), ASPECT-Imaging (Shoham, Israel) and Cubresa (Winnipeg, Canada). The first three manufactures offer complete MR-PET systems for sequential and/or simultaneous studies. All three systems have cryogen-free magnets so that the operating costs are reduced. In contrast, Cubresa has developed an MR-compatible PET insert, which can be operated in standard clinically available MR-scanner, but also standalone.

References

1. T. Beyer, D. W. Townsend and T. Brun, *J. Nucl. Med.*, 2000, **41**, 1369.
2. D. W. Townsend, *J. Nucl. Med.*, 2001, **42**, 533.
3. G. Antoch, F. M. Vogt, L. S. Freudenberg, F. Nazaradeh, S. C. Goehde, J. Barkhausen, G. Dahmen, A. Bockisch, J. F. Debatin and S. G. Ruehm, *J. Am. Med. Assoc.*, 2003, **290**, 3199.

4. R. Bar-Shalom, N. Yefremov, L. Guralnik, D. Gaitini, A. Frenkel, A. Kuten, H. Altman, Z. Keidar and O. Israel, *J. Nucl. Med.*, 2003, **44**, 1200.
5. W. A. Weber and R. Figlin, *J. Nucl. Med.*, 2007, **48**(Suppl 1), 36S.
6. T. C. Kwee and R. M. Kwee, *Eur. Radiol.*, 2009, **19**, 731.
7. B. H. Hasegawa, E. L. Gingold, S. M. Reilly, S. C. Liew and C. E. Cann, *Proc. SPIE*, 1990, **1231**, 50.
8. D. Delbeke, H. Schöder, W. H. Martin and R. L. Wahl, *Semin. Nucl. Med.*, 2009, **39**, 308.
9. A. Bockisch, L. S. Freudenberg, D. Schmidt and T. Kuwert, *Semin. Nucl. Med.*, 2009, **39**, 276.
10. R. J. Hicks and M. S. Hofman, *Nat. Rev. Clin. Oncol.*, 2012, **9**, 712.
11. U. Pietrzyk, K. Herholz and W. D. Heiss, *J. Comput. Assist. Tomogr.*, 1990, **14**, 51.
12. C. Burger and A. Buck, *J. Nucl. Med.*, 1997, **38**, 1818.
13. Y. Shao, S. R. Cherry, K. Farahani, K. Meadors, S. Siegel, R. W. Silverman and P. K. Marsden, *Phys. Med. Biol.*, 1997, **42**, 1965.
14. K. Farahani, R. Slates, Y. Shao, R. Silverman and S. R. Cherry, *J. Magn. Reson. Imaging*, 1999, **9**, 497.
15. R. R. Raylman, S. Majewski, S. K. Lemieux, S. S. Velan, B. Kross, V. Popov, M. F. Smith, A. G. Weisenberger, C. Zorn and G. D. Marano, *Phys. Med. Biol.*, 2006, **51**, 371.
16. J. E. Mackewn, D. Strul, G. Charles-Edwards, J. J. Totman, W. A. Hallett and P. Halsted, *IEEE NSS-MIC Conf. Record*, 2008, 4662.
17. J. E. Mackewn, P. Halsted, G. Charles-Edwards, R. Page, J. J. Totman and K. Sunassee, *IEEE Trans. Nucl. Sci.*, 2010, **57**, 1052.
18. A. J. Lucas, R. C. Hawkes, R. E. Ansorge, G. B. Williams, R. E. Nutt, J. C. Clark, T. D. Fryer and T. A. Carpenter, *Technol. Cancer Res. Treat.*, 2006, **5**, 33.
19. B. Pichler, G. Boening, M. Rafecas and S. I. Ziegler, *IEEE Trans. Nucl. Sci.*, 1999, **46**, 298.
20. B. J. Pichler, M. S. Judenhofer, C. Catana, J. H. Walton, M. Kneilling, R. E. Nutt, S. B. Siegel, C. D. Claussen and S. R. Cherry, *J. Nucl. Med.*, 2006, **47**, 639.
21. C. Catana, Y. Wu, M. S. Judenhofer, J. Qi, B. J. Pichler and S. R. Cherry, *J. Nucl. Med.*, 2006, **47**, 1968.
22. M. S. Judenhofer, C. Catana, B. K. Swann, S. B. Siegel, W. I. Jung, R. E. Nutt, S. R. Cherry, C. D. Claussen and B. J. Pichler, *Radiology*, 2007, **244**, 807.
23. D. Schlyer, P. Vaska, D. Tomasi, C. Woody, S. H. Maramraju and S. Southekal, *IEEE NSS-MIC Conf. Record*, 2007, 3256.
24. K. Nagy, G. Nemeth, P. Major, Z. Nyitrai, D. Máthé, B. Gulyás, C. Halldin, J. Lantos, D. Völgyes and I. Bagaméry, *J. Nucl. Med.*, 2011, **52**(Suppl 1), 259.
25. M. Schmand, Z. Burbar, J. Corbeil, N. Zhang, C. Michel, L. Byars, L. Eriksson, R. Grazioso, M. Martin, A. Moor, J. Camp, V. Matschl, R. Ladebeck, W. Renz, H. Fischer, K. Jattke, G. Schnur, N. Rietsch, B. Bendriem and W. D. Heiss, *J. Nucl. Med.*, 2007, **48**(Suppl. 2), 45P.

26. H. P. Schlemmer, B. Pichler, M. Schmand, Z. Burbar, C. Michel, R. Ladebeck, K. Jattke, D. Townsend, C. Nahmias, P. K. Jacob, W. D. Heiss and C. D. Claussen, *Radiology*, 2008, **248**, 1028.
27. H. Herzog, K. J. Langen, C. Weirich, E. Rota Kops, J. Kaffanke, L. Tellmann, J. Scheins, I. Neuner, G. Stoffels, K. Fischer, L. Caldeira, H. H. Coenen and N. J. Shah, *Nuklearmedizin*, 2011, **50**, 74.
28. A. Kolb, H. F. Wehrl, M. Hofmann, M. S. Judenhofer, L. Eriksson, R. Ladebeck, M. P. Lichy, L. Byars, C. Michel, H. P. Schlemmer, M. Schmand, C. D. Claussen, V. Sossi and B. J. Pichler, *Eur. Radiol.*, 2012, **22**, 1776.
29. H. Zaidi, N. Ojha, M. Morich, J. Griesmer, Z. Hu, P. Maniawski, O. Rati, D. Izquierdo-Garcia, Z. A. Fayad and L. Shao, *Phys. Med. Biol.*, 2011, **56**, 3091.
30. H. Herzog and J. van den Hoff, *Q. J. Nucl. Med. Mol. Imaging*, 2012, **56**, 247.
31. G. Delso, S. Fürst, B. Jakoby, R. Ladebeck, C. Ganter, S. G. Nekolla, M. Schwaiger and S. I. Ziegler, *J. Nucl. Med.*, 2011, **52**, 1914.
32. A. N. Otte, J. Barral, B. Dolgoshein, J. Hose, S. Klemin, E. Lorenz, R. Mirzoyan, E. Popova and M. Teshima, *Nucl. Instrum. Methods Phys. Res., Sect. A*, 2005, **545**, 705.
33. G. Llosa, N. Belcari, M. G. Bisogni, G. Collazuol, A. Del Guerra, S. Marcatili, S. Moehrs and C. Piemonte, *IEEE NSS-MIC Conf. Record*, 2007, 3220.
34. E. Roncali and S. R. Cherry, Application of silicon photomultipliers to positron emission tomography, *Ann. Biomed. Eng.*, 2011, **39**, 1358.
35. https://www.philips.de/healthcare/product/HC882380/ingenuity-tf-pet-mr-petmrsystem, last accessed October 2017.
36. https://www.healthcare.siemens.de/magnetic-resonance-imaging/mr-pet-scanner/biograph-mmr, last accessed October 2017.
37. http://www3.gehealthcare.com/en/products/categories/magnetic_resonance_imaging/3-0t/signa_pet-mr#geUnleash, last accessed October 2017.
38. C. S. Levin, S. H. Maramraju, M. M. Khalighi, T. W. Deller, G. Delso and F. Jansen, *IEEE Trans. Med. Imag.*, 2016, **35**, 1907.
39. V. Schulz, B. Weissler, P. Gebhardt, T. Solf, C. W. Lerche and P. Fischer, *IEEE NSS-MIC Conf. Record*, 2011, 4467.

CHAPTER 10

MR-PET Instrumentation

C. W. LERCHE*[a], J. FELDER[a], C.-H. CHOI[a] AND N. J. SHAH[a,b,c]

[a]Forschungszentrum Jülich, Institute of Neuroscience and Medicine - 4, Wilhelm-Johnen-Straße 1, 52428 Jülich, Germany; [b]JARA-BRAIN – Translational Medicine, Aachen, Germany; [c]Department of Neurology, RWTH Aachen University, Pauwelsstr. 30, 52074 Aachen, Germany
*E-mail: c.lerche@fz-juelich.de

10.1 Mutual Interferences Between MR and PET

Positron emission tomography (PET) imaging allows the observation of the metabolic and physiologic processes of organs in their normal and diseased states. However, its ability to show anatomic detail is limited. Therefore, the complementary use of PET with computed tomography (CT) or magnetic resonance imaging (MRI) is becoming common practice and has largely been facilitated by the development of dedicated programs, which enable the combined display and analysis of images from both modalities. The advantages of the combined analysis have led to the development of sequentially combined PET/CT scanners and to the abandonment of PET-only scanners. When compared to CT, MRI also offers a large number of contrast mechanisms for functional and physiological imaging, thus allowing for single exams delivering metabolic, functional and structural data when using hybrid MR-PET devices.[1] Moreover, the realisation of truly simultaneous acquisitions is more difficult to achieve in the case of the combination

New Developments in NMR No. 19
Hybrid MR-PET Imaging: Systems, Methods and Applications
Edited by N. Jon Shah
© The Royal Society of Chemistry 2019
Published by the Royal Society of Chemistry, www.rsc.org

of PET and CT, since both modalities use high-energy photons as the information carrier and both modalities also use scintillation detectors for detecting them. Furthermore, although it is possible to discriminate the photons by their energy, CT requires a photon flux, which is several orders of magnitude higher than for PET imaging and which will saturate the PET detector making it insensitive to the annihilation radiation. In contrast, MRI uses radiofrequency (RF) for excitation and for the detection of the signal from the nuclei; this can be more easily combined and mutual interferences can be better minimised. Nevertheless, the strong static magnetic field, the switched magnetic gradient fields, the use of high-power RF transmission and the extraordinarily high sensitivity of MR for the detection of RF signals require very careful design when combining MR with PET in a simultaneous device.

One major problem in the development of a single, integrated MR-PET device is the strong B_0 magnetic field, which not only impedes the use of classical photomultiplier tubes (PMTs) for photo detection, but also prohibits the use of many essential electronic components from being arranged close to the imaging FOV of the MR system. The reasons for this are manifold: the B_0 field may lead to the displacement of loose components; some components become ineffective, for example, ferrites, which are used in many electronic devices, have a saturation flux density of ≈0.3 T; and many plugs, integrated circuit (IC) sockets and pins of electronic components contain ferromagnetic materials. MRI not only requires a strong magnetic field but it must also be extremely homogeneous with deviations of only a few parts per million (ppm) over the whole MRI FOV. MRI images already become significantly distorted by field variations of $\delta B_0 \approx 0.1\ \text{ppm} \times B_0$. Thus, the image distortion effect that a magnetic component of volume V_D causes at a distance r can be estimated by evaluating its magnetic dipole strength (eqn (10.1))

$$H_D \approx \frac{B_r}{\mu} \frac{V_D}{2\pi r^3} = \frac{B_r}{\mu} \frac{m_D}{2\pi \rho_D r^3} \tag{10.1}$$

where B_r is the saturation remanence of the material, ρ_D is its density, μ is the permeability and m_D is the total mass corresponding to the volume.[2] For instance, a volume of 0.1 mm^3 ferrite causes a field variation of ≈0.1 ppm in a 3 T scanner at ≈2.5 cm distance. However, paramagnetic and diamagnetic materials can also affect the MR image. Therefore, potentially all the materials required for the construction of the PET system inside the magnet bore can lead to MR image degradation. This can be especially relevant for gamma shielding materials, which are often used in PET imaging devices to protect the scintillation detectors from annihilation radiation emitted outside the PET FOV, and also for the scintillation crystals, which are key components for PET and have a considerable volume in order to stop the gamma photons. To overcome this problem, it is preferable to use materials with only a small susceptibility mismatch, compared to the

Table 10.1 Volume magnetic susceptibility of some selected materials and their corresponding function in PET devices or use in phantoms.[3–5] (a) Bismuth germanate, (b) lutetium orthosilicate, (c) gadolinium orthosilicate.

Material	Function	$\chi\,[10^{-6}]$	Material	Function	$\chi\,[10^{-6}]$
CsI	Scintillator	−18.0	Lead	γ-Shielding	−15.8
BGO[a]	Scintillator	−19.0	Tungsten	γ-Shielding	78.0
LSO[b]	Scintillator	−21.7	Copper	Electronics	−9.7
GSO[c]	Scintillator	9530	Silicon	Electronics	−4.2
Fused silica	Light guide	−11.28	Nickel	Electronics	600

surrounding medium.[3,4] To meet MR compatibility criteria while designing a PET device, only materials with a susceptibility mismatch smaller than 10 ppm compared to water are recommended for use.[5] The volume magnetic susceptibility of water at 37 °C is -9.05×10^{-6} and for human tissues it ranges from -11.0×10^{-6} to -7.0×10^{-6}. The susceptibility, χ, for selected shielding materials, scintillation crystals, and other materials used in PET devices are summarised in Table 10.1. While many typical materials fulfil the condition mentioned above, some of them still cannot be used for building a combined MR-PET device, *e.g.* the common GSO scintillator and nickel, which is often used in small amounts in connectors and pins of electronic components.

In addition to magnetic compatibility, there are further limitations on the materials that can be used. Table 10.1 suggests that lead, which is an inexpensive and efficient shield, could be used in hybrid MR-PET devices. However, lead is a good conductor, and when exposed to the switching gradient fields, eddy-currents will be induced on the surface of the lead. For materials with homogeneous magnetisation $M = \frac{\chi}{\mu}B$ and a thickness, d, of the conductive material considerably thinner than the corresponding skin depth and frequency, f, of the electromagnetic radiation, the power dissipation due to eddy-currents can be approximated by eqn (10.2)

$$P = \frac{\pi^2 M^2 d^2 f^2}{6\sigma\rho} = \frac{\pi^2 M^2 d^3 f^2 A}{6\sigma m} \tag{10.2}$$

where ρ is the density of the material, σ its resistivity, and A the surface area.[6] This power dissipation can lead not only to heating of the γ-shields, but also appears in other large conductive structures like large copper planes, *e.g.* power supply and ground layers of printed circuit boards (PCBs) in the PET data acquisition electronics. In addition to heating effects and large induced voltages, a second important eddy-current related issue, which may lead to the failure of electronic components, is the creation of eddy-current induced counter fields that are opposed to its cause (Lenz's law). These are especially relevant for the gradient system of the MRI system, since eddy-current fields can be sufficiently strong to distort the applied gradient fields. As the gradient fields are used for the spatial encoding of the MR receive signals, their distortion leads to image distortion and, potentially, other MR image artefacts.

As discussed in Chapter 2, RF transmit and receive coils are required for MRI imaging. While the transmit coil emits RF pulses with typical peak powers of few kW in humans, the RF receive coil needs to detect very small signals. Consequently, special care has to be taken to avoid noise pick up through the receive coils. Since MRI operates in an RF shielded environment, both the PET sub-system and the MRI sub-system need to be built in a way to avoid generating electromagnetic interferences in the MR frequency bands.

10.2 MR Compatible PET Detector Technology

During the development of a single integrated MR-PET device, a major problem encountered was that the photo detectors commonly used in PET, the PMTs, cannot be operated in the high magnetic fields required for MRI (see Section 6.2.2). In contrast, silicon has a susceptibility of -4.2×10^{-6} and is therefore intrinsically MR compatible. Consequently, silicon-based photo detectors, such as avalanche photo diodes (APDs) or silicon photomultiplier (SiPMs), can also be manufactured to be MR-compatible if sufficient care is taken to only employ materials for encapsulation and pins that have magnetic susceptibilities close to that of water. SiPMs, in particular, have additional advantages compared to PMTs, such as higher intrinsic timing resolution and considerably reduced dimensions.

Similarly, all electronic components, such as ICs resistors, capacitors, connectors, *etc.* should be carefully selected with regard to the MR compatibility of the materials used in these components. Precise attention is required concerning the connectors and pins of components used in PET systems, as these are often ferromagnetic, however, specialised suppliers offer MR compatible alternatives. Close care is also required for the design of the printed circuit boards. As explained in Section 10.1, the switching gradient fields may induce eddy-currents on large copper planes, such as common ground and power supply planes. Eddy-currents may also be induced in loops within both the signal paths and the power supply paths. In all cases, the induced voltages can cause additional noise on signal and power lines, leading to altered or falsely detected signals in the PET electronics. Most often, this reduces PET energy resolution and/or causes an altered PET signal count rate, both of which affect the quantifiability of the PET images.

In order to improve the mutual electrical compatibility, at least the readout electronics, and most often the whole PET detector modules, are enclosed by an RF shield, which normally is integrated in the PET module housing. RF shields can be realised by depositing either structured or solid conductive layers, which short-circuit the E-field component of the RF radiation.[7] Care has to be taken, since large conductive planes might cause eddy-currents

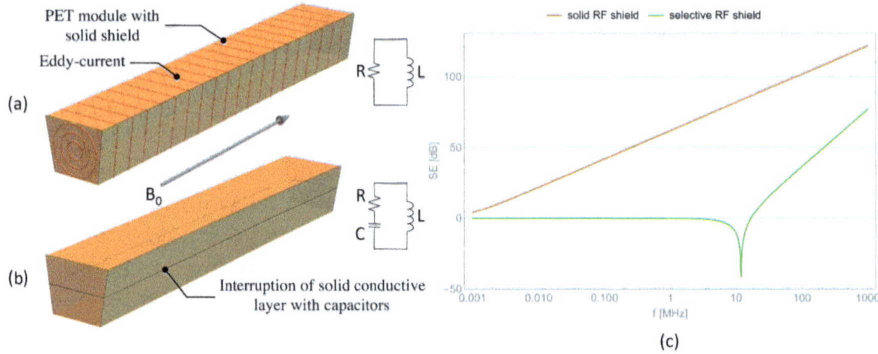

Figure 10.1 Two possible realisations of RF shielding for MR compatible PET modules. (a) PET module with solid, continuous, and thin conductive layer (LR-low pass). Large eddy-currents can occur owing to the continuously high conductivity along the circumference of the PET module and the orthogonal orientation of gradient fields (parallel to the B_0 field) and the cross-section of the module. (b) Incorporating slits, which are parallel to the B_0 field and which are bridged by high value capacitors, can prevent the eddy-currents (LC-low pass). (c) Exemplary shielding effectiveness (SE) of a solid RF shield and a selective RF shield.

that can distort the gradient fields and therefore can cause artefacts in the MR image. The simplest RF shielding can be achieved by depositing a thin, closed conductive layer on the PET module housing. However, in this case, the fast-varying gradient fields, whose field vectors are parallel to the B_0 field vector, will induce large eddy-currents on the PET module surface (Figure 10.1(a)). Arranging interruptions in the conductive layer parallel to the B_0 vector can efficiently reduce these eddy-currents. However, these interruptions will act as a dipole antenna when irradiated with a plane wave of wavelength λ.[8] The shielding effectiveness of a single slot of length l is given approximately by eqn (10.3)

$$\mathrm{SE} = 20\log\left[\frac{\lambda}{l}\right]\mathrm{d}B \tag{10.3}$$

RF radiation penetration through these slits can be avoided by connecting both conductive half-surfaces with capacitors (Figure 10.1(b)). The shield then acts like a low pass filter which blocks RF radiation close to the Larmor frequency, but which only minimally affects the gradient fields.[9] Alternatively, carbon fibre reinforced plastic (CFRP) can be used, since it effectively blocks RF radiation and also minimises distortion of the gradient fields.[10] In all the cases discussed above, both the RF penetration from the MR transmit coil into to the PET module, and the pick-up of electromagnetic radiation, from the PET electronics inside the module, by the sensitive MR receive coil, is suppressed. It should be mentioned that the frequency of the RF emissions

Table 10.2 Approximate Larmor frequencies, f_0, in MHz for most relevant nuclei and field strengths.

	Field strength			
Nuclei	1.5 T	3 T	7 T	9.4 T
^1H	64	123	297	400
^7Li	25	48	115	155
^{13}C	16	32	75	101
^{17}O	7	17	40	54
^{19}F	60	116	280	376
^{23}Na	17	33	79	106
^{31}P	26	50	120	162

by the MR transmit coil depends on the field strength and the imaged nuclei through the relation $f_0 = \frac{\gamma}{2\pi} B_0$, where γ is the gyromagnetic ratio of the nuclei being imaged. The most important nucleus for MRI is the proton ^1H. It has the highest abundance in human tissue and also the highest gyromagnetic ratio—resulting in the highest MR signal of all MR visible nuclei. Depending on the field strength and the target nuclei, the Larmor frequencies can be very different (Table 10.2). This has to be considered when designing frequency-selective RF shields for MR compatible modules.

10.3 MR-PET System Architecture

In general, there are three general ways in which MR and PET can be incorporated into a combined device. The most straightforward approach is to combine the PET scanner and the MRI scanner sequentially (Figure 10.2(a)). In this case, both scanners are aligned along their system axes, and since their field-of-views (FOVs) do not overlap, simultaneous MR-PET imaging is not possible. However, this allows the optimisation of both imaging modalities to a maximum while mutual interference is kept to a minimum. Furthermore, if the distance between both modalities is sufficiently large, it is also possible to use PMTs as photodetectors, although additional shielding against the magnetic fringe field may be required. Most MR scanners for humans have to be installed inside a Faraday cage to avoid the RF emissions from the multiple and omnipresent electronic devices, present in a normal environment, to be picked up by the sensitive receive coil of the MR device. The sequential MR-PET configuration may be realised either by installing the PET inside the Faraday cage or in a separate room and combining the scanner through a shuttle bed. In the case that the PET system is installed inside the Faraday cage, the PET device itself has to be carefully shielded to minimise the emission of RF noise.

In another possible configuration, the PET scanner is built as an insert for the MR scanner (Figure 10.2(b)). This allows alignment of both FOVs and simultaneous measurement, however, achieving MR compatibility is

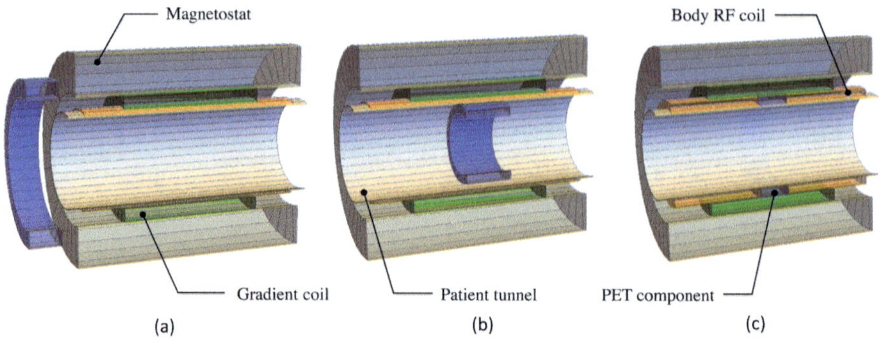

Figure 10.2 Possible system configurations of hybrid MR-PET devices. (a) Sequential dual modality device where the PET ring is installed in front of the MRI scanner. (b) Insert configuration. The PET subsystem rests on the patient tunnel and is aligned concentrically with the MR magnet; simultaneous operation possible but restricted FOV. (c) Fully integrated device, which allows simultaneous MR-PET acquisitions and large FOVs.

more difficult. The PET detectors and readout electronic is fully exposed to the static magnetic fields and the switching gradient fields. Also, all components from the PET device are very close to the MRI FOV and even very small amounts of incompatible materials can cause significant artefacts. PET inserts for a standard MRI scanner will necessarily have a reduced patient port because the detectors and associated electronics occupy additional space in the already confined gantry. A wide bore MRI scanner would allow for whole body PET inserts but no combined wide-bore MR-PET has been built to date. This is potentially due to the costs associated with wide bore magnets. Finally, the requirement of an RF shield for the PET modules makes them, in general, opaque to the radiation from the integrated body coil, which is a large combined transmit and receive coil for scout scans and body imaging applications that requires a large FOV. Thus, a dedicated joined transmit and receive coil that fits inside the PET insert is required. These coils must fulfil the special requirements that are explained in the next section.

The PET sub-system can also be fully integrated into the MR system (Figure 10.2(c)). This allows the simultaneous acquisition with both modalities for large FOVs and has the advantage that the body coil can also be used for MR imaging. The PET subsystem is embedded in the body coil and is, therefore, closer to the gradient coil, which can reach high temperatures. Consequently, special care has to be taken to assure good thermal isolation. All system design considerations mentioned for the insert solution, also apply to the fully integrated MR-PET device. In addition, the latter two design approaches require the use of silicon-based photo detectors for the PET subsystem, and as a consequence of their higher sensitivity to temperature variations, the liquid cooling of the PET

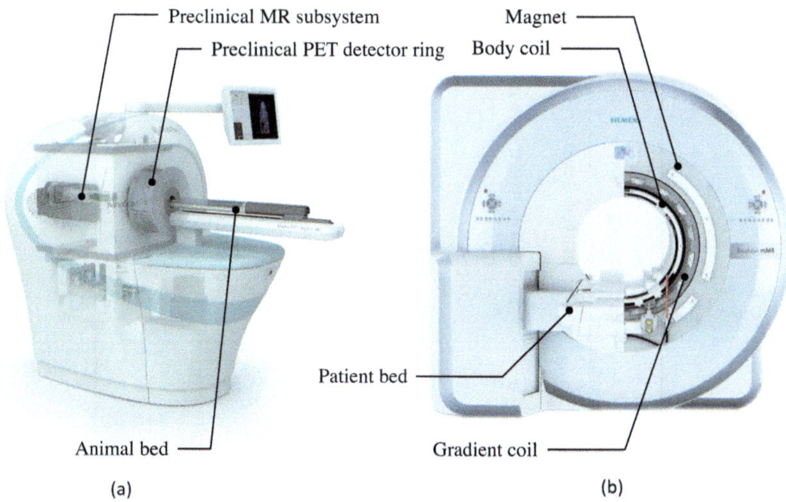

Figure 10.3 (a) Preclinical in-line 1 T MR-PET system nanoScan with permanent magnet from Mediso. (b) Integrated human 3 T MR-PET system from Siemens Healthcare GmbH (the PET detector ring is not visible in the image of the Siemens system.). Image A provided by Mediso Medical Imaging Systems. Image B provided by Siemens Healthcare GmbH.

module electronics is also preferable. A typical commercially available, fully integrated human MR-PET scanner and a smaller preclinical product are shown in Figure 10.3.

Currently, only two MR-PET systems are commercially available for use on humans. Both are fully integrated, whole-body devices at 3 T field strength. Previously, Philips offered a sequential MR-PET system, but it was recently discontinued. Several companies, including Mediso and Bruker Biospin, offer a preclinical MR-PET device at different field strengths. For preclinical applications, sequential and insert devices are available. The full integration of preclinical PET subsystems into a preclinical MRI scanner is hindered by the additional problem that, in general, these MR devices have a very small inner bore diameter (see Chapter 18). Even if sufficiently small PET module technology can be built, the scintillation detectors will be much closer to the imaged object, leading to larger parallax errors and requiring depth of interaction detection. Table 10.3 summarises the general properties of several commercially available hybrid MR-PET devices.

10.4 Hybrid MR-PET Coils

Hybrid systems, configured for consecutive measurements, resemble individual MR and PET systems placed in close vicinity, and hence do not require special RF coil technology. For a general introduction to RF coils and technology the reader is referred to Chapter 2 and the references therein.

Table 10.3 Performance parameters of commercially available MR-PET scanners.[10,11,12]

	System	Mediso Nanoscan	Siemens mMR	GE Signa
PET subsystem	Configuration and field strength	Preclinical in-line 3 T	Integrated human 3 T	Integrated human 3 T
	Type	Superconducting, dry magnet (no liquid helium)	Superconducting, with liquid helium	Superconducting, with liquid helium
	Bore size with gradients in	100 mm	60 cm	60 cm
	Gradient strengths	>650 mT m^{-1}	45 mT m^{-1} @ 200 T m^{-1} s^{-1}	44 mT m^{-1} @ 200 T m^{-1} s^{-1}
MR subsystem	Axial FOV	100 mm	258 mm	250 mm
	Transaxial FOV	120 m	594 mm	600 mm
	Number of crystals/module, crystal properties	39 × 39 – LYSO:Ce, 1.12 × 1.12 × 13 mm	LSO:Ce, 4 × 4 × 20 mm	LYSO:Ce, 4 × 5.3 × 25 mm
	Spatial resolution @ centre according to NEMA	1.25 mm	4.6 mm	4.4 mm
	NEC	850 kcps @ 60 MBq (mouse)	180 kcps	218 kcps
	Temporal resolution	1.2 ns	5.9 ns (coincidence window)	400 ps
	Energy resolution	19%	14.5%	10.5%
	Abs. sensitivity @ center, in 250–750 keV energy window	>8%	14.1 cps kBq^{-1}	23.3 cps kBq^{-1}

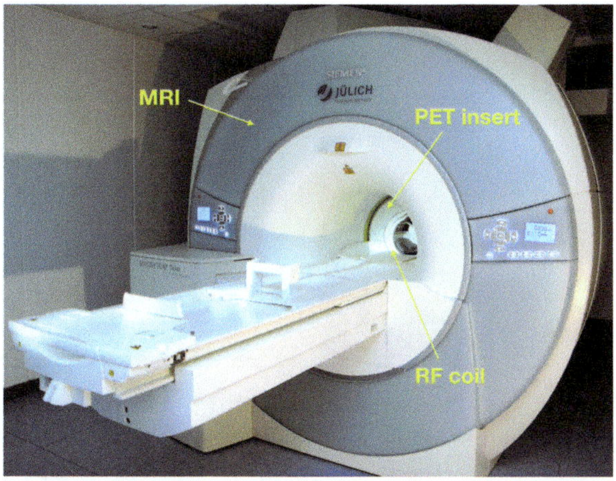

Figure 10.4 Overview of a human hybrid MR-PET system.

Figure 10.4 shows an MR-PET system, capable of simultaneous measurement, with the PET ring co-located inside the MRI magnet. In this configuration, the MRI coils are either placed inside the PET system within the PET acquisition FOV,[13] between the MR gradient coil and the PET modules[11,14] or integrated with the PET modules.[15] In the first case, any material used in the MR coil will contribute to the additional absorption and scatter of the gamma-quanta, which need to be minimised to optimise PET performance. The other way around, MR-compatible PET detector modules are likely—although not mandatorily[16]—to be completely enclosed within a copper or aluminium shield in order to eliminate interferences between the two modalities. Thus, the RF coil environment is closely confined with a highly conductive shielding material. In the other cases—when placing the RF coil outside the PET modules or integrating both into a single unit—the issues of PET attenuation/scatter through the MR coil are avoided. However, the MR coil still operates in a modified environment with increased coupling to metallic components, which needs to be considered to avoid B_1 field distortions and resonance frequency shifts.

MRI coils consist of a plastic casing that surrounds electronic components. While there are a number of PCBs—which may include the quadrature hybrid, transmit/receive switch, low-noise amplifier, and detuning circuitry among others—the coil elements themselves are usually copper or aluminium traces separated by lumped elements, such as capacitors (non-magnetic). The relatively dense materials (PCBs, lumped elements, coaxial cables, plugs, *etc.*) lead to attenuation and scattering of the annihilation photons.[17] Therefore, one optimisation criterion is where to place the components geometrically: high-density PCBs are placed outside the PET FOV and, because of out-of-FOV scatter, ideally at a certain distance from the PET FOV. Preferably only a minimum number of coil tracks should be located within the PET FOV. An example of a PET-MR coil is the

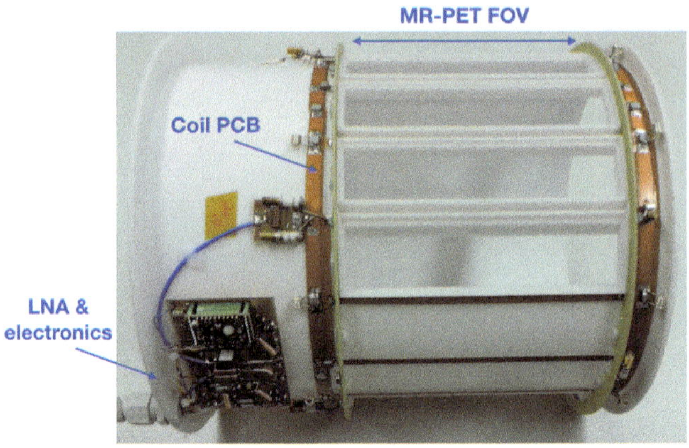

Figure 10.5 Custom made MR-PET coil.

high-pass birdcage coil, as shown in Figure 10.5. The high-pass design only requires capacitors on the two end-rings and not on the rungs of the resonator structure. Thus, the only remaining structures in the PET FOV are the PCB tracks and the coils plastic casing.

The PCB tracks can be constructed using a thin flexible PCB in combination with an (equally) thin copper or aluminium track. The standard thickness of the copper on the PCBs is around 35 μm, which represents a compromise between minimising the electrical resistance of the coil by increasing the conductor cross-section and PET attenuation. Given the limited thickness of the PCBs and tracks, the effects on gamma absorption and scatter are extremely small. As can be computed from Beer–Lambert's law, the attenuation of a thin, single layer PCB, with a thickness below 0.5 mm is less than 1% for γ-energies of 511 keV. Because of the theoretically lower absorption of aluminium in comparison to copper, at almost equal electrical conductivity, the use of Al as a conductor material has been widely investigated.[17,18] However, no consensus can be found in the literature and the choice of material remains a trade-off between gamma attenuation and signal-to-noise ratio (SNR) degradation in MRI caused by a slightly increased electrical resistance.

Finally, for PET compatibility the coil case also needs to be considered. Delso *et al.*[19] reported that a case made from a low-density material, with limited thickness, and ideally a flat surface should be employed in the PET FOV. Otherwise, the coil case might account for as much as a 15% reduction in count rate. The choice of the case material is a difficult task, as it needs to fulfil several requirements at the same time. It should be low-density for low gamma radiation attenuation and must be sturdy enough to support the mechanical stability of the coil structure. An issue that is emphasised by the impact, force and heating tests imposed for medical devices by the medical device directive.[20] Finally, the material must be an electrical insulator that sufficiently insulates the high voltages present in the transmit RF pulses. The properties listed for the case material also hold for the case varnish—with the addition that the varnish needs to be biocompatible, since it comes into direct contact with the patient's skin.

Geometric optimisation of the coil former has been proposed. This is in order to avoid streaking artefacts in the PET image reconstruction, which arise at air-material boundaries of the coil case[21] and at high attenuation along certain trajectories.[19] In this respect, a closed cylindrical coil case seems to be optimal. However, this might reduce patient comfort—especially when being applied to close fitting head coils. Moreover, functional MRI frequently requires the installation of a mirror system on the MR coil so that the patient has an unobstructed view of the task presentation screen.

Another important aspect in MR-PET coil design is the interaction of the RF field with the metallic shielding of the PET detector ring. If properly designed to be transparent for field gradients but presenting a closed shield for RF fields, the PET shield can be used as an RF screen for the MR coil.[7] Since many volume coils employ RF screens, the MR performance penalty is minor—the mirror currents on the RF screen reduce the field generated inside the coil and thus require an elevated drive power to achieve the desired flip angle. However, the positioning of the coil inside the PET cassette needs to be highly reproducible in this case, as even a slight offset will result in a severe impedance mismatch of the coil. An alternative is to add an additional RF screen to the coil itself. This removes the necessity of stringent repositioning accuracy and enables the coil to be operated without the PET cassettes present. However, this comes at the cost of increasing gamma attenuation through another metallisation layer.

10.5 Mutual Interferences with Other Modalities and Equipment

MR-PET measurements may be combined with further imaging modalities, especially with electroencephalography (EEG), electrocardiography (ECG), electrooculography (EOG), and electromyography (EMG), with EEG being the most relevant for brain imaging. EEG records the electrical signals, called brain waves, resulting from the changing action potentials of the activated neurons that are close to the surface of the brain, mainly from neurons in the cerebral cortex. The brain waves are grouped into delta-waves (1–5 Hz), theta-waves (4–7 Hz), alpha-waves (8–13 Hz), beta-waves (14–30 Hz), and gamma-waves (14–30 Hz). For data acquisition, electrodes, which are integrated in a soft cap, are placed on the scalp under special standardised conditions and at well-defined positions (Figure 10.6(a)). Brain potentials are very small (50–100 μV) and need to be amplified with dedicated, differential preamplifier having a gain of up to 10^6. For concurrent MR-EEG acquisitions, the interconnections between the EEG electrodes and those preamplifiers have to be routed through the MR patient bore, where they are exposed to the

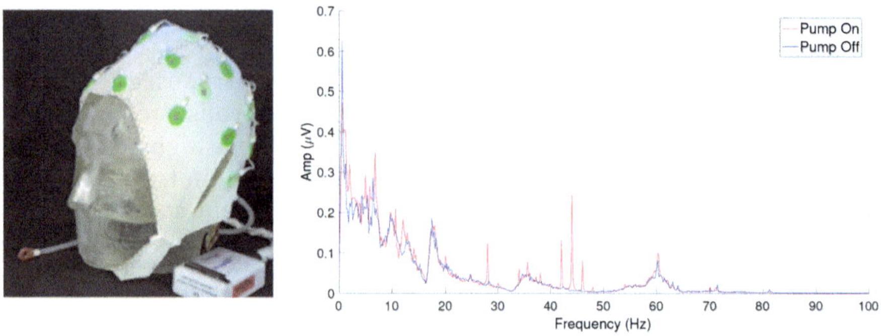

(a) (b)

Figure 10.6 (a) MR-compatible EEG cap on a head phantom for dedicated MR-PET brain imaging.[23] (b) Frequency spectrum of an EEG acquisition (Loudness Dependence of Auditory Evoked Potential) with (red) and without (blue) cryogenic pump operation. EEG spectra from the C_z channel (one of the EEG channels) is shown.

static magnetic field B_0 and the switching gradient fields. When the MR system acquires data, the gradient fields constantly vary and induce currents in the loops built by the connections between EEG electrodes and preamplifier, leading to very strong artefacts in the EEG data. Removing these artefacts is non-trivial, as their amplitude normally significantly exceeds the amplitudes from EEG signals, and it contains components in the EEG frequency range. However, to reduce this artefact, Allen *et al.* have proposed subtracting an averaged artefact followed by adaptive noise cancellation.[22] In the case that the EEG acquisition is combined with MR-PET imaging, possible attenuation of the γ-radiation from the PET tracer also has to be considered. However, it can be shown that the EEG electrodes (small Ag-AgCl metal electrodes embedded in a plastic housing made of polycarbonate and acrylate) only cause a minor attenuation with only a weak effect on the reconstructed emission images.[24]

Mechanical vibrations can also cause additional artefacts in the EEG acquisition. Most modern MRI scanners use a superconducting magnet for the generation of the B_0 field. The temperature of the superconducting magnet is cooled by a liquid helium (*ca.* 4.2 K) and the temperature is constantly maintained by a cryocooler coupled to a compressor. This cryocooler causes faint vibrations with relative, periodic variations of the static B_0 field, with respect to the loops created by the EEG signal cables, and therefore induces noise in the EEG data. These artefacts have characteristic frequencies correlated with the type of the coldhead and the mechanical design of the MR. Also, this artefact contains components in the EEG frequency range, but they are of much lower amplitude compared to the gradient switching artefact (Figure 10.6(b)).

fMRI also often requires stimulation of the human auditory, visual, or motor cortex. As is the case for the EEG cap, it has to be assured that no

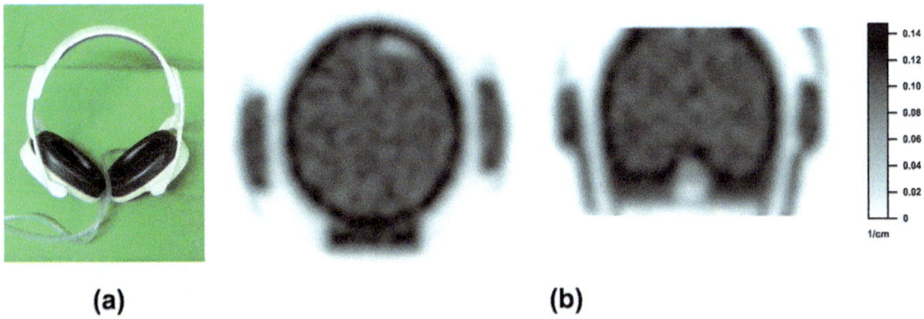

Figure 10.7 (a) MR-compatible headphones. (b) Transaxial and coronal attenuation map of a head phantom with the same headphones (acquired on a PET scanner with rotating ^{68}Ge emission sources for transmission imaging at 511 keV). The scale to the right shows the linear attenuation coefficients of the attenuation maps.

ferromagnetic materials are used in the construction of the stimulation devices, or if they are used, that these components are placed sufficiently far away from the MRI FOV. Furthermore, it is necessary to suppress RF emissions and excessive eddy-currents In addition, any component that is to be placed in the PET FOV, must not attenuate the γ-radiation, or an attenuation map for correcting the attenuation must be provided. Figure 10.7(a) shows MR compatible headphones for ear protection, which can also be used for stimulation of the auditory cortex and are constructed exclusively from plastic. The attenuation maps of the headphones placed on a head phantom for dedicated MR-PET brain imaging show a considerable amount of attenuation from the ear cups. Correcting PET artefacts caused by the attenuation is hardly possible, since the exact position of the headphones with respect to head can neither be derived from the MR images nor from the PET emission image. Instead of the headphones shown, small, air driven in-ear plugs should be used. For any additional equipment that might be required, *e.g.* ergometer, infusion pumps for contrast agents, TV screens and small mirrors for stimulation during fMRI acquisitions, blood sampler for measuring the PET tracer concentration in the blood, *etc.* all aspects of MR and PET compatibility mentioned above always have to be considered.

References

1. H. Herzog and C. Lerche, *PET Clinics*, 2016, **11**(2), 95.
2. D. W. Harberts and M. van Helvoort, *IEEE Int. Symp. Electromag. Comp., Conf. Rec.*, 2015, **1**, 1463.
3. S. Yamamoto, K. Kuroda and M. Senda, *IEEE Trans. Nucl. Sci.*, 2003, **50**(5), 1683.
4. D. Strul, D. Cash, S. F. Keevil, P. Halsted, S. C. R. Williams and P. K. Marsden, *IEEE Trans. Nucl. Sci.*, 2003, **50**(1), 60–69.
5. J. F. Schenck, *Med. Phys.*, 1996, **23**(6), 815.

6. G. Bertotti, *IEEE Trans. Magn.*, 1988, **24**(1), 621.
7. B. J. Lee, R. D. Watkins, C. M. Chang and C. S. Levin, *Magn. Reson. Med.*, 2018, **79**(3), 1745.
8. C. H. Kraft, *IEEE Int. Symp. Electromag. Comp., Conf. Rec.*, 1994, **1**, 73.
9. A. Berneking, R. Trinchero, Y. Ha, F. Finster, P. Cerello, C. Lerche and N. J. Shah, *IEEE Trans. Nucl. Sci.*, 2017, **64**(5), 1118.
10. D. D. L. Chung, *Carbon*, 2001, **39**(2), 279.
11. G. Delso, S. Fürst, B. Jakoby, R. Ladebeck, C. Ganter, S. G. Nekolla, M. Schwaiger and S. I. Ziegler, *J. Nucl. Med.*, 2011, **52**(12), 1914.
12. A. M. Grant, T. W. Deller, M. M. Khalighi, S. H. Maramraju, G. Delso and C. S. Levin, *Med. Phys.*, 2016, **43**(5), 2334.
13. C. Catana, *Magn. Reson. Imaging Clinics*, 2017, **25**(2), 231.
14. A. M. Grant, B. J. Lee, C. M. Chang and C. S. Levin, *Med. Phys.*, 2017, **44**(1), 112.
15. M. S. H. Akram, T. Obata, M. Suga, F. Nishikido, E. Yoshida, K. Saito and T. Yamaya, *J. Magn. Reson.*, 2017, **283**, 62.
16. N. Omidvari, G. Topping, J. Cabello, S. Paul, M. Schwaiger and S. I. Ziegler, *Phys. Med. Biol.*, 2018, **63**(9), 095002.
17. C. Y. Sander, B. Keil, D. B. Chonde, B. R. Rosen, C. Catana and L. L. Wald, *Magn. Reson. Med.*, 2015, **73**(6), 2363.
18. V. Rieke, A. Ganguly, B. L. Daniel, G. Scott, J. M. Pauly, R. Fahrig, N. J. Pelcs and K. Butts, *Magn. Reson. Med.*, 2005, **53**(6), 1409.
19. G. Delso, A. Martinez-Möller, R. A. Bundschuh, R. Ladebeck, Y. Candidus, D. Faul and S. I. Ziegler, *Phys. Med. Biol.*, 2010, **55**(15), 4361.
20. IEC, *Medical Electrical Equipment - Part 1: General Requirements for Basic Safety and Essential Performance*, IEC, 2012.
21. C. W. Lerche, T. Kaltsas, L. Caldeira, J. Scheins, E. Rota Kops, L. Tellmann, U. Pietrzyk, H. Herzog and N. J. Shah, *Phys. Med. Biol.*, 2018, **63**(3), 035039.
22. P. J. Allen, O. Josephs and R. Turner, *NeuroImage*, 2000, **12**(2), 230.
23. H. Iida, Y. Hori, K. Ishida, E. Imabayashi, H. Matsuda, M. Takahashi and S. Iguchi, *Ann. Nucl. Med.*, 2013, **27**(1), 25–36.
24. R. Rajkumar, E. Rota Kops, J. Mauler, L. Tellmann, C. Lerche, H. Herzog and I. Neuner, *PLoS One*, 2017, **12**(9), e0184743.

Section II

Special aspects of Data Corrections in MR-PET

CHAPTER 11

MR-based Corrections for Quantitative PET Image

L. CALDEIRA*[a], E. ROTA KOPS[a], N. A. DA SILVA[b] AND H. HERZOG[a]

[a]Forschungszentrum Jülich, Institute of Neuroscience and Medicine – 4, Wilhelm-Johnen-Straße 1, 52428, Jülich, Germany; [b]Hospital da Luz Learning Health, Hospital da Luz, Rua Carlos Alberto da Mota Pinto, Edifício Amoreiras Square 17 – 9º, 1070-313 Lisbon, Portugal
*E-mail: l.caldeira@fz-juelich.de

11.1 Introduction

In this chapter, we address positron emission tomography (PET) corrections that use magnetic resonance (MR) information, namely attenuation correction (AC), partial volume correction (PVC) and arterial input function (AIF). AC is mandatory in PET measurements and represents a major challenge in MR-PET, when compared to PET-only or PET-CT scanners. PVC and AIF are two challenges in PET that benefit greatly from MR information.

11.2 MR-Based AC

One of the major challenges for quantitating brain PET images when physiological or pharmacokinetic processes have to be analysed is related to photon interactions with biologic tissue, *i.e.* photons are 'lost' by a photoelectric

New Developments in NMR No. 19
Hybrid MR-PET Imaging: Systems, Methods and Applications
Edited by N. Jon Shah
© The Royal Society of Chemistry 2019
Published by the Royal Society of Chemistry, www.rsc.org

effect or Compton scattering. This phenomenon of attenuation is one dominant source of biased images if no correction is applied.[1] The consequence of not applying attenuation correction would be incorrect qualitative visual interpretation and erroneous quantitative evaluation. Thus, it is important to understand the physical processes that underlie photon attenuation and also the different methods that are used to correct this phenomenon.

The process of how the 511 keV annihilation photons result from the decay of positron emitting nuclides was described in Chapter 5. This chapter presents the physical and methodological basis of attenuation correction and summarises state-of-the-art developments in algorithms used to derive the attenuation map enabling accurate attenuation correction of brain PET data.

11.2.1 The Attenuation Process

Photon attenuation is the loss of photons emitted from a radioactive source due to their interaction with the tissue they traverse. This stochastic interaction process includes Compton scattering and photoelectric absorption, depending on the electron density of the material. Compton scattering[2] leads to a deviation in the original trajectory of the photon with an inevitable partial energy loss (a portion of the photon energy is transferred to the nearby atomic electrons it interacts with). Photoelectric absorption leads to a complete energy loss and total disappearance of the photon; the entire photon energy is transferred to the nearby atomic electron. Within the soft tissue, which comprises the largest part of the human body, two ranges can be defined where each of the two interactions dominates: photoelectric absorption below 20 keV and Compton scattering above 20 keV. Since approximately 99.5% of all interactions at 511 keV in soft tissue tend to be Compton scattering, the photoelectric absorption can be neglected in the context of PET.

Regardless of the type of interaction, photon attenuation is often considered to be the most important physical factor degrading PET image quality and quantification accuracy. While photons are travelling along a straight line in tissue (see Chapter 5) only a fraction of them arrive at distance L without scattering or absorption. The survival fraction is a path integral as follows

$$\mathrm{SF} = \mathrm{e}^{-\int_L \mu(l)\mathrm{d}l} \tag{11.1}$$

where μ is the spatially-dependent linear attenuation coefficient of the tissue in units of cm^{-1}. In other words, eqn (11.1) expresses the probability, p, that the photon will not interact with the nearby electrons while passing along L. The attenuation coefficient, μ, depends on the energy of the photons and on the electron density of the tissue. The attenuation is essentially proportional to the electron density due to the Compton and photoelectric effects. Thus, a higher tissue density results in larger attenuation coefficients,

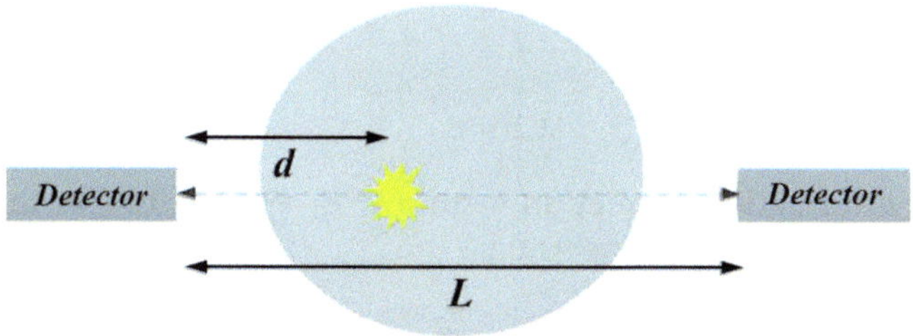

Figure 11.1 Coincidence detection in an attenuating object with the distance L relevant for attenuation.

μ, and consequently in lower survival fractions. For instance, the 511 keV attenuation coefficients of PET for important tissues are: lung 0.018–0.027 cm^{-1}, adipose tissue 0.086–0.093 cm^{-1}, soft/brain tissue 0.094–0.100 cm^{-1}, spongious bone 0.110 cm^{-1} and cortical bone 0.130–0.172 cm^{-1}.

A special advantage of PET with regard to photon attenuation is that the survival fraction (SF) of the biological tissue along the given distance L between two detectors, called the line-of-response (LOR), does not depend on the exact location of the annihilation event along this line, but rather on the total distance travelled by both annihilation photons between the two detectors (Figure 11.1).

The independence of the annihilation position is based on the fact that the total probability, p, for the photons to be detected in coincidence by the detectors is equal to the product of the single probabilities, p_d and p_{L-d}, related to the distances d and $(L-d)$, respectively. Thus, it will result in

$$p = p_d \cdot p_{L-d} = \mathrm{e}^{-\int_0^L \mu(l)\mathrm{d}l} \tag{11.2}$$

Since the attenuation fraction is $(1-p)$, the photon attenuation along a given LOR is also independent of the emission point.

Based on eqn (11.2), the number of photons emitted along each LOR and measured by the detectors, corresponding to the measured activity distribution A_{meas} for that LOR, can be written as

$$\int_0^L A_{\mathrm{meas}} = \int_0^L A \cdot \mathrm{e}^{-\int_0^L \mu(l)\mathrm{d}l} \tag{11.3}$$

where A is the actual activity of the LOR concerned, multiplied by the probability of non-attenuated photons. Thus, the actual activity, A, is given by rewriting eqn (11.3) as (11.4)

$$\int_0^L A = \int_0^L A_{\mathrm{meas}} \cdot \mathrm{e}^{\int_0^L \mu(l)\mathrm{d}l} = \int_0^L A_{\mathrm{meas}} \cdot \mathrm{ACF} \tag{11.4}$$

with eqn (11.5)

$$\mathrm{ACF} = \mathrm{e}^{\int_0^L \mu(l)\mathrm{d}l} \tag{11.5}$$

which is the attenuation correction factor (ACF) responsible for the attenuation correction of the LOR concerned.

11.2.2 The Attenuation Correction Process

The aim of the attenuation correction process is to obtain the individual ACFs for each LOR. This is possible by directly measuring the ACFs, or indirectly by measuring or calculating a map of the attenuation coefficients, μ, of the considered object with a subsequent calculation of the ACF.

11.2.2.1 Standalone PET Scanners

In the case of standalone PET scanners, *e.g.* the ECAT Exact HR+ scanner (Siemens, Erlangen, Germany), the ACFs can be directly measured.

In such scanners, one or more positron-emitting rod sources rotate between object and detectors. During such a transmission measurement, *T*, (Figure 11.2) the counts attenuated by the considered object are measured along the LORs. A similar measurement, but without any object in FOV (*blank*

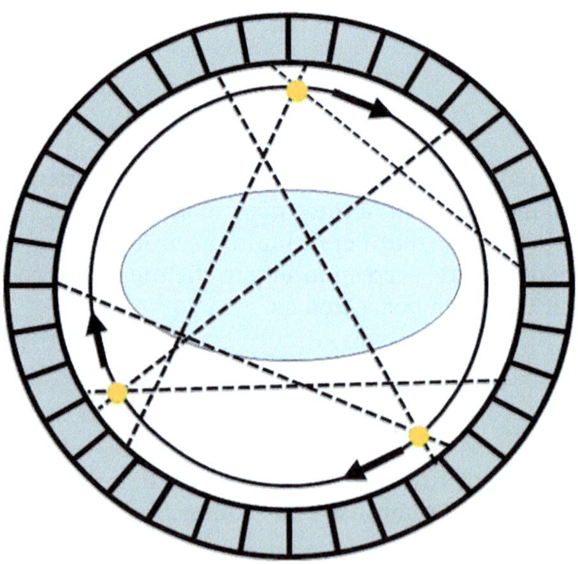

Figure 11.2 Principle of the transmission scanning with three rotating sources (yellow dots on their orbit) and examples of coincidence lines.

measurement B), is carried out by measuring non-attenuated count values. Thus, considering eqn (11.4) we get eqn (11.6)

$$B = T \cdot e^{\int_0^L \mu(l)\mathrm{d}l} = T \cdot \mathrm{ACF} \rightarrow B/T = \mathrm{ACF} \tag{11.6}$$

The ratio B/T represents an unbiased (but noisy) estimator of the true ACFs, which can be integrated directly in the reconstruction process (Section 7.1.2.2). Although this is the only direct measure of the attenuation coefficients, the transmission-based AC method suffers from inaccuracies, such as the propagation of noise[3] and subject position differences between transmission and emission scans. The latter is a consequence of the transmission and emission scans being performed sequentially.

11.2.2.2 PET-CT Scanners

A different situation pertains for PET-CT scanners. In a PET-CT scan, the CT measurement essentially corresponds to a transmission scan (Figure 11.3), where the radiation originating from a rotating X-ray tube is recorded by detector elements being arranged on the opposite side of the patient's body. However, such measurement delivers tomographic images describing the radiodensity (the relative inability of electromagnetic radiation, particularly X-rays, to pass through a particular material) of the investigated object in

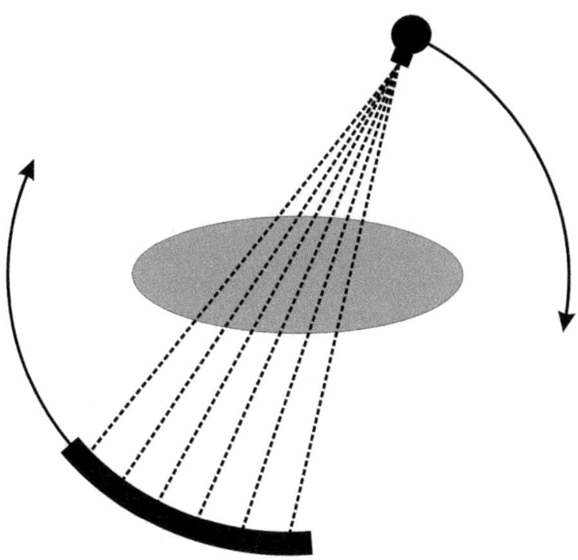

Figure 11.3 Principles of a CT scan. A source, emitting X-rays, and a detector assembly rotate around the body together.

physical units, *i.e.* the Hounsfield Unit (HU). By definition,[4] the HU value in a voxel with average linear attenuation coefficient, μ, is given by HU = 1000 × $(\mu - \mu_{water})/(\mu_{water} - \mu_{air})$ where μ_{water} = 0 HU and μ_{air} = −1000 HU are the linear attenuation coefficients of water and air, respectively.

Consequently, an extrapolation from the HU values to the PET attenuation coefficients μ is mandatory.[5,6] Furthermore, the CT measurements are based on a spectrum of photons from 30–140 keV. These photon energies are significantly lower than the 511 keV photons used in PET and, therefore, a conversion is unavoidable. To make the conversion accurate, one should consider two crucial facts: (1) the X-rays have a broad (polychromatic) energy spectrum and only reflect the average attenuation coefficient of a material at different energies; and (2) the energy scaling is material-dependent and thus behaves in a different way in materials with low-Z (soft tissue) or high-Z (cortical bone) values. The reason for the former is the well-known beam-hardening effect in CT imaging. This effect causes the CT number of the same tissue to be smaller when it is deeper inside the patient's body. Consequently, this results in attenuation values that are not absolute and thus require conversion. For non-uniform objects, point (2) indicates that simple scaling of the CT images may result in attenuation maps at the energy of interest with a large bias.

Once the PET attenuation coefficients, μ, are available in the form of an attenuation image (also known as attenuation map), the line integrals along all possible LOR trajectories are calculated according to eqn (11.5). In summary, this method provides low-noise ACFs and is therefore considered the current ground truth of attenuation correction.

11.2.2.3 MR-PET Scanners

In MR-PET scanners, the calculation of the required μ-map of the subject encounters new limits. None of the methods described above are applicable in MR-PET hybrid scanners, mainly due to space restrictions or technical constraints.[7] A straightforward approach to solve this problem is to derive the attenuation data from acquired MR images, preferably from anatomical images typically acquired during MR investigations. Unfortunately, image intensities in MRI are related to tissue proton density, relaxation times and a whole host of sequence parameters and have no relationship to attenuation coefficients valid for photon radiation at 511 keV. As such, new methods to compute attenuation maps from MR information have been developed. Due to the growing number of hybrid MR-PET systems installed worldwide, manufacturers and research groups have proposed numerous AC methods. A major problem of MR-based AC is the recognition of bone. Anatomical MR images acquired routinely show no, or a very low, signal for air and bone in spite of their extremely different attenuation properties arising from their different electron density. The suggested AC methods mainly differ in the way they image/recognise bone, and also the soft tissue and air regions. In whole-body MR-PET, an accurate definition of bone may not be so important if PET images are only evaluated qualitatively. However, in MR-PET of the brain, PET images are both qualitatively and quantitatively disturbed if the different tissue types and especially bone are not properly defined or even neglected.[8] The following overview primarily deals with MR-based AC in the case of brain studies.

MR-based AC methods can be categorised into three classes based on the techniques applied to create the attenuation map. First, one can list the segmentation-based methods, which sub-classify the various tissues within the MR image and assign AC values to each of these tissue classes. While in the past these methods assumed uniform attenuation coefficients, nowadays they are moving to continuous values by applying probabilistic approaches. Second, there are several alternatives of template- or atlas-based methods. These mainly deform or adapt an attenuation template/atlas to an individual patient's MR image to obtain a corresponding attenuation map. Methods using machine-learning algorithms can be also listed under this second group as they require atlas information for the learning process. The third class of methods exploits PET emission data together with anatomical information from MR images to compute the attenuation maps. In recent years, significant progress has been made in the field of MR-PET AC. A detailed overview of early approaches for MR-based AC can be found in these articles.[9–11]

Segmentation-based Methods: The ability to identify the number and the type of tissue segments largely depends on the MR sequences used. The regions are labelled on the basis of their MRI signal intensity or their anatomical location (*e.g.* the shape of the lungs can be identified) and pre-defined attenuation coefficients for each considered tissue are assigned to the identified regions. The main advantage of segmentation methods is that they directly adopt the individual MR images without going the diversion over a general template/atlas. Thus, these methods account for physiological inter-subject and pathology related variability in brain anatomy. However, segmentation methods also have limitations and consequently classification errors. In general, cortical bone can now be reliably recognised, but areas of complex air–bone interfaces, such as the paranasal sinuses, are challenging.[12] Furthermore, several components of the MR-PET scanners, such as the table and the radiofrequency coils, do not produce an MR signal. Thus, for these structures, an attenuation map cannot be obtained by MR segmentation and additional procedures have to be applied.[13] Finally, linear attenuation coefficients vary from subject to subject, thus diminishing the significance of inter-subject invariability.

A large number of approaches have been published which derive attenuation maps purely from MR images. Some of these are based on segmentation of anatomical T1 weighted MR images[14–17] or dual ultra-short time Echo (UTE) images[12,18–21] into multiple tissue classes (*e.g.* air, brain tissue, adipose tissue and bone). A detailed simulation study by Keereman *et al.*[22] has determined the number of tissue classes that need to be segmented in order to have acceptable errors in the emission reconstruction; these are air, (lung, in case of whole-body MR-PET), soft tissue, spongious bone and cortical bone.

Unfortunately, a differentiation between different bone structures is presently the most difficult. Nevertheless, the UTE sequence is able to image the bone, although with poor spatial resolution in the axial direction. The dual-UTE sequence acquires data at two different echo times (*e.g.* TE_{UTE1}: 0.07 ms and TE_{UTE2}: 2.46 ms); at the extremely short echo time of the first echo, TE_{UTE1}, the T_2^* signals from the bone have not decayed, thus yielding an image of bone. At longer times, such as TE_{UTE2}, the signal from bone no longer exists, resulting in image intensities equal to those of the air. The segmentation methods based on these two UTE images principally estimate the tissue class of the bone by calculating the difference of the logarithms between the UTE images, resulting in an R_2 map (R_2 defines the inverse of the spin–spin relaxation time T_2).[23] A newer method is based on a regular four-compartment segmentation from a Dixon sequence[24] to obtain a realistic bone attenuation map. Other approaches are based on special combinations of MRI sequences as, *e.g.*, short echo time (STE)/Dixon[25] or UTE/Dixon.[12] In addition, improved UTE sequences have been proposed, such as zero TE (ZTE),[26,27] which applies an even shorter echo time, giving optimal image features for the recognition of cranial bone structures.

Atlas/Template-Based Methods: The first step of the atlas/template-based methods is to create an atlas or a single template of MR images, which are largely similar to those acquired during MR-PET scans; the corresponding CT images or attenuation maps complete the approach. The basic approach of these methods is to find similarities between the patient's MR image and the atlas/template MR images; the corresponding similarities are then transferred/applied to the co-registered CT images or attenuation maps. In this way, an indirectly generated CT (pseudo-CT) or an attenuation map matching the patient is generated. The pseudo-CT is finally converted into an attenuation map using a bilinear transformation.[5,6] The potential advantage of this technique is that the atlas/template can incorporate much more structural details than the segmented and labelled approaches described above. In particular, bone is automatically included. Currently, some of the atlas/template-based methods are the best performing methods; however, limitations also exist. When severe brain alterations or deformations are present, the process of similarity identification is not able to capture them exactly, resulting in failures of the attenuation maps. Furthermore, the application of these methods to children, for example, would require the additional acquisition of CT scans in order to create a new atlas or templates, raising ethical issues due to unnecessary radiation exposure.

The several methods currently published differ in: (1) how the atlas/templates are generated; (2) the combination of MR images and attenuation maps accounts for the atlas/template; and (3) how the similarities between the patient's MR image and the atlas/template MR images are calculated. The template can be a simple[28,29] or a complex[30] average of a large number of MR and CT/transmission image pairs, while the atlas consists of a large number of single MR and CT image pairs.[31,32] Most of the atlas/templates include different basic types of images, as *e.g.* pairs of CT and dual UTE[33,34] or CT and T1 weighted MR images.[31,35] An exception, whereby the template consists of an attenuation map—generated from eight PET transmission images—and a T_1-weighted MR image was proposed by Rota Kops *et al.*[28] In some approaches, the UTE images are coupled with data from Dixon sequences[36] or T_2-weighted images.[37] The pseudo-CT images or the attenuation maps are finally derived by using different user-defined methods. These methods are based on manifold approaches, which again combine different MR sequences and their images and include various probabilistic measures, pattern recognition, or machine learning approaches.[15,31,33,38] The method presented by Izquierdo-Garcia *et al.*[30] works on a template-based approach; the template is created using preliminary segmented T_1-weighted images to improve the registration process. Atlas or template-based methods generally result in high-quality μ-maps, with significant similarity to typical CT-based μ-maps for the identification of bone and air.[39] However, some of them require a long computational time (*e.g.* up to 1 h) and are therefore not practical for use in routine clinical practice.

11.2.2.4 Emission-based Methods

The final set of methods to be covered here are based on the estimation of the attenuation information directly from the emission data. These being the attenuation coefficients estimated with simultaneous maximum likelihood reconstruction of emission and attenuation coefficients (MLAA).[40] Due to the statistical variability of the emission data, it is not normally possible to uniquely estimate the activity values and attenuation correction factors as these often suffer from cross talk artefacts, where the features of the emission data propagate into the attenuation map and *vice versa*. However, with the advent of time-of-flight (TOF) MR-PET scanners, it became possible to use MR anatomical information for the TOF-MLAA estimation of attenuation coefficients. This shows that attenuation correction factors can now be uniquely determined up to a constant scaling factor.[41,42] Furthermore, the noise and cross-talk artefacts were considerably reduced. These AC methods are particularly promising for deriving patient-specific attenuation maps. To improve the robustness of the MLAA algorithm, the estimation of attenuation maps over different anatomical regions was also proposed.[43] However, the long computational time and the limited applicability to only TOF systems are highly limiting factors. Nevertheless, another interesting application of the MLAA method is its use to estimate the attenuation of flexible hardware components used in the MR-PET scanners,[44] which is a particular challenge in whole-body measurements.

11.2.2.5 Final Considerations

The challenge of generating a reliable attenuation map for attenuation correction in MR-PET scanners remains a current topic of research. This is in spite of all the approaches presented above and also despite the statement given in Bailey *et al.*[45] that such issue can be put to rest because adequately addressed solutions have been found for the majority of clinical situations.

Most of the methods cited above presented comparisons of whole brain features; they did not consider tumour patients; that is the really challenging task. In fact, due to the presence of abnormal anatomy, the methods based on atlas or template information might fail or, alternatively, the data analysis and comparison of such patient groups cannot be performed with the generally used atlas of volumes-of-interest (VOIs) and means over the whole brain. Instead, a patient-by-patient analysis considering every single tumour has to be performed. Ladefoged *et al.*[46] presented such results, which are very promising but still require further investigations.

11.3 Partial Volume Correction

The spatial resolution of PET scanners is one of the limiting factors in the absolute quantification of, for example, blood flow and metabolism in small anatomical structures, like the cerebral cortex or blood vessels. The limited spatial resolution causes a cross-contamination of activity from neighbouring tissues.[47] One consequence of this is a partial loss of the signal in structures with high activity (spill-out) together with the increase of signal in regions with low activity (spill-in). These effects are usually more visible in smaller structures or regions with high contrast. The partial volume (PV) effect includes, therefore, not only spill-in and spill-out effects but also tissue fraction effects. Tissue fraction is a result of the finite voxel size, which means each voxel can contain a mixture of tissues. The measured PET signal, in this case, represents a mean activity concentration of two or more tissues and can be confounded with a 'new' structure or incorrectly regarded as a consequence of the resolution of the system. This effect is not always considered during correction as it affects images to a smaller degree. In clinical use, the question often arises, whether a decrease in the PET signal corresponds to a lower tissue accumulation or is a consequence of a partial volume effect or a combination of both. This is why it is so important to correct for PV effects.

11.3.1 Partial Volume Effect (PVE)

After the reconstruction of a PET image, the measured activity concentration is proportional to the real concentration (kBq cm^{-3}) in the object. Phantom studies and calculations with simulated data have been used to characterise and correct PVE. Here, we show two simple examples, a one-dimensional and a three-dimensional example. In Figure 11.4, different one-dimensional signals are shown at different resolutions (FWHM = 2 mm, 4 mm, and 8 mm) and different conclusions can be drawn from this example:

- The smaller the object the lower the intensity appears, although the objects have the same initial intensity (Figure 11.4(a) and (b)). Thus, the ratio of the measured activity concentration in the objects and the real activity concentration is a function of the object size.[48] This ratio is called the recovery coefficient (RC).

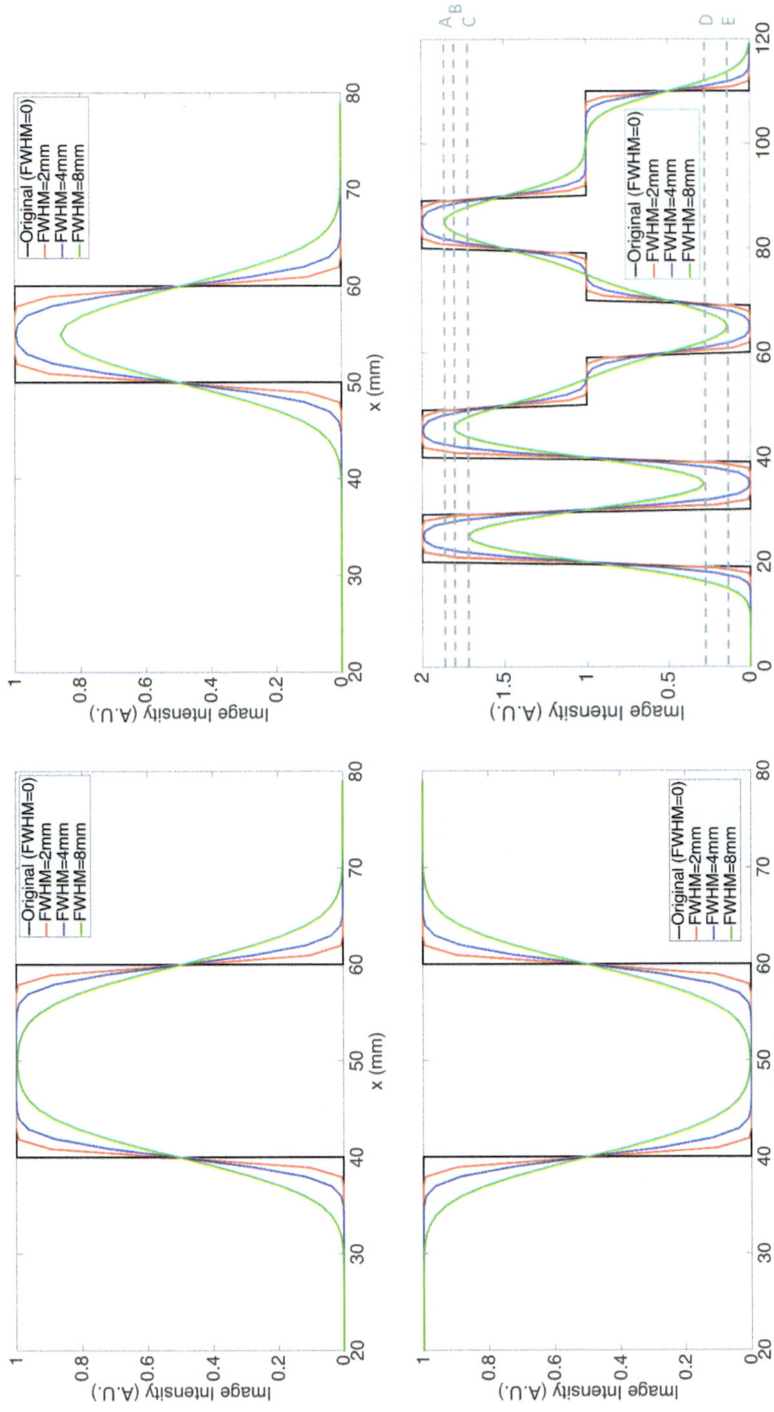

Figure 11.4 Effect of different resolutions in one-dimensional signals.

- The PV effect is not only a decrease of activity but can also be an increase of activity in certain areas due to spill-in effect (Figure 11.4(c)).
- As a rule of thumb, the intensity of the signal is not preserved if the object is smaller than twice the resolution of the scanner (*i.e.* the FWHM).
- The spatial position of the object and its relation to neighbouring radioactivity has an important influence in PVE. Thus, objects in areas of very different activity concentrations are much more influenced by the PV effect than objects in areas with activity concentrations similar to their own. For example, in Figure 11.4(d) at 8 mm resolution, points A, B, and C have different image intensities, although originally, they all had the same image intensity (value of 2). The same happens for points D and E.

A simulated three-dimensional brain phantom was created (Figure 11.5(a) and (b)) in which all voxels of the grey matter (GM) were set to a value of 1, the voxels of the white matter (WM) to a value of 0.25 and the remaining tissue to zero. The histogram can be seen in Figure 11.5(c). The simulated data set was blurred with a three-dimensional Gaussian filter (isotropic) with FWHM of 2 mm, 4 mm and 8 mm. Figures 11.5(d), (e), (g), (h), (j), (k) show the visual effect of the filter on the images while the histograms in Figure 11.5(f), (i), (l) demonstrate how the intensity distribution changes as a function of the FWHM of the filter. The GM values decrease significantly, while WM values increase mainly at its borders with GM, but several WM voxels still preserve their original values due to the large dimension of WM when compared with the resolution. Two profiles in transverse and coronal direction are shown and the spill-over effect is clearly visible (Figure 11.6). The profiles show how complex PVE can be. Finally, there is no clear distinction between grey and white matter in the case of FWHM of 8 mm.

11.3.2 Partial Volume Correction (PVC)

11.3.2.1 Empirical Methods

Since the end of the 1970s, several techniques have been developed to correct PVE. The first methods developed described the PV empirically, that is, phantom measurements were performed to understand the effects.[48,49] These are empirical methods because they do not use knowledge of the system resolution. The correction of the PVE can then be achieved by calculating the recovery coefficient (RC), which is the ratio of measured activity over the real activity. To be able to implement these methods, phantom measurements are necessary using a cylindrical phantom with several spheres or tubes of different diameters. However, it should be noted that when using tubes, the PVE in the axial direction cannot be corrected; consequently, the use of spheres is preferred. In both cases, the spheres/tubes are filled with an ^{18}F solution of equivalent activity concentration. In order to simulate different activity contrasts,

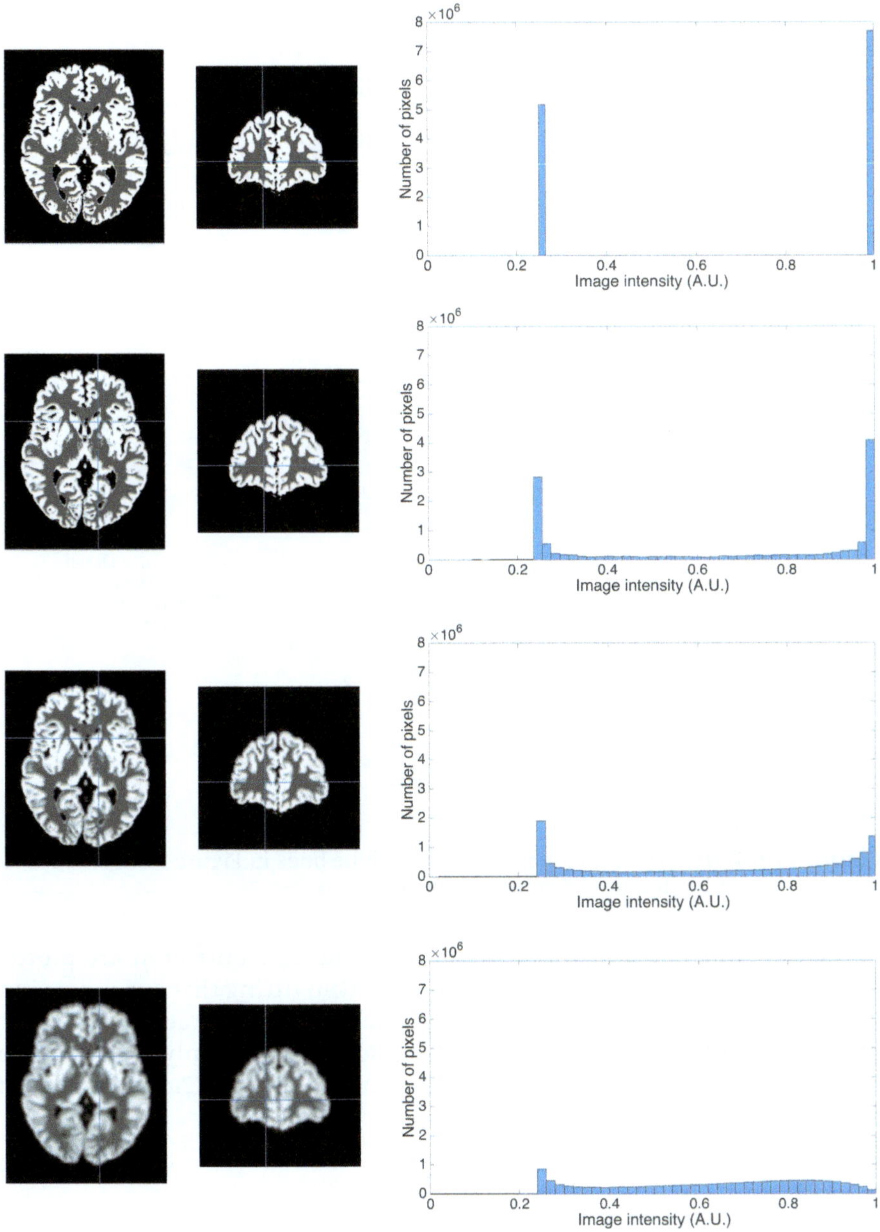

Figure 11.5 Effect of different resolutions in a brain image and corresponding histogram.

the measurement must be repeated several times with different activity ratios between the main cylindrical compartment and the spheres/tubes inside of it (*e.g.* 10 : 1, 5 : 1, *etc.*). In conjunction with this, the parameters of standard patient measurements for acquisition and reconstruction should also be used for phantom data processing. The ratios between the

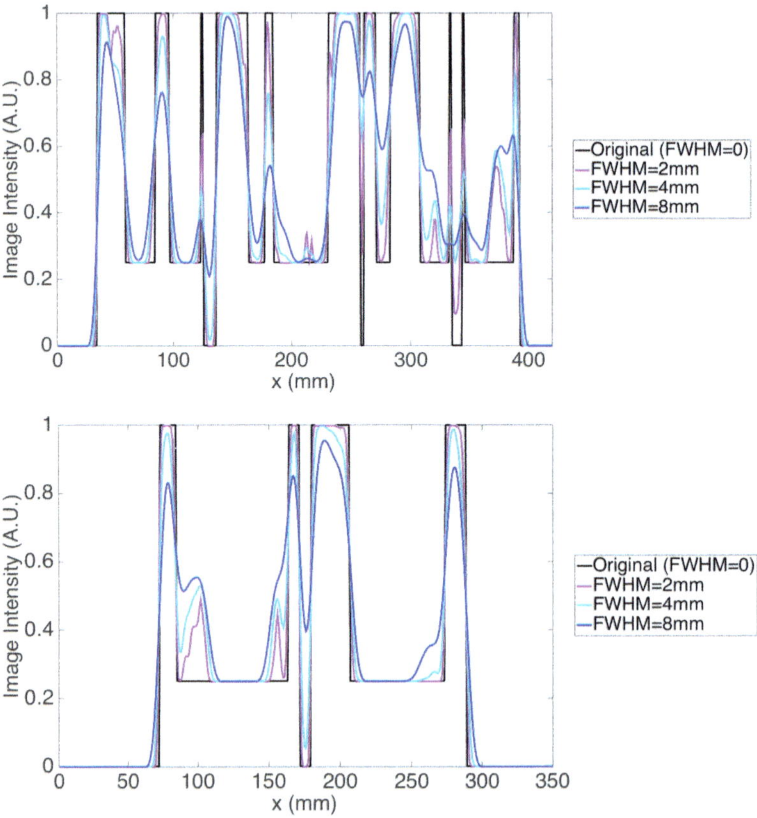

Figure 11.6 Transverse and coronal profiles (blue lines in Figure 11.5) of the brain phantom at different resolutions.

measured concentrations and the known true concentration are plotted as a function of the diameters and an algorithm fitting this curve provides the mathematical tool with which to calculate the correct RC for every diameter and for each activity contrast. However, this only represents an approximation of the measured environment and does not provide reliable quantitative values.

11.3.2.2 Deconvolution Methods

Subsequently, correction methods have been developed which consider the resolution of the scanner for PVC and this is typically described by the point spread function (PSF) (see Chapter 5). In order to describe the PSF, a Gaussian curve is usually used. Methods to determine the PSF of a PET scanner have been reviewed by Rahmim *et al.*[50] It is important to describe the PSF in three dimensions and, if possible, as a function of the distance to the centre of the PET scanner, since resolution in a PET scanner is usually spatially-variant. Mathematically, the PVE can be described in image I as the convolution of the original activity distribution O with the PSF (eqn (11.7))

$$I = \text{PSF} \otimes O \tag{11.7}$$

The convolution can be seen as a filter operation. Some PVC methods try to reverse the effect of the PV by inverting the problem stated in the equation above. However, these methods suffer from noise amplification when trying to recover information at high frequencies. Iterative deconvolution methods tackle this issue and show promising results.[51]

11.3.2.3 Anatomically-Guided Methods

Several methods use regions of interest to correct for PVE. It is possible to derive these regions directly from PET images, but this can be quite laborious. Therefore, anatomical information acquired by MRI was introduced in PVC methods so that high-frequency image information was available. Although the use of anatomical information to derive regions-of-interest presents difficulties relating to accurate alignment and, in some cases, segmentation, it is possible to overcome these issues as accurate alignment is achievable with simultaneous MR-PET acquisition.[52] Segmentation algorithms can separate classes on a voxel-by-voxel basis (including background), which can be used for PVC. Therefore, not only can grey matter, white matter and CSF be classified, but also tumour tissue, if, for example, it shows intensity values on the MR image which are different to the neighbouring tissue or blood vessels. Different MR images can be used for classification (T_1, T_2 and PD weighted images).[53] Segmentation can be performed before or after registration with the PET data set.

In the following section, PVC methods using anatomical information are described. The first methods were developed to correct one ROI, usually the grey matter,[48] by calculating an RC, which accounts for the decrease or increase of activity in the ROI. The RC is calculated by applying the PSF to an image of that ROI with homogeneous activity (eqn (11.8) and (11.9))

$$\text{ROI}_{\text{Image}} = \begin{cases} 0 \text{ outside ROI} \\ 1 \text{ inside ROI} \end{cases} \tag{11.8}$$

$$\text{Filtered ROI}_{\text{Image}} = \text{PSF} \otimes \text{ROI}_{\text{Image}} \tag{11.9}$$

The RC can be calculated as the ratio of activity inside the ROI of the convoluted/filtered image to the original image. In this case, spill-in effects are not taken into consideration.

An alternative method uses the same procedure described before but considers more ROIs within the brain. The image with homogeneous activity in each single ROI is filtered; this is done as many times as there are ROIs, and finally, for each filtered image, all ROIs are analysed. Consequently, this method considers not only spill-out, but also spill-in. After the application of the filter, the ROI being considered suffers from spill-out, but the other ROIs suffer from spill-in. All these RCs are put into a matrix, the so-called geometric transfer matrix (GTM), describing all possible cross-correlations.[54–56] The GTM matrix is then (pseudo)-inverted and used to calculate the correct activity values for each ROI.

The methods described so far do not provided a final corrected image, but rather corrected values for each ROI. Therefore, these kinds of methods were further extended to be applied voxel-wise, thus finally providing a corrected image.[57] Nowadays, the PVC is usually based on fully voxel-wise data processing.[47] The procedure is as previously described, but instead of calculating an RC for the whole ROI, an RC for each GM voxel is calculated by dividing the filtered image by the original image after subtraction of the filtered WM image (eqn (11.10))

$$\mathrm{RC}_{\mathrm{Image}} = \frac{\mathrm{filtered\,ROI}_{\mathrm{Image}}}{\mathrm{ROI}_{\mathrm{Image}}};\ \mathrm{Image}_{\mathrm{new}} = \frac{\mathrm{Image}}{\mathrm{RC}_{\mathrm{Image}}}(\mathrm{inside\,ROI}) \tag{11.10}$$

Although this method does not consider several regions, the background activity can be considered (Müller-Gärtner Method (MGM), as described by Müller-Gärtner *et al.*[58]). For example, to correct the GM, first the WM is considered by subtracting the activity (eqn (11.11)

$$\mathrm{Image}_{\mathrm{new}} = \mathrm{Image} - \mathrm{WM}_{\mathrm{avg}} \times \mathrm{filtered\ WM}_{\mathrm{Image}} \tag{11.11}$$

The GM ROI is then corrected as described previously in eqn (11.10). The filtered $\mathrm{WM}_{\mathrm{image}}$ is the filtered image of the binary or probabilistic image of WM. In this method, the average of the region $\mathrm{WM}_{\mathrm{avg}}$ has to be estimated before, either with GTM-modified MGM (mMGM)[59] or with a WM ROI chosen in a region without PVE. Furthermore, eqn (11.11) can be extended for several background regions. Although each voxel of the ROI can be individually corrected, only the target region, in this case the GM, is corrected. The multi-target correction (MTC) proposes to iteratively go through all regions,[59] considering one ROI as the target region in each iteration. Another possibility to extend this method to all ROIs is to create a $\mathrm{ROI}_{\mathrm{image}}$ with all ROIs (eqn (11.12))

$$\mathrm{ROI}_{\mathrm{Image}} = \begin{cases} 0 & \mathrm{outside\,ROIs} \\ \mathrm{WM}_{\mathrm{avg}} & \mathrm{inside\,WM\,ROI} \\ \mathrm{GM}_{\mathrm{avg}} & \mathrm{inside\,GM\,ROI} \end{cases} \tag{11.12}$$

Table 11.1 Anatomically-guided methods.

	Single region		Multi-region	
Region-based Voxel-wise:	RC	No spill-in	GTM	Spill-in
	Voxelwise RC	No spill-in	Yang	Spill-in
	MGM	Spill-in from background regions	RBV	Spill-in, GTM initialisation
	mMGM	Spill-in from background regions, GTM initialisation		
Iterative			MTC	Spill-in, GTM initialisation, extension of mMGM
			Iterative Yang	Spill-in, extension of Yang

This method is called the Yang method[60] and is an extension of eqn (11.8). The method works as described in eqn (11.10), but all ROIs are corrected simultaneously. Due to this, some consideration has to be given to the activity values in different ROIs, in this case, WM_{avg} and GM_{avg}. However, one can assume different activities for each ROI based on previous knowledge, *e.g.* GM : WM is usually set to 4 : 1. A further possibility is to take the average of the ROI directly from the image, although due to PV effects, this can be a very rough estimation. Another group[61] proposed the use of the GTM results to initialise the image ROI_{Image} in the so-called region-based voxel-wise (RBV) method. Finally, the Yang method can also be performed iteratively until convergence (iterative Yang); in this case, the GTM initialisation is not used, but the values of the image directly.[61]

The anatomically-guided methods are summarised in Table 11.1.

11.3.2.4 Reconstruction Methods

As an alternative to the image-based methods described thus far, PVE can be directly corrected during the PET data reconstruction. This is advantageous as it also takes the noise properties into consideration. During reconstruction, the PSF can be included in the system matrix (Section 7.1.2.2).[62] This reduces PVE, but on the other hand, the quantitative recovery is limited[63] and 'Gibbs artefacts' (also known as ring artefacts) can appear in the reconstructed images.[64] The reconstruction algorithms 'maximum *a posteriori*' (MAP) are an alternative to obtaining reconstructed PET images with a reduced level of noise and a better convergence. MAP reconstruction can correct for PVE and prevent Gibbs artefacts by imposing smoothness within tissues.[63] In 1997, Lipinski *et al.*[65] showed a better delineation of activity uptake borders using the so-called boundaries method. This method allows smoothness within tissues and avoids blurring among tissues by defining anatomical boundaries where smoothing is forbidden. This is considered true for brain imaging. However, a very good agreement between anatomical boundaries (usually from MRI or CT) and the functional boundaries in PET is required. Instead of defining boundaries where smoothing is forbidden, an alternative method was proposed, which does not require explicit segmentation.[66] This method, known as the Bowsher method, uses the intensity similarities in the MRI image to decide which voxels should be used in the neighbourhood.[67] Other approaches to include anatomical information have also been proposed.[54,68–73]

Despite these various approaches, there is still work needed to show the effectiveness of the MAP approach in correcting PVE.[63] Further work also needs to be done in order to understand the influence of mismatched anatomical information caused by mis-registration or functional/anatomical differences. These approaches have been mostly applied to brain studies, where mismatched anatomical information is minimal.

11.3.3 Applications

The correction of the PVE may play a relevant role in clinical diagnosis and in therapy monitoring for a higher accuracy and an exact analysis of quantitative parameters. A variety of methods for PVC are used in the context of neurological questions and some are supported by CT or MR images with high spatial resolution.[61,74,75] In recent years, the PVC has been increasingly used in the quantification of PET brain studies, because the activity concentration in the grey matter is underestimated as a consequence of its limited thickness. Methods allowing absolute quantification of the PET signals have been developed to quantify blood flow, metabolism, and receptor density. In the area of neuro-oncology, the PVC methods developed for PET brain studies can be applied using anatomical information from high resolution morphological imaging (CT, MR) and could, theoretically, be utilised for brain tumours, given that the tumour extent is exactly derived using MR or CT. However, this is not straightforward because the extent of malignant disease is often not exactly delineated in CT or MR images.[76] In order to meet the demand for greater diagnostic precision, the quantitative analysis of PET studies in oncological diagnosis is gaining significance. These PVC methods will have to be validated from the clinical point-of-view and should provide semi-quantitative and quantitative parameters for the judgment of lesions related to recurring, metastatic disease in oncology.

More complete reviews of PV correction methods are available by Erlandsson *et al.*[63,77]

11.4 Image-derived Input Function

Knowledge of the input function is mandatory if PET is to deliver fully quantitative data of metabolic or physiological functions, such as cerebral glucose consumption or cerebral blood flow (see Chapter 8). The input function is the time–activity curve of the injected activity in the blood, which presents a peak shape at the beginning directly after the bolus injection and decreases slowly with time. In contrast to the invasive procedure of arterial blood sampling for obtaining the input function, an image derived input function (IDIF), in which the blood information is acquired from the dynamic PET data, is an attractive non-invasive alternative. Consequently, to track the dynamics of the IDIF, PET data need to be reconstructed with higher sampling frequency at early times, that is, the time frames have to be very short. The IDIF approach was first validated for large blood pools, such as the heart (left ventricle)[78] and aortic segments.[79] Because of the large volume of these vascular structures, the partial volume effect (PVE, Section 11.3) has low or even no influence. In brain PET imaging, these vascular structures are out of the field-of-view and cannot be used to extract the IDIF. As an alternative, the intra-cranial arteries need to be considered. However, because of the small diameter of these structures (*e.g.* 5 mm for internal carotid arteries[80]) the PVE becomes relevant. This results in an underestimated input function, which generates quantification errors. In order to obtain an IDIF, three steps should be considered:[81] (1) identify the time–activity curve (TAC) relative to the blood pool; (2) correct for PVE; and (3) correct for plasma to whole-blood ratio and metabolites (when required) (Section 8.2).

11.4.1 IDIF in Non-simultaneous MR-PET

Initial efforts to obtain an IDIF used PET data[82] exclusively. However, using only the reduced anatomical information from PET for the identification of the blood pool, which is a critical step in obtaining an IDIF, can be challenging. In order to overcome this limitation, several authors explored the dynamic information from PET to extract the blood dynamics using cluster analysis[83] or independent component analysis.[84,85] Furthermore, methods based exclusively on PET data can be suboptimal due to the difficulty of performing partial volume correction (PVC) without anatomical information.[86]

Theoretically, MRI should be the reference method as its high anatomical detail enables the overall carotid diameter and the entire length of the vessel to be obtained, which is very important in PVC. However, in non-simultaneous MR-PET, errors related to mis-registration were reported and hampered the use of MR information on PET data. These errors are the consequence of the used co-registration methods which are based on the brain structures localised above the carotid arteries. Even when advanced co-registration algorithms are used, the elastic properties of the vessels, which are not rigid structures, can hamper the use of the data due to the different head position between MR and PET measurements.[87–89] To reduce co-registration errors, methods using immobilisation devices were explored.[81,86] However, these procedures proved uncomfortable for the patients and increased the workflow time.

Obtaining the IDIF corrected for PVE is challenging. In order to consider PVE, different PVC methods can be applied, either in post-processing[84,85,87,88]

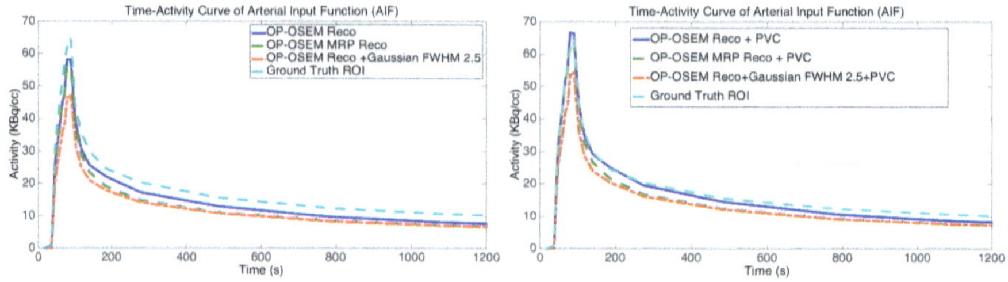

Figure 11.7 IDIF using different reconstruction methods before (left) and after PVC (right). Reproduced from ref. 74 with permission, Copyright 2015 IEEE.

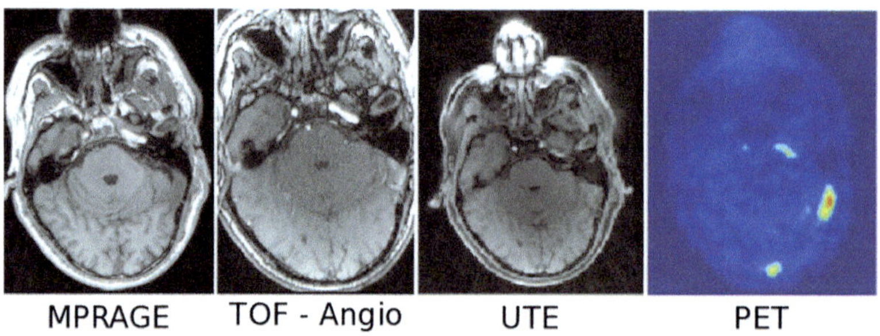

Figure 11.8 MR-PET contrasts from vessels: MPRAGE, TOF-Angiography, UTE and averaged PET image from early times.

or in reconstruction[88] with or without MR information (Figure 11.7).[74] Moreover, choosing a limited number of hottest pixels from a pre-defined region can also contribute to a better IDIF estimation.[87,90,91] In addition, late blood samples can also be used to scale the IDIF and correct for residual PVE. This was shown to be a robust method in different radiotracers.[87–92] Depending on the radiotracer, the exclusive use of imaging data is limited by the presence of radiometabolites. In the case of radiotracers generating only a few metabolites, the use of a single blood sample to correct for this effect and scale the IDIF is an attractive solution. In the case of tracers with many radiometabolites, other solutions may be considered and validated for each tracer.

11.4.2 IDIF in Simultaneous MR-PET

The introduction of hybrid MR-PET systems offers new possibilities to investigate IDIF by avoiding the necessity of using immobilisation devices or complex co-registration algorithms, resulting in a reduction in workflow and increased subject comfort. In addition, it also affords the possibility of exploring different image contrast to define the vessel regions, which must be precise to avoid signal contamination from adjacent tissue. In Figure 11.8, different contrasts/resolutions of MR images allow the identification of vessels (*e.g.* MPRAGE, TOF-Angiography and UTE). PET can also be used to identify the vessels based on the tracer dynamics (*e.g.* summed PET image from early times). In Figure 11.8, the differences in resolution between MR and PET data are also clear. MRI offers different dedicated methods to image the vessels where accurate vessel segmentation can be obtained, for example MR angiography with or without contrast agent (Chapter 3). Furthermore, using the high anatomical details from different brain structures also provides the possibility to apply PVC methods based on different anatomical regions defined on MRI.[63]

Image resolution must be carefully considered when investigating IDIF using MR-PET. MR images are often acquired with non-isotropic resolution and slice gap, whereas PET data are acquired with nearly isotropic resolution in the internal carotid artery (ICA) region. (Note that resolution and pixel size are different concepts. Resolution is the FWHM of the point spread function while pixel size is related to the reconstruction matrix. One should also consider that the resolution in PET is not isotropic due to DOI effects. Nevertheless, once carotid arteries are in the centre of the FOV this nearly isotropic assumption holds.) This can introduce errors in IDIF. Consequently, setting up and optimisation sequences to investigate IDIF must consider those variables. In addition, image distortions can also affect the MR-signal and should be evaluated to avoid errors in IDIF.

In simultaneous imaging, errors arising from the elastic properties are minimised; nevertheless, small misalignments can still occur. In order to consider these, MR vessel images can be registered to an early summed PET image where the vessels are still highlighted (Figure 11.8). Moreover, the registration can be performed for each vessel individually (*e.g.* left and right carotid) in order to obtain a better IDIF.[93]

The concept of the input function is also commonly used in the quantification of dynamic MRI following contrast agent injection, either in dynamic susceptibility contrast (DSC) or dynamic contrast enhancement (DCE) (Chapter 3). In the MR-PET context, the information from both modalities is acquired simultaneously and can be used synergically. In this way, the higher temporal resolution from MRI (1.5–5 s—note that the temporal resolution in MRI is related to the number of slices acquired and consequently to the volume coverage) can be explored to measure the rapid changes during the first bolus passage. Each modality allows information about the arterial input function (AIF) to be obtained but with different sampling frequencies—MRI allows better sampling of the AIF peak (Figure 11.9). Consequently, PET IDIF can be derived using both MR and PET dynamic information. The initial efforts to convert MR dynamic information in PET IDIF were tested in small animals and, thereafter, translated to human applications.[94–98]

Despite its associated benefits, using dynamic MR data also introduce challenges, which must be considered. For example, in DSC, calibration measurements in whole blood have shown a quadratic relationship between image contrast changes and contrast agent concentration for the input function measured with MRI.[99] This non-linear relationship in DSC MR might be a drawback not present in PET, where a linear relationship exists. Moreover,

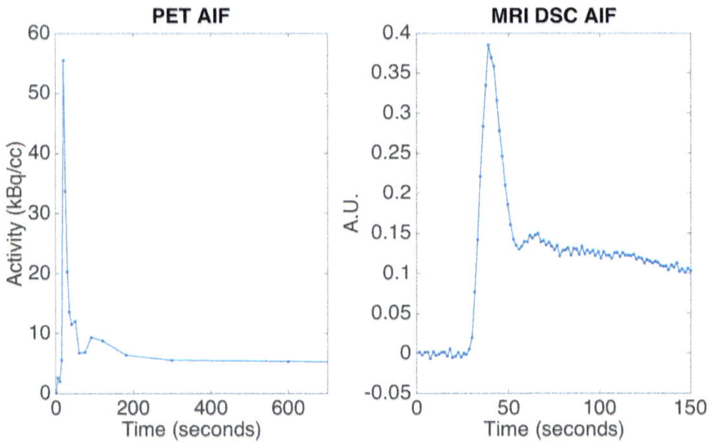

Figure 11.9 Dynamic MR-PET AIFs. On the left, AIF extracted from dynamic PET data. On the right, AIF extracted from dynamic MRI using DSC.

differences between injection rates in PET and MR can also produce errors. Such details must be considered in order to reduce uncertainties.[100]

11.4.3 Alternative Method

Dynamic PET data can also be explored to simultaneously estimate the input function and kinetic parameters in the so-called simultaneous estimation (SIME) method. In this approach, the input function is modelled analytically, and its parameters are estimated simultaneously with the kinetic parameters. The input function is often scaled with a blood sample in order to guarantee quantitative results. This is an attractive but complex method, which was further extended to the use of dynamic MRI data to improve the peak estimation and reduce kinetic parameter errors.[98]

11.4.4 Evaluation Methods

11.4.4.1 Metrics

The validation of IDIF methodologies is important and is often performed against the gold standard of arterial blood sampling and sometimes arterialised venous blood samples.[101] In this context, two metrics are often used: area under the curve (AUC) ratio of the estimated and ground truth curves, and kinetic model parameters/parametric images. AUC can be evaluated as the integral over the curve at different periods of the examination (*e.g.* peak and tail AUC ratios).[87,90–92] Kinetic model parameters/parametric images resulting from the use of IDIF and arterial blood sampling can be compared. This is a tracer dependent procedure, since the kinetic models applied are tracer dependent. Graphical methods using linear regression (*e.g.* Patlak plot or Logan plot) are often preferred due to their higher stability (compared to non-linear regression) regarding noise level and parameter identification.

Different radiotracers present different dynamics, and consequently, IDIF methods cannot be directly translated between tracers and always require a re-validation. For example, a method that works for [^{18}F]-FDG cannot be directly translated to [^{11}C]-flumazenil due to the different dynamics of the tracers. This influences the spill-over and spill-in effects corrected with PVC.

11.4.4.2 Uncertainties

The use of arterial blood sampling is also associated with uncertainties, which should be considered in the data evaluation. The first uncertainty is related to the blood sampling location. The sampling is often performed at the radial artery, which is far from the place of interest, *e.g.* the brain. Consequently, dispersion and delay can introduce quantification errors. The second uncertainty is related to problems with the cannulation, such as the coagulation of the blood in the tubes. A further uncertainty concerns possible stickiness of the tracer to the tube, which must be corrected to avoid an overestimation of the arterial blood input curve. Finally, the sampling strategies used might not match the PET framing scheme. The mentioned aspects highlight the limitations of arterial blood sampling and must be considered in IDIF validation.

In an MR-PET context, the position of the internal carotid arteries at the base of the skull can also be a crucial issue for IDIF due to attenuation uncertainties. This critical location can result in quantification errors which may influence the PVC methods. Consequently, MR-PET provides advantages and disadvantages, which must be critically considered when developing IDIF methodologies.

References

1. P. Kinahan, D. Townsend, T. Beyer and D. Sashin, *Med. Phys.*, 1998, **25**, 2046.
2. T. G. Turkington, *Semin. Nucl. Med.*, 2000, 255.
3. M. N. Wernick and J. N. Aarsvold, *Emission Tomography: The Fundamentals of PET and SPECT*, Academic Press, 2004.
4. R. A. Brooks, *J. Comput. Assist. Tomogr.*, 1977, **1**, 487.
5. C. Burger, G. Goerres, S. Schoenes, A. Buck, A. Lonn and G. Von Schulthess, *Eur. J. Nucl. Med. Mol. Imaging*, 2002, **29**, 922.
6. J. P. Carney, D. W. Townsend, V. Rappoport and B. Bendriem, *Med. Phys.*, 2006, **33**, 976.
7. S. Vandenberghe and P. K. Marsden, *Phys. Med. Biol.*, 2015, **60**, R115.
8. F. L. Andersen, C. N. Ladefoged, T. Beyer, S. H. Keller, A. E. Hansen, L. Højgaard, A. Kjær, I. Law and S. Holm, *NeuroImage*, 2014, **84**, 206.

9. I. Bezrukov, F. Mantlik, H. Schmidt, B. Schölkopf and B. J. Pichler, *Semin. Nucl. Med.*, 2013, 45.
10. A. Mehranian, H. Arabi and H. Zaidi, *Med. Phys.*, 2016, **43**, 1130.
11. Y. Chen and H. An, *Magn. Reson. Imaging Clin.*, 2017, **25**, 245.
12. Y. Berker, J. Franke, A. Salomon, M. Palmowski, H. C. Donker, Y. Temur, F. M. Mottaghy, C. Kuhl, D. Izquierdo-Garcia and Z. A. Fayad, *J. Nucl. Med.*, 2012, **53**, 796.
13. M. Eldib, J. Bini, D. D. Faul, N. Oesingmann, C. Tsoumpas and Z. A. Fayad, *PET Clinics*, 2016, **11**, 151.
14. H. Zaidi, M. L. Montandon and D. O. Slosman, *Med. Phys.*, 2003, **30**, 937.
15. G. Wagenknecht, E. R. Kops, L. Tellmann and H. Herzog, *Nuclear Science Symposium Conference Record (NSS/MIC)*, 2009 IEEE, 2009.
16. J. Teuho, J. Linden, J. Johansson, J. Tuisku, T. Tuokkola and M. Teräs, *IEEE Trans. Nucl. Sci.*, 2016, **63**, 2452.
17. F. Liu, H. Jang, R. Kijowski, T. Bradshaw and A. B. McMillan, *Radiology*, 2017, 170700.
18. J. Cabello, M. Lukas, S. Förster, T. Pyka, S. G. Nekolla and S. I. Ziegler, *J. Nucl. Med.*, 2015, **56**, 423.
19. C. Catana, A. van der Kouwe, T. Benner, C. J. Michel, M. Hamm, M. Fenchel, B. Fischl, B. Rosen, M. Schmand and A. G. Sorensen, *J. Nucl. Med.*, 2010, **51**, 1431.
20. M. R. Juttukonda, B. G. Mersereau, Y. Chen, Y. Su, B. G. Rubin, T. L. Benzinger, D. S. Lalush and H. An, *NeuroImage*, 2015, **112**, 160.
21. C. N. Ladefoged, D. Benoit, I. Law, S. Holm, A. Kjær, L. Højgaard, A. E. Hansen and F. L. Andersen, *Phys. Med. Biol.*, 2015, **60**, 8047.
22. V. Keereman, R. V. Holen, P. Mollet and S. Vandenberghe, *Med. Phys.*, 2011, **38**, 6010.
23. V. Keereman, Y. Fierens, T. Broux, Y. De Deene, M. Lonneux and S. Vandenberghe, *J. Nucl. Med.*, 2010, **51**, 812.
24. T. Koesters, K. P. Friedman, M. Fenchel, Y. Zhan, G. Hermosillo, J. Babb, I. O. Jelescu, D. Faul, F. E. Boada and T. M. Shepherd, *J. Nucl. Med.*, 2016, **57**, 918.
25. P. Khateri, H. S. Rad, A. H. Jafari, A. F. Kazerooni, A. Akbarzadeh, M. S. Moghadam, A. Aryan, P. Ghafarian and M. R. Ay, *Mol. Imaging Biol.*, 2015, **17**, 884.
26. F. Wiesinger, L. I. Sacolick, A. Menini, S. S. Kaushik, S. Ahn, P. Veit-Haibach, G. Delso and D. D. Shanbhag, *Magn. Reson. Med.*, 2016, **75**, 107.
27. G. Delso, F. Wiesinger, L. I. Sacolick, S. S. Kaushik, D. D. Shanbhag, M. Hüllner and P. Veit-Haibach, *J. Nucl. Med.*, 2015, **56**, 417.
28. E. Rota Kops, H. Hautzel, H. Herzog, G. Antoch and N. J. Shah, *IEEE Trans. Nucl. Sci.*, 2015, **62**, 2115.
29. S. Wollenweber, S. Ambwani, G. Delso, A. Lonn, R. Mullick, F. Wiesinger, Z. Piti, A. Tari, G. Novak and M. Fidrich, *IEEE Trans. Nucl. Sci.*, 2013, **60**, 3383.
30. D. Izquierdo-Garcia, A. E. Hansen, S. Förster, D. Benoit, S. Schachoff, S. Fürst, K. T. Chen, D. B. Chonde and C. Catana, *J. Nucl. Med.*, 2014, **55**, 1825.

31. N. Burgos, M. J. Cardoso, K. Thielemans, M. Modat, S. Pedemonte, J. Dickson, A. Barnes, R. Ahmed, C. J. Mahoney and J. M. Schott, *IEEE Trans. Med. Imaging*, 2014, **33**, 2332.
32. A. Torrado-Carvajal, J. L. Herraiz, E. Alcain, A. S. Montemayor, L. Garcia-Cañamaque, J. A. Hernandez-Tamames, Y. Rozenholc and N. Malpica, *J. Nucl. Med.*, 2016, **57**, 136.
33. A. Santos Ribeiro, E. Rota Kops, H. Herzog and P. Almeida, *Nucl. Instrum. Methods Phys. Res., Sect. A*, 2014, **734**, 166.
34. S. Roy, W.-T. Wang, A. Carass, J. L. Prince, J. A. Butman and D. L. Pham, *J. Nucl. Med.*, 2014, **55**, 2071.
35. T. Sekine, A. Buck, G. Delso, E. E. ter Voert, M. Huellner, P. Veit-Haibach and G. Warnock, *J. Nucl. Med.*, 2016, **57**, 215.
36. B. K. Navalpakkam, H. Braun, T. Kuwert and H. H. Quick, *Invest. Radiol.*, 2013, **48**, 323.
37. A. Johansson, M. Karlsson and T. Nyholm, *Med. Phys.*, 2011, **38**, 2708.
38. M. Hofmann, F. Steinke, V. Scheel, G. Charpiat, J. Farquhar, P. Aschoff, M. Brady, B. Schölkopf and B. J. Pichler, *J. Nucl. Med.*, 2008, **49**, 1875.
39. A. Johansson, M. Karlsson, J. Yu, T. Asklund and T. Nyholm, *Med. Phys.*, 2012, **39**, 3283.
40. J. Nuyts, P. Dupont, S. Stroobants, R. Benninck, L. Mortelmans and P. Suetens, *IEEE Trans. Med. Imaging*, 1999, **18**, 393.
41. M. Defrise, A. Rezaei and J. Nuyts, *Phys. Med. Biol.*, 2012, **57**, 885.
42. A. Salomon, A. Goedicke and T. Aach, *10th Int. Meeting on Fully 3D Image Reconstruction in Radiology and Nuclear Medicine Beijing*, China, 2009.
43. A. Salomon, A. Goedicke, B. Schweizer, T. Aach and V. Schulz, *IEEE Trans. Med. Imaging*, 2011, **30**, 804.
44. T. Heußer, C. M. Rank, Y. Berker, M. T. Freitag and M. Kachelrieß, *EJNMMI Phys.*, 2017, **4**, 12.
45. D. Bailey, B. Pichler, B. Gückel, G. Antoch, H. Barthel, Z. Bhujwalla, S. Biskup, S. Biswal, M. Bitzer and R. Boellaard, Summary Report of the 6th International Workshop on PET/MRI, *Mol. Imaging Biol.*, 2017, 1.
46. C. N. Ladefoged, F. L. Andersen, A. Kjær, L. Højgaard and I. Law, *Front. Neurosci.*, 2017, **11**, 453.
47. E. Rota Kops and B. Krause, in *PET in Clinical Oncology*, Springer, 2000, p. 33.
48. E. J. Hoffman, S.-C. Huang and M. E. Phelps, *J. Comput. Assist. Tomogr.*, 1979, **3**, 299.
49. R. M. Kessler, J. R. Ellis Jr and M. Eden, *J. Comput. Assist. Tomogr.*, 1984, **8**, 514.
50. A. Rahmim, J. Qi and V. Sossi, *Med. Phys.*, 2013, **40**.
51. J. Tohka and A. Reilhac, *NeuroImage*, 2008, **39**, 1570.
52. H. Herzog, K. Langen, C. Weirich, E. Rota Kops, J. Kaffanke, L. Tellmann, J. Scheins, I. Neuner, G. Stoffels and K. Fischer, *Nuklearmedizin*, 2011, **50**, 74.
53. K. Held, E. Rota Kops, B. J. Krause, W. M. Wells, R. Kikinis and H.-W. Muller-Gartner, *IEEE Trans. Med. Imaging*, 1997, **16**, 878.

54. E. Henze, S.-C. Huang, O. Ratib, E. Hoffman, M. E. Phelps and H. R. Schelbert, *J. Nucl. Med.*, 1983, **24**, 987.
55. P. Herrero, J. Markham and S. R. Bergmann, *J. Comput. Assist. Tomogr.*, 1989, **13**, 862.
56. O. G. Rousset, Y. Ma and A. C. Evans, *J. Nucl. Med.*, 1998, **39**, 904.
57. T. O. Videen, J. S. Perlmutter, M. A. Mintun and M. E. Raichle, *J. Cereb. Blood Flow Metab.*, 1988, **8**, 662.
58. H. W. Müller-Gärtner, J. M. Links, J. L. Prince, R. N. Bryan, E. McVeigh, J. P. Leal, C. Davatzikos and J. J. Frost, *J. Cereb. Blood Flow Metab.*, 1992, **12**, 571.
59. K. Erlandsson, A. Wong, R. Van Heertum, J. Mann and R. Parsey, *NeuroImage*, 2006, **31**, T84.
60. J. Yang, S. Huang, M. Mega, K. Lin, A. Toga, G. Small and M. Phelps, *IEEE Trans. Nucl. Sci.*, 1996, **43**, 3322.
61. B. A. Thomas, K. Erlandsson, M. Modat, L. Thurfjell, R. Vandenberghe, S. Ourselin and B. F. Hutton, *Eur. J. Nucl. Med. Mol. Imaging*, 2011, **38**, 1104.
62. A. J. Reader, P. J. Julyan, H. Williams, D. L. Hastings and J. Zweit, *IEEE Trans. Nucl. Sci.*, 2003, **50**, 1392.
63. K. Erlandsson, I. Buvat, P. H. Pretorius, B. A. Thomas and B. F. Hutton, *Phys. Med. Biol.*, 2012, **57**, R119.
64. S. Tong, A. M. Alessio, K. Thielemans, C. Stearns, S. Ross and P. E. Kinahan, *Nuclear Science Symposium Conference Record (NSS/MIC)*, 2010 IEEE, 2010.
65. B. Lipinski, H. Herzog, E. Rota Kops, W. Oberschelp and H. Muller-Gartner, *IEEE Trans. Med. Imaging*, 1997, **16**, 129.
66. J. E. Bowsher, V. E. Johnson, T. G. Turkington, R. J. Jaszczak, C. Floyd and R. E. Coleman, *IEEE Trans. Med. Imaging*, 1996, **15**, 673.
67. L. Caldeira, J. Scheins, P. Almeida and H. Herzog, *Nucl. Instrum. Methods Phys. Res., Sect. A*, 2013, **702**, 141.
68. G. Gindi, M. Lee, A. Rangarajan and I. G. Zubal, *IEEE Trans. Med. Imaging*, 1993, **12**, 670.
69. R. Leahy and X. Yan, *Information Processing in Medical Imaging*, 1991, 105.
70. X. Ouyang, W. H. Wong, V. E. Johnson, X. Hu and C.-T. Chen, *IEEE Trans. Med. Imaging*, 1994, **13**, 627.
71. K. Baete, J. Nuyts, W. Van Paesschen, P. Suetens and P. Dupont, *IEEE Trans. Med. Imaging*, 2004, **23**, 510.
72. S. Somayajula, C. Panagiotou, A. Rangarajan, Q. Li, S. R. Arridge and R. M. Leahy, *IEEE Trans. Med. Imaging*, 2011, **30**, 537.
73. K. Vunckx, A. Atre, K. Baete, A. Reilhac, C. M. Deroose, K. Van Laere and J. Nuyts, *IEEE Trans. Med. Imaging*, 2012, **31**, 599.
74. L. L. Caldeira, N. da Silva, J. J. Scheins, M. E. Gaens and N. J. Shah, *IEEE Trans. Nucl. Sci.*, 2015, **62**, 1725.
75. Y. Rong, I. Vernaleken, O. H. Winz, A. Goedicke, F. M. Mottaghy and E. Rota Kops, *Z. Med. Phys.*, 2015, **25**, 230.

76. K.-J. Langen, N. Galldiks, E. Hattingen and N. J. Shah, *Nat. Rev. Neurol.*, 2017, **13**, 279.
77. K. Erlandsson, J. Dickson, S. Arridge, D. Atkinson, S. Ourselin and B. F. Hutton, *PET Clinics*, 2016, **11**, 161.
78. Y. Choi, R. A. Hawkins, S.-C. Huang, S. S. Gambhir, R. C. Brunken, M. E. Phelps and H. R. Schelbert, *J. Nucl. Med.*, 1991, **32**, 733.
79. A. P. Van Der Weerdt, L. J. Klein, R. Boellaard, C. A. Visser, F. C. Visser and A. A. Lammertsma, *J. Nucl. Med.*, 2001, **42**, 1622.
80. J. Krejza, M. Arkuszewski, S. E. Kasner, J. Weigele, A. Ustymowicz, R. W. Hurst, B. L. Cucchiara and S. R. Messe, *Stroke*, 2006, **37**, 1103.
81. P. Zanotti-Fregonara, E. M. Fadaili, R. Maroy, C. Comtat, A. Souloumiac, S. Jan, M.-J. Ribeiro, V. Gaura, A. Bar-Hen and R. Trébossen, *J. Cereb. Blood Flow Metab.*, 2009, **29**, 1825.
82. K. Chen, D. Bandy, E. Reiman, S.-C. Huang, M. Lawson, D. Feng, L.-s. Yun and A. Palant, *J. Cereb. Blood Flow Metab.*, 1998, **18**, 716.
83. M. Liptrot, K. H. Adams, L. Martiny, L. H. Pinborg, M. N. Lonsdale, N. V. Olsen, S. Holm, C. Svarer and G. M. Knudsen, *NeuroImage*, 2004, **21**, 483.
84. M. Naganawa, Y. Kimura, K. Ishii, K. Oda, K. Ishiwata and A. Matani, *IEEE Trans. Biomed. Eng.*, 2005, **52**, 201.
85. K.-H. Su, L.-C. Wu, R.-S. Liu, S.-J. Wang and J.-C. Chen, *Nucl. Med. Commun.*, 2005, **26**, 995.
86. J.-E. Litton, *J. Comput. Assist. Tomogr.*, 1997, **21**, 907.
87. J. E. Mourik, M. Lubberink, U. M. Klumpers, E. F. Comans, A. A. Lammertsma and R. Boellaard, *NeuroImage*, 2008, **39**, 1041.
88. P. Zanotti-Fregonara, J.-S. Liow, M. Fujita, E. Dusch, S. S. Zoghbi, E. Luong, R. Boellaard, V. W. Pike, C. Comtat and R. B. Innis, *PLoS One*, 2011, **6**, e17056.
89. E. K. Fung, B. Planeta-Wilson, T. Mulnix and R. E. Carson, *Nuclear Science Symposium Conference Record (NSS/MIC)*, 2009 IEEE, 2009.
90. J. E. Mourik, F. H. van Velden, M. Lubberink, R. W. Kloet, B. N. van Berckel, A. A. Lammertsma and R. Boellaard, *NeuroImage*, 2008, **43**, 676.
91. J. E. Mourik, M. Lubberink, A. Schuitemaker, N. Tolboom, B. N. van Berckel, A. A. Lammertsma and R. Boellaard, *Eur. J. Nucl. Med. Mol. Imaging*, 2009, **36**, 463.
92. P. Zanotti-Fregonara, K. Chen, J.-S. Liow, M. Fujita and R. B. Innis, *J. Cereb. Blood Flow Metab.*, 2011, **31**, 1986.
93. E. K. Fung and R. E. Carson, *Phys. Med. Biol.*, 2013, **58**, 1903.
94. E. Evans, G. Buonincontri, D. Izquierdo, C. Methner, R. C. Hawkes, R. E. Ansorge, T. Kreig, T. A. Carpenter and S. J. Sawiak, *EJNMMI Phys.*, 2014, 84.
95. E. Evans, S. J. Sawiak, A. O. Ward, G. Buonincontri, R. C. Hawkes and T. A. Carpenter, *Nucl. Instrum. Methods Phys. Res., Sect. A*, 2014, **734**, 137.
96. E. Poulin, R. Lebel, E. Croteau, M. Blanchette, L. Tremblay, R. Lecomte and M. Lepage, *Magn. Reson. Med.*, 2013, **69**, 781.
97. M. A. Richard, J. P. Fouquet, R. Lebel and M. Lepage, *PET Clinics*, 2016, **11**, 193.

98. H. Sari, K. Erlandsson, K. Thielemans, D. Atkinson, S. Ourselin, S. Arridge and B. F. Hutton, *IEEE Trans. Nucl. Sci.*, 2015, **62**, 612.
99. F. Calamante, *Prog. Nucl. Magn. Reson. Spectrosc.*, 2013, **74**, 1.
100. H. Sari, K. Erlandsson, A. Barnes, D. Atkinson, S. Arridge, S. Ourselin and B. Hutton, *EJNMMI Phys.*, 2014, 54.
101. R. W. Parks, D. A. Loewenstein, K. L. Dodrill, W. W. Barker, F. Yoshii, J. Y. Chang, A. Emran, A. Apicella, W. A. Sheramata and R. Duara, *J. Clin. Exp. Neuropsychol.*, 1988, **10**, 565.

CHAPTER 12

Motion Correction in Brain MR-PET

R. P. BUSCHBECK*[a], L. CALDEIRA[a], J. SCHEINS[a], L. TELLMANN[a] AND N. J. SHAH[a,b,c]

[a]Forschungszentrum Jülich, Institute of Neuroscience and Medicine – 4, Wilhelm-Johnen-Straße 1, 52428 Jülich, Germany; [b]JARA-BRAIN – Translational Medicine, Aachen, Germany; [c]Department of Neurology, RWTH Aachen University, Pauwelsstr. 30, 52074 Aachen, Germany
*E-mail: r.buschbeck@fz-juelich.de

12.1 Introduction

In conventional magnetic resonance imaging (MRI) and positron emission tomography (PET) reconstructions (see Chapters 1 and 7, respectively), it is implicitly assumed that scan data are acquired while the imaged object lies absolutely motionless in the scanner. However, *in vivo*, this is not the case as there will always be a certain amount of motion, regardless of how compliant the subject might be. This includes voluntary body motion, *e.g.* due to discomfort during the typically long measurements, as well as (semi)-involuntary motion, *e.g.* from the subject's heartbeat, blood flow, respiration or convulsions. In general, motion issues are especially pronounced in infants, elderly subjects and patients suffering from diseases affecting the motor system, *e.g.* Parkinson's disease.

New Developments in NMR No. 19
Hybrid MR-PET Imaging: Systems, Methods and Applications
Edited by N. Jon Shah
© The Royal Society of Chemistry 2019
Published by the Royal Society of Chemistry, www.rsc.org

In the case of intra-scan motion during an MR-PET measurement, the quality of the acquired data is detrimentally affected.[1] Consequently, image artefacts can appear up to the point where the images become non-diagnostic (Figure 12.1), which is not only bothersome but also extremely costly in the clinic.[2] The appearance and the severity of the motion artefacts that arise depends on the type of motion (translational, rotational, periodic, elastic, inelastic *etc.*), the timing and the extent of subject motion, as well as the post-processing and reconstruction techniques. Additionally, in MRI, the sequence type also impacts the proneness to motion. While in PET, motion causes mainly image blurring and inaccurate quantification, *e.g.* due to the mismatch of the attenuation maps,[1] in MRI there is a large variety of possible artefacts including, but not limited to, blurring, ghosting, signal dropouts and distortions.[3]

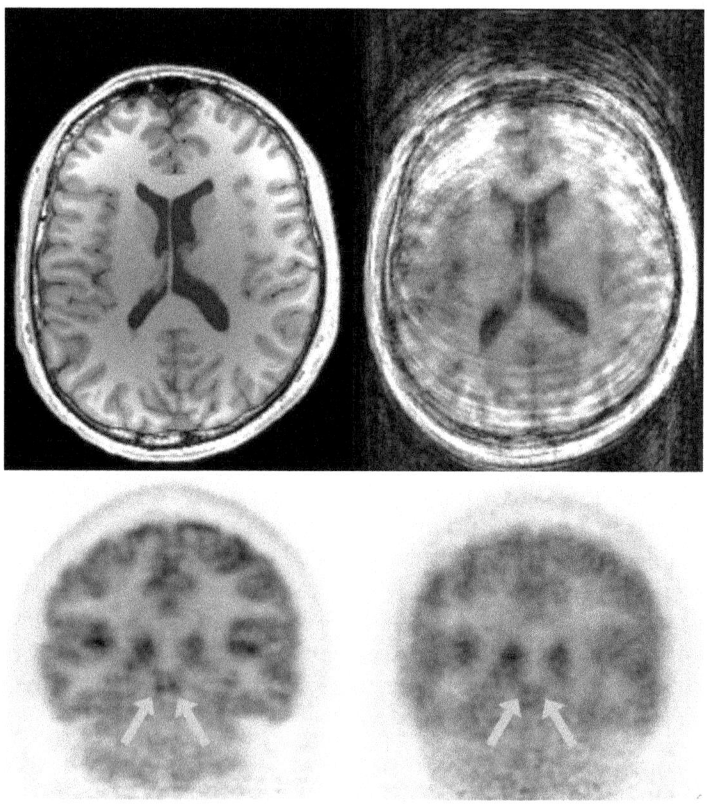

Figure 12.1 Comparison of MRI and PET images without (left) and with motion artefacts (right). Motion can cause a large variety of possible artefacts in MRI not all of which necessarily shown here. In PET the main type of motion artefact is blurring and incorrect quantification. The arrows mark the *superior colliculi* where tracer uptake is visible only in the motion corrected images.

In order to reduce or eliminate the negative effects of motion, a wide range of motion correction techniques have been developed for both MRI and PET. Most of these methods require information on the motion that happens over the time course of the scan. The motion information can be used to either correct the data acquisition itself in real time (prospective correction) or to correct the acquired data after the measurement has finished (retrospective correction). In the beginning, these techniques were only used in MRI and PET separately. However, since the introduction of hybrid MR-PET systems, synergy effects can be exploited, *e.g.* by using the same motion tracking method (see Section 12.2) for corrections in both modalities (see Section 12.3).

The importance of motion correction methods is expected to significantly increase in the future as MRI and PET strive for higher resolutions and quantitative precision. The higher the desired signal accuracy is, the more even the slightest motion can significantly degrade the image quality and call for appropriate corrections.

In the following sections, several methods for motion detection and tracking, as well as motion correction techniques for both imaging modalities, are discussed. In this chapter, we focus mainly on the estimation and correction of rigid-body motion, *i.e.* translations and rotations only, as is the most prominent type of motion in brain MR-PET. It is noted though, that especially in whole-body MR-PET, there are additional correction methods, which for instance, address elastic motion effects.[1]

12.2 Motion Detection and Tracking

In the following section, different motion detection and tracking techniques are discussed. The section is structured by the way in which the motion information is obtained: external device-based, MR-based and/or PET-based.

12.2.1 External Device-based

A popular method for measuring intra-scan motion in MRI or PET scans is with external devices that operate independently of the scanner, acquiring the motion information without making use of the imaging data. In order for a certain external device to be suited for use in MR-PET, it needs to be MR-compatible and preferably have a minimal (or at least known) absorption coefficient for 511 keV photons. In addition, it has to be sufficiently small to fit into the narrow space inside the PET detector and the MRI head coil when used for brain imaging, for example. Several motion detection and tracking methods will be listed below, not all of which have been applied to hybrid MR-PET thus far.

One of the most popular approaches for motion tracking in medical imaging is to use optical cameras. Despite current efforts to enable the filming of the subject's head directly, most techniques still require a marker, *e.g.* an encoded chequerboard, to be attached to the head.[3] The camera images are used to calculate the motion information and, using markers, very high accuracies on the order of 0.01° and 0.01 mm have been achieved.[4] In reality, these accuracies might be extremely challenging to attain due to scanner vibrations *etc.* It is important to note, that these optical systems require an unobstructed line of sight to the subject, which, due to the additional equipment in the scanner bore, might be even more challenging in hybrid MR-PET than in MRI or PET alone. This is especially the case for brain PET inserts where the space between the PET insert and the RF coil is very tight as is the space between the coil and the subject.

In MRI, subject motion has also been measured using RF-based methods.[3] Here, small additional pickup coils with very localised sensitivities are rigidly attached to the subject's head. The RF signal that is received in the presence of a magnetic gradient can be used to encode the current spatial position of the coil.

If cardiac or respiratory phases, rather than bulk motion, are of interest for gating, these can be measured with respiratory belts, pulse oximeters or ECG electrodes.[1] Note that although the cardiorespiratory motion happens mostly in the abdomen, it was shown in studies that brain imaging can also benefit significantly from gating, since abdominal motion effects propagate into the head.[5] This is especially so at ultra-high fields where the rising and falling abdominal cavity, and the associated magnetic susceptibility changes, cause particularly strong periodic changes in the strength of the magnetic field experienced in the brain.

External devices can provide very accurate motion information in (near) real time while interfering minimally, if at all, with the imaging process. However, since the devices or markers often need to be attached to the subject, the feasibility of their use in clinical environments is not without restrictions. Also, depending on the technique, cross-calibration between the device and the scanner coordinate system, as well as a synchronisation between the clocks, can be necessary for good motion correction results.

12.2.2 MR-based

In MR-based motion detection and tracking, the motion information is obtained using only the available MRI scanner hardware. Since no devices such as cameras or markers are required, the clinical feasibility of MR-based methods is considered to be superior as compared to external device-based methods.[6] However, compromises potentially have to be made in terms of accuracy, latency and MRI sequence interference.

In the following, two categories of MR-based motion detection and tracking techniques are discussed, including their respective merits and shortcomings: navigator-based and image-based motion estimation.

12.2.2.1 Navigator-based Motion Information

Motion information can be obtained from so-called navigator scans in which parts of *k*-space are repeatedly sampled. This is due to the fact that, according to the Fourier shift and rotation theorems, subject translations in real space manifest as phase shifts while rotations manifest as magnitude rotations in the *k*-space data, respectively. The number of measurable degrees of freedom in the motion parameters depends on the dimensionality of the navigator. The available trajectories in navigator scans range from 1D line navigators[7] to 2D orbital navigators[8] to 3D spherical[9,10] or cloverleaf navigators.[11] Only the 3D case allows for full rigid-body motion estimation with six degrees of freedom (three rotation axes, three translation directions). In order to calculate the motion parameters, a reference navigator, which is typically acquired at the beginning of the scan, is compared to target navigators acquired at a later time point. The translation and rotation estimates are calculated from the navigator data *via* fitting algorithms,[9] pattern matching algorithms[9] or spherical harmonic analyses.[12]

Usually, navigators are acquired as additional scans in an MRI sequence, which can prolong the total scan time. It is noted though, that very often there are dead times in MR sequences, *e.g.* in IR sequences such as MP-RAGE (see Chapter 4) that can be exploited for navigator acquisitions without affecting the overall acquisition time. However, minimal changes to the SNR of the parent sequence are unavoidable. That is, the RF pulses of the navigator scans influence the steady-state behaviour of the parent sequence (see Chapter 1), which is why they should be executed with as few excitation pulses as possible and using low flip angles. The achievable accuracies of navigators for translations and rotations have been reported to be on the order of fractions of millimetres and degrees, respectively.[9]

A special case of navigator motion estimation is performed in sequences with auto-navigation techniques where the sequence inherently provides navigator data without additional navigator scans like *e.g.* PROPELLER[13] or EPIK.[14] In PROPELLER, for instance, rectangular partitions of *k*-space rotating about the origin are sampled. This sampling scheme results in the centre of *k*-space being frequently resampled, *i.e.* oversampled, during the measurement. While the data from the *k*-space centre are later used for the reconstruction of the image, initially they can also be used to estimate subject motion either directly from the *k*-space data or from corresponding low-resolution images (see next paragraph).

FID navigators are another type of navigator, which has been recently investigated, in which the FID signal behaviour (see Chapter 1) is measured.[15] Their biggest merit is that they can be acquired a lot faster than other navigators and, hence, interfere less with the parent sequence. However, to date, FID navigators can only be used for the detection and not the quantification of motion; ongoing work suggests that quantification might also be feasible in the near future.

12.2.2.2 Image-based Motion Information

Motion information can be obtained from reconstructed images if multiple consecutive volumes (or at least parts of them) are acquired during a scanning session, as is inherently the case in fMRI or DCE-MRI.[16–19] The motion information is calculated through the registration of target images to a reference image.

If the parent sequence does not acquire multiple volumes, these so-called image navigators can also be obtained from redundant data in the parent sequence, as in PROPELLER,[13] or from dedicated scans acquiring typically low-resolution image navigators.[20] While the latter approach might increase the scan time if the image navigators cannot be acquired in waiting times of the sequence, the former approaches do not change the original measurement at all. Image navigators can provide motion information with similar accuracies as *k*-space navigators. However, if the navigator data cannot be obtained from the redundant data of the parent sequence, the typical durations of full-image navigators are significantly longer than for *k*-space navigators with corresponding dimensionality.

Compared to external device-based motion information, the temporal resolution of the MR-based information is typically lower because the information has to be calculated from data that are only available after certain time intervals. However, navigator-based methods are advantageous in that they are built into the sequence and run without operator input and are thus 'invisible' to the patient/subject. These factors have to be considered when opting for a tracking method to perform motion correction in MR-PET.

12.2.3 PET-based

The practical advantage of detecting and tracking motion using PET only is that no additional devices or data are required, which is desirable in terms of clinical feasibility. Analogous to MRI, motion information can be obtained both from PET raw data and images. In both types of methods, PET coincidence data are first binned into multiple time intervals (frames) and the motion is then measured between individual frames. The framing scheme can be regular (equidistant intervals) or irregular (unequal intervals) and is usually defined at the user's discretion. However, regardless of the chosen framing scheme, one common shortcoming of all PET-based motion tracking methods is that they cannot measure intra-frame motion, but only inter-frame motion, which lowers the temporal resolution considerably.

12.2.3.1 Image-based Motion Information

Image-based methods in PET can be used to detect or estimate motion.[21] In these methods, an image is reconstructed for each frame to form a series of images. Several algorithms can be used to estimate motion based on this series of images, most notably optical flow and registration algorithms.

Optical flow algorithms assume that the same image intensity should remain in the same voxel, otherwise there must have been motion.[22] Motion is described in this case as a velocity field, *i.e.* there is a spatial displacement in a short interval of time (between two consecutive images), which can be defined as a vector. Therefore, for each voxel in the image, a vector describes the motion to a voxel in the next image. These vectors have to follow optical flow constraints, such as continuity and smoothness.[22,23] Note that optical flow algorithms are mainly used in whole-body imaging where elastic deformations are relevant.

Registration algorithms assume that a geometrical transformation occurred between consecutive images. In this case, a reference image has to be defined and an iterative process finds the most likely transformation according to a given similarity metric. Most research regarding registration methods investigates the appropriate similarity metric and target image for a given application.[24]

A simple method for only detecting motion (without quantifying it) is defining octants in an image and calculating the sum of coincidences in each octant. If the sums change too abruptly between time frames, it is considered to be motion.[25] One of the issues in image-based methods is whether corrections such as attenuation correction, and consequently scatter correction, should be applied, as they depend on the position of the structures.[24,25] The spatial mismatch between the attenuation map, usually acquired at the beginning of the PET scan, and the emission map can introduce artefacts, especially if the movements are large, which leads to errors in motion estimation.[26] Traditionally, attenuation correction has been performed before estimating the motion, but recently, researchers have shown that not applying it gives a better motion estimation.[24,27] After the motion estimation, the attenuation correction needs to be applied with the now available motion information. Alternatively, the reconstruction and motion estimation/correction can be performed iteratively until convergence.[24,28]

12.2.3.2 Raw Data-based Motion Information

Raw data-based methods use LOR projection data which are sorted according to projection angles (see also Section 7.1) into so-called sinograms to detect motion. However, motion quantification is very limited due to the usually low count statistics. Since these techniques do not require a reconstruction beforehand, they can be less computationally intensive and, hence, faster. They are often called 'data-driven' because they do not make assumptions about the motion. One of these techniques is principal component analysis (PCA). PCA is a generic method that produces a series of orthogonal basis functions that describe the variations in the data.[25] Using the three largest principal components, motion can be successfully detected[29] PCA can be used not only in sinograms but can also be applied to images.[25] PCA-based methods only detect motion, but do not estimate it. Another crude method consists of calculating the centre of mass of sinogram data, but for low count data (low SNR) this is only reasonably accurate for axial displacements.[30]

There are some methods in between these two, *i.e.* images are calculated from the sinograms without using iterative reconstruction, but simply using a back-projection operator[25] and then image-based methods are used to estimate motion.

One of the issues in all PET-based motion estimations is how long the frames should be. Usually, frames are defined arbitrarily by the user, based on the consideration of the trade-off between image quality (SNR) and temporal resolution. Longer frames have more coincidences and correspondingly better statistics but do not take account of intra-frame motion, which is usually assumed to be minimal. Such frames provide reconstructed images with higher quality, making the motion estimation, in principle, more accurate, but with lower temporal resolution. Shorter frames have a better temporal resolution, but more noise, which can lead to erroneous motion estimates. To address the noise, strategies such as filtering, and regularisation can be used.[31] In image-based methods, shorter frames increase the total time of reconstruction, because more images have to be reconstructed. Because of this, raw data-based methods can use higher temporal resolution (~2 s frames) compared to image-based methods (~5 s frames). The feasible temporal resolutions also depend on the type of positron emitter. To take advantage of both methods, instead of defining a framing scheme arbitrarily, one can first detect motion with a high temporal resolution from sinogram data and afterwards define an appropriate framing scheme to optimally estimate motion from the image data.

In the case of more complex motion, *e.g.* elastic and cyclic cardiorespiratory movements, which is more relevant in whole-body MR-PET, a motion model is often defined. When using a motion model, MR and PET information can be both used to find the best model parameters.[32,33]

12.3 Motion Correction Techniques

The following section discusses different motion correction techniques that are available in MRI and PET. For optimal MR-PET results, the corrections need to be applied to both modalities separately before combining the respective images. Note that, when using data from different modalities, it is essential to check the temporal and spatial alignments between the systems beforehand, since the coordinate systems do not necessarily coincide, and the clocks of each system are also independent.[21]

12.3.1 MRI

If motion occurs during an MRI scan and is not corrected, the complex data in *k*-space can become inaccurate, spatially misplaced, phase-shifted or combinations thereof[34] leading to significant artefacts as shown in Figure 12.1. Motion correction techniques in MRI aim at undoing or preventing these errors and can be subdivided into two categories: retrospective and prospective approaches.

12.3.1.1 Retrospective Motion Correction

In retrospective correction techniques the motion information is used to correct flawed *k*-space data after the scan is completed, *i.e.* during reconstruction or post-processing.[3] The measurement itself is executed

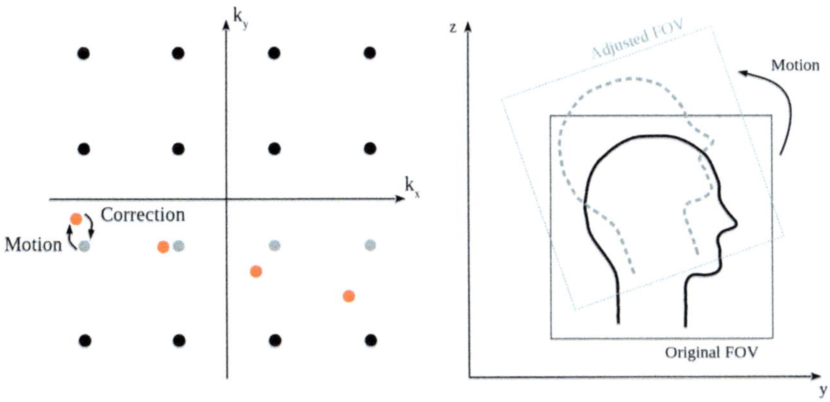

Figure 12.2 Schematic representation of the principles of retrospective (left) and prospective (right) motion correction techniques. Retrospective corrections aim at correcting image or *k*-space data after the acquisition is completed. In the diagram on the left, the black dots represent *k*-space sampling points that were correctly acquired. However, due to subject motion, some points might have been acquired at wrong positions (red) leading to an inhomogeneously filled *k*-space, which is problematic for the reconstruction. Using retrospective motion correction techniques, the intended data points on the regular grid (grey) are recovered. Prospective corrections aim at correcting the data acquisition itself during the measurement, for instance, by adjusting the FOV in case of motion as shown in the right depiction.

irrespective of any potential occurrence of subject motion during the scan. Based on the motion information, the *k*-space data are modified or regenerated in order to undo the effects caused by motion[35] as exemplified in Figure 12.2. If cardiorespiratory gating was performed, the *k*-space data might as well be retrospectively selected, regrouped, reordered and correlated based on the stored ECG or respiration data.[36] The effectiveness of the correction depends on many factors: the spatial accuracy and temporal resolution as well as the dimensionality of the motion information, the sequence details, the course of motion during the scan, coil characteristics *etc.* This often makes the optimal choice of retrospective correction techniques dependent on the specific case.

Other retrospective corrections, which do not necessarily require any motion information, are 'autofocusing'[37,38] and deep learning techniques.[39] 'Autofocusing' techniques iteratively correct flawed images through the optimisation of a cost function representing a measure for the image quality. The *k*-space raw data are deliberately altered according to a motion model until the cost function reaches a minimum. In deep learning techniques, artificial neural networks are trained to identify and correct motion artefacts.[39] In both techniques, special care needs to be taken so that no pathology is mistakenly taken for a motion artefact and removed from the image.

Retrospective motion corrections have been shown to greatly increase the quality of images from motion-affected data. However, retrospective methods can only partly correct motion artefacts, as some consequences of motion, *e.g.* through-plane motion in 2D sequences or spin history effects, cannot be corrected retrospectively. Hence, it is desirable to acquire the image data as flawlessly as possible in the first place, requiring a prospective correction, which is discussed below.

12.3.1.2 Prospective Motion Correction

In contrast to retrospective methods, in prospective motion correction, the motion information is used to correct the MRI measurement itself in real time while the scan is still running. The goal is to acquire data that are unaffected by any motion that occurs during the measurement.

For this purpose, the motion information can be used to discard and reacquire corrupted acquisitions where motion was detected.[20] Also, it can be used to constantly adjust the FOV of the measurement[40] such that the position of the head with respect to the FOV remains constant (Figure 12.2). Moreover, if gating is performed, the starting points of the MRI acquisitions can be 'triggered' such that the images are consistently acquired in a fixed position of the cardiorespiratory cycle.[36] Since motion also affects the B_0 field homogeneity, methods have been developed to perform real time shimming based on the motion information acquired.[11]

While some prospective techniques cause a prolongation of the scan time, this drawback is often significantly overcompensated by the benefit of having artefact-free images in the end. In general, prospective motion corrections can be computationally challenging, because the motion parameters have to be calculated, *e.g.* from navigators or external tracking devices, in real time with as little latency (ideally <0.1 s) and as accurately as possible. Thereafter, they have to be fed back into the scanner immediately to adjust the pulse sequence parameters.

Prospective and retrospective correction techniques can also be used jointly. In this case, the retrospective techniques try correcting the residual artefacts, which were not eradicated by the prospective correction. An extensive review of prospective motion correction in MRI can be found in ref. 40.

12.3.2 PET

Subject motion causes the detected LORs to become inconsistently attributed to spatial positions leading to significant blurring, incorrect quantification and additional noise in PET images. In this context, the importance of appropriate PET data corrections is again highlighted. As described in Chapter 7, conventional data corrections are mandatory to obtain artefact-free, quantitative images, even in the stationary case without motion. In the case of subject motion, these data corrections become motion-variant and dependent on the specific motion track of the subject during the whole measurement (see Figure 7.10). Additionally, as soon as motion-tracking parameters for the measurement are available (see Section 12.2), various correction methods allow the elimination of motion artefacts (Figure 12.1, bottom). All motion correction methods differ in terms of effort relating to the interplay with conventional data corrections and flexibility. In the simple case of rigid-body motion, *e.g.* motion of the whole head, two general approaches are feasible to appropriately compensate motion. These are the multiple acquisition frame (MAF) method[41,42] and the direct LOR-based motion compensation method.[19,21]

The well-established MAF method applies image reconstruction to subsequent frames. As the motion parameters for each frame are already known, only the attenuation map of the subject needs to be adjusted accordingly, *i.e.* the attenuation map is transformed into the matching position and then frame-dependent attenuation correction factors are calculated (see Section 7.2.1). This implies only minimal modifications of the processing workflow in comparison to a standard reconstruction without motion compensation. Normalisation, random correction and scatter correction can be applied as usual without further motion-related modifications of the workflow. In a final step, the obtained images of all reconstructed frames are transformed, according to the frame-dependent motion information, into a specific reference position. A drawback of the MAF method is that intra-frame motion is neglected. Thus, in situations involving substantial subject motion, intra-frame motion can only be minimised by splitting up frames into multiple shorter frames, which all require separate costly reconstructions.

Alternatively, methods to perform motion correction directly at the level of LORs have also been proposed.[19] Such LOR-based motion correction methods provide more flexibility with respect to the intended framing scheme because intra-frame motion can be appropriately considered at the cost of even higher computational effort. That means, for each detected position of the subject, all measured prompt events of the list mode data (see Section 5.9) have to be appropriately transformed before sorting them into the projection space domain. In contrast to the MAF method, the normalisation needs special attention in LOR-based motion correction because the spatial correlation between scanner LORs and LORs through the subject becomes time-variant and dependent on the observed motion (see Figure 7.10). Furthermore, the complete set of normalisation factors, N_j, need to be transformed and superposed accordingly for all different positions of the subject in order to obtain appropriate normalisation for all LORs. The costly calculation of effective normalisation factors is mandatory to realise an artefact-free image reconstruction.[21]

In general, the above methods allow rigid motion to be compensated and can achieve the same level of quantification as in the case without motion (Figure 12.1). However, any such motion correction method requires the acquisition of appropriate motion information with high spatial and temporal accuracy (Section 12.2). This can be challenging when using MR-based motion information since during the PET scan it is typical that several different MRI sequences are executed with small pauses in between, *e.g.* MP-RAGE, FLAIR, fMRI *etc.* All of these sequences have different characteristics in terms of the motion information that can be obtained from them, which makes it difficult to apply uninterrupted and consistent motion correction in PET. This problem has not yet been satisfactorily solved and requires further research effort. One possible option would be to acquire the same kind of navigator repeatedly throughout the different MRI sequences in order to obtain consistent motion information.

Note that, in contrast to MRI, all motion correction techniques in PET are applied retrospectively. A prospective correction is neither necessary nor possible.

References

1. C. Catana, *Semin. Nucl. Med.*, 2015, **45**, 212.
2. J. B. Andre, B. W. Bresnahan, M. Mossa-Basha, M. N. Hoff, C. P. Smith, Y. Anzai and W. A. Cohen, *J. Am. Coll. Radiol.*, 2015, **12**, 689.
3. F. Godenschweger, U. Kägebein, D. Stucht, U. Yarach, A. Sciarra, R. Yakupov, F. Lüsebrink, P. Schulze and O. Speck, *Phys. Med. Biol.*, 2016, **61**, R32.
4. J. Maclaren, B. S. R. Armstrong, R. T. Barrows, K. A. Danishad, T. Ernst, C. L. Foster, K. Gumus, M. Herbst, I. Y. Kadashevich, T. P. Kusik, Q. Li, C. Lovell-Smith, T. Prieto, P. Schulze, O. Speck, D. Stucht and M. Zaitsev, *PLoS One*, 2012, **7**, e48088.
5. S. Skare and J. L. Andersson, *Magn. Reson. Imaging*, 2001, **19**, 1125.
6. J. Maclaren, O. Speck, D. Stucht, P. Schulze, J. Hennig and M. Zaitsev, *Magn. Reson. Med.*, 2010, **63**, 162.
7. R. J. Ordidge, J. A. Helpern, Z. Qing, R. A. Knight and V. Nagesh, *Magn. Reson. Imaging*, 1994, **12**, 455.
8. Z. W. Fu, Y. Wang, R. C. Grimm, P. J. Rossman, J. P. Felmlee, S. J. Riederer and R. L. Ehman, *Magn. Reson. Med.*, 1995, **34**, 746.
9. E. B. Welch, A. Manduca, R. C. Grimm, H. A. Ward and C. R. Jack Jr, *Magn. Reson. Med.*, 2002, **47**, 32.
10. D. W. Petrie, A. F. Costa, A. Takahashi, Y.-F. Yen and M. Drangova, *Magn. Reson. Med.*, 2005, **53**, 1080.
11. A. J. Van der Kouwe, T. Benner and A. M. Dale, *Magn. Reson. Med.*, 2006, **56**, 1019.
12. A. F. Costa, Y.-F. Yen and M. Drangova, *Magn. Reson. Imaging*, 2010, **28**, 185.
13. J. G. Pipe, *Magn. Reson. Med.*, 1999, **42**, 963.
14. R. G. Nunes, P. Jezzard, T. E. J. Behrens and S. Clare, *Magn. Reson. Med.*, 2005, **53**, 1474.
15. T. Kober, J. P. Marques, R. Gruetter and G. Krueger, *Magn. Reson. Med.*, 2011, **66**, 135.
16. S. Thesen, O. Heid, E. Mueller and L. R. Schad, *Magn. Reson. Med.*, 2000, **44**, 457.

17. G. H. Glover, T.-Q. Li and D. Ress, *Magn. Reson. Med.*, 2000, **44**, 162.
18. N. White, C. Roddey, A. Shankaranarayanan, E. Han, D. Rettmann, J. Santos, J. Kuperman and A. Dale, *Magn. Reson. Med.*, 2010, **63**, 91.
19. M. G. Ullisch, J. J. Scheins, C. Weirich, E. R. Kops, A. Celik, L. Tellmann, T. Stöcker, H. Herzog and N. J. Shah, *PLoS One*, 2012, **7**, e48149.
20. M. D. Tisdall, A. T. Hess, M. Reuter, E. M. Meintjes, B. Fischl and A. J. van der Kouwe, *Magn. Reson. Med.*, 2012, **68**, 389.
21. C. Catana, T. Benner, A. van der Kouwe, L. Byars, M. Hamm, D. B. Chonde, C. J. Michel, G. El Fakhri, M. Schmand and A. G. Sorensen, *J. Nucl. Med.*, 2011, **52**, 154.
22. D. Fortun, P. Bouthemy and C. Kervrann, *Comput. Vis. Image Understanding*, 2015, **134**, 1.
23. M. Dawood, F. Buther, X. Jiang and K. P. Schafers, *IEEE Trans. Med. Imaging*, 2008, **27**, 1164.
24. N. Costes, A. Dagher, K. Larcher, A. C. Evans, D. L. Collins and A. Reilhac, *Neuroimage*, 2009, **47**, 1496.
25. P. Schleyer, J. Dunn, S. Reeves, S. Brownings, P. Marsden and K. Thielemans, *Phys. Med. Biol.*, 2015, **60**, 6441.
26. F. Sforazzini, Z. Chen, J. Baran, J. Bradley, A. Carey, N. J. Shah and G. Egan, in *Biomedical Imaging (ISBI 2017), 2017 IEEE 14th International Symposium on*, IEEE, 2017, p. 231.
27. J. M. Mukherjee, C. Lindsay, A. Mukherjee, P. Olivier, L. Shao, M. A. King and R. Licho, *Med. Phys.*, 2016, **43**, 2443.
28. J. Jiao, A. Bousse, K. Thielemans, N. Burgos, P. S. Weston, J. M. Schott, D. Atkinson, S. R. Arridge, B. F. Hutton, P. Markiewicz and S. Ourselin, *IEEE Trans. Med. Imaging*, 2017, **36**, 203.
29. K. Thielemans, P. Schleyer, J. Dunn, P. K. Marsden and R. M. Manjeshwar, in *Nuclear Science Symposium and Medical Imaging Conference (NSS/MIC), 2013 IEEE*, IEEE, 2013, p. 1.
30. P. Markiewicz, K. Thielemans, J. Schott, D. Atkinson, S. Arridge, B. Hutton and S. Ourselin, *Phys. Med. Biol.*, 2016, **61**, N322.
31. W. Bai and M. Brady, *Phys. Med. Biol.*, 2009, **54**, 2719.
32. M. Fieseler, F. Gigengack, X. Jiang and K. P. Schäfers, *Biomed. Eng. Online*, 2014, **13**, S2.
33. R. Manber, K. Thielemans, B. F. Hutton, A. Barnes, S. Ourselin, S. Arridge, C. O'Meara, S. Wan and D. Atkinson, *J. Nucl. Med.*, 2015, **56**, 890.
34. M. Zaitsev, J. Maclaren and M. Herbst, *J. Magn. Reson. Imaging*, 2015, **42**, 887.
35. R. Bammer, M. Aksoy and C. Liu, *Magn. Reson. Med.*, 2007, **57**, 90.
36. P. Lanzer, C. Barta, E. Botvinick, H. Wiesendanger, G. Modin and C. Higgins, *Radiology*, 1985, **155**, 681.
37. D. Atkinson, D. L. Hill, P. N. Stoyle, P. E. Summers and S. F. Keevil, *IEEE Trans. Med. Imaging*, 1997, **16**, 903.
38. D. Atkinson, D. L. Hill, P. N. Stoyle, P. E. Summers, S. Clare, R. Bowtell and S. F. Keevil, *Magn. Reson. Med.*, 1999, **41**, 163.

39. K. Pawar, Z. Chen, N. J. Shah and G. Egan, Abstract No. 1174 in *Proceedings of the Joint Annual Meeting ISMRM -ESMRM, Paris, France*, 2018.
40. J. Maclaren, M. Herbst, O. Speck and M. Zaitsev, *Magn. Reson. Med.*, 2013, **69**, 621.
41. Y. Picard and C. J. Thompson, *IEEE Trans. Med. Imaging*, 1997, **16**, 137.
42. R. R. Fulton, S. R. Meikle, S. Eberl, J. Pfeiffer and C. J. Constable, *IEEE Trans. Nucl. Sci.*, 2002, **49**, 116.

Section III

Special considerations in MR-PET

CHAPTER 13

MR-PET Measurement

K.-J. LANGEN[a,b], I. NEUNER[a,c,d], N. MATELA[f], E. ROTA KOPS*[a], W. WORTHOFF[a] AND N. J. SHAH[a,d,e]

[a]Forschungszentrum Jülich, Institute of Neuroscience and Medicine – 4, Wilhelm-Johnen-Straße 1, 52428 Jülich, Germany; [b]Department of Nuclear Medicine, RWTH Aachen University Hospital, Aachen, Germany; [c]Department of Psychiatry, Psychotherapy and Psychosomatics, RWTH Aachen University, Aachen, Germany; [d]JARA – BRAIN – Translational Medicine, Germany; [e]Department of Neurology, RWTH Aachen University Hospital, Aachen, Germany; [f]Facul. de Ciências, IBEB - Inst. de Biofísica e Eng. Biomédica, Univ. de Lisboa, Lisbon, Portugal
*E-mail: e.rota.kops@fz-juelich.de

13.1 Introduction

The increasing use of multimodal magnetic resonance-positron emission tomography (MR-PET) hybrid imaging raises several complex issues. These relate not only to radiation protection and dose considerations in PET, and MR safety with respect to MR imaging, but also to issues concerning combined patient handling, measurement workflow, data acquisition and logistical efforts. In addition to these practical and technological challenges, one must also bear in mind that the consideration of ethical issues can also alter when these two modalities are applied simultaneously. Furthermore, rather

New Developments in NMR No. 19
Hybrid MR-PET Imaging: Systems, Methods and Applications
Edited by N. Jon Shah
© The Royal Society of Chemistry 2019
Published by the Royal Society of Chemistry, www.rsc.org

than being confined to one specialisation, radiologists and nuclear medicine physicians are required to work in cooperation, considering the needs of both specialisms.

In this chapter, we aim to analyse these themes to highlight factors of consideration in the simultaneous use of MR-PET for brain measurements. Information relating to whole-body measurements can be found in Chapter 15.

13.2 Patient Handling and Measurement Workflows

Several aspects of a PET/CT measurement are also common to MR-PET measurements. These relate to general information about the patient, *e.g.*, medical history, chronic diseases, dorsal or neck pains and also general information concerning the measurement, *e.g.*, scan duration and patient positioning, although the latter is only valid in the case of whole-body scans. For brain measurements, the positioning of the patient within the MR-PET field-of-view (FOV) is fixed and only depends on the extent of the PET FOV within the MR FOV. PET tracer dependent constrains are also common for both the modalities of PET/CT and MR-PET. For example, in the case of a standard ^{18}F-FDG PET examination, patients must fast for approximately 4 h before scanning and their blood glucose level is taken shortly before scanning. However, when combined with an MR scan, additional information is required and is collected by the physician during the explanatory meeting with the patient. The scope of this information relates not only to metallic implants and pace makers (Section 13.5) but also to issues concerning claustrophobia and possible allergic reactions to MR contrast agents.

13.2.1 Patient Preparation

As the choice of radiotracer to be injected is directly related to the patient's suspected disease/diagnosis or the aims of the research study, the measurement requirements for different subjects are necessarily different. In neurological and neuro-oncological cases, when ^{18}F-FDG or ^{18}F-FET is applied, the patient is required to fast for around 4 h before the measurement. In addition, in the case of ^{18}F-FDG, the blood glucose level should be in the standard range (130 mg dL^{-1}). During the medical consultation, the information sheet concerning MR and PET measurements is discussed, completed, and finally signed by the patient. After a venous access is established, the patient is injected with an amount of radiotracer calculated ad hoc, depending on their weight. In a clinical environment, the measurement usually begins 20–30 min following injection of the tracer and the patient is required to wait for some time in a quiet and often darkened room. Before the patient is positioned on the scanner bed, all metal objects are removed from their person and placed in a box (Section 13.5). During a brain scan, the patient is positioned on their back on the scanner bed with their shoulders tight to the head coil. Earplugs, rather than earphones, are used, as earphones are not PET compatible due to attenuation influence. Small pillows, left and right of the head, serve to avoid head movements. Before the upper part of the head coil is closed, a laser marker is projected at the bridge of the nose, fixing the head position. Finally, the patient is given a bell in their hand and is instructed to press it in the eventuality that they feel discomfort. The patient is then moved into the bore.

In some centres, where clinicians and researchers are working together, the PET scan may start simultaneously with the injection of the radiotracer. In this way, radiotracer uptake can be followed over the whole time and special data and image analyses can be performed. However, starting a PET scan simultaneously with an MR scan is only possible if an RF lock between the MR scanner room and outside is available.

13.2.2 Image Acquisition and Protocols

Once the patient is lying in the bore, PET and MR acquisition can begin. The PET acquisition times, as well as the required MR sequences, are dependent on the radiotracer injected and the medical/research requirements. Although each centre is free to decide on its own acquisition time and sequences based on previous literature or research aims, official guidelines and standards are available for PET (*e.g.*, for Germany: https://www.awmf.org/leitlinien/aktuelle-leitlinien/ll-liste/deutsche-gesellschaft-fuer-nuklearmedizin-e-v.html; Boellaard *et al.*;[1] for the USA: http://www.snmmi.org/ClinicalPractice/). The standards help clinicians choose the best procedure for obtaining a high-quality examination. Furthermore, by following the official guidelines, levels of reproducibility are improved, increasing the applicability of clinical research among multiple institutions. Also, the exchange of results and experiences among clinicians working in different centres becomes possible. While the patient is in the bore, PET acquisition only runs in list mode, allowing any frame scaling to be chosen by sorting the single detected coincidences with a user-defined protocol. Following completion of the measurement, the PET data are sorted and reconstructed according to the given protocol, while the MR images are, in general, already available. After reconstruction of the PET data, co-registration with the MR images enables further analysis and image processing.

13.3 Ethical Issues

13.3.1 Introduction

In general, scientific studies in humans using MRI-PET are subject to review and approval by an official ethics committee and all studies have to be conducted according to the principles of the Helsinki Declaration. The scientific use of radiopharmaceuticals for PET investigations in humans requires additional approval by government agencies in most countries. This concerns not only protection against the harmful effects of ionising radiation but also the general regulations for the use of pharmaceutical products in humans, which vary from country to country. Another ethical aspect in biomedical research is data protection. In particular, this relates to the prevention of the misuse of personal data belonging to the patients and volunteers participating in scientific studies. Furthermore, incidental findings can occur in scientific investigations, which may have a lasting effect on the patient's life and need to be handled with care.

13.3.2 Risks of Magnetic Fields

The greatest risk potential in MRI comes from metal parts in the body (*e.g.* implants) that could move or heat-up during the MRI scan. Consequently, patients or volunteers are clearly informed of these risks and are questioned about all the usual forms of implants. Despite this risk, and considering other appropriate exclusion criteria according to our current knowledge, MRI is harmless. This is also the case in high-field MRI where over 10 000 measurements worldwide (7 T and above) have been taken with no serious side effects reported.[2–4] Minor side effects of high-field MRI may include slight dizziness when entering the magnet at 7 T MRI. In a small percentage of people, other mild subjective discomfort such as the appearance of flashes of light, unsteady gait after magnetic field exposure, flushing or drowsiness may occur. Furthermore, the patient may experience abnormal sensations during the measurement due to the activation of peripheral nerve fibres, *e.g.* a tingling in the hands. However, these effects are harmless to health. Magnetic ingredients of tattoos, however, may cause skin burns during MRI scans. Tattoo inks used in Germany since the 1990s usually contain few to no magnetic ingredients and are suitable for MRI. If the situation is unclear, tattoos can be an exclusion criterion for MRI. At present, there is no known risk of MR brain scanning of a pregnant woman to the developing foetus for scanning at 4 T or less. However, pregnancy generally precludes the use of radioactive substances with PET.

13.3.3 Radiation Issues

The use of radioactive tracers and ionising radiation may increase the risk of cancer. Therefore, in all investigations involving the exposure of humans to radiation, a careful assessment of the radiation dose to the patient should be carried out. The associated risk should then be weighed against the benefit for the patient, in the case of a diagnostic scan, or for society, in the case of a scientific study. Consideration should be given to the age of the subjects proposed for the investigation. In particular, persons under 18 years of age should not be involved except when problems specific to their age groups are under investigation. Whenever possible, volunteers should be aged over 50 years. The numbers of individuals participating in a research project should be restricted to the minimum necessary to obtain the information required. Consideration should be given to the risks to an individual who is involved in a series of research investigations and steps should be taken to prevent individuals repeatedly taking part in research projects, thus leading to a substantial cumulated radiation dose. The possibility of early pregnancy should always be borne in mind if women of childbearing age must be used as subjects. Furthermore, pregnant women and mothers who are breastfeeding must not be involved in any project except where problems specific to their health are under investigation and alternative techniques, which do not involve ionising radiation, have been considered and rejected. In addition, when employees who work with radiation are considered as possible volunteers for research, the researcher must ensure that, as volunteers, they are aware of the additional risk arising from their exposure to radiation at work. As applicable to any subject participating in a study, the legitimate right of the employee to refuse participation in the research project must be respected. For research projects involving the use of ionising radiation, the International Commission on Radiological Protection (ICRP) and the World Health Organization (WHO) recommend the use of categories of risk arranged according to the radiation dose and estimated level of risk to be received by the subject. The risk may then be evaluated with regard to the expected benefit.

13.3.4 Data Protection

In the European Union, the General Data Protection Regulation (GDPR) has replaced previous legislation in 2018.[5] The GDPR law has created a number of core rights for the protection of individual data that also concerns biomedical research. Important aspects are the written informed consent of the subject, information about data transfer, processing and access to personal data and other supplementary information. Furthermore, subjects have the right to have their personal data rectified if their data is incorrect or incomplete and to have their personal data erased (unless in special circumstances). All data must be anonymised, stored for 10 years and then deleted.

13.3.5 Incidental Findings

During the course of a research study, medical information concerning a participant, but not relating to the principle reason for the study, may also be discovered. Since no patient–physician relationship is established in a research study, the recording and analysis of data is solely based on scientific criteria and does not aim at an individual clinical diagnosis. The study subject must be informed of this incidental finding and give consent in advance to receiving such information about incidental findings in order to preclude ethical and legal conflicts. The researcher has the obligation not to inflict any harm on a subject, in respect of their communication.

13.4 Radiation Protection and Dose Considerations

In addition to the aforementioned diagnostic benefits associated with MR-PET, MR-PET also presents a further key benefit when compared to PET/CT: a significant reduction in the amount of radiation exposure to the patient. This reduction occurs mainly as a result of replacing the CT component of the modality with MR to provide the source of anatomical information. As a consequence of the As Low As Reasonably Achievable (ALARA) principle, the reduction of radiation exposure should be a permanent goal in all radiological examinations but is of paramount importance for paediatric patients and for patients requiring repeated radiological examinations. The possibility of reducing doses to these patients is contributing to the clinical adoption of MR-PET as a standard diagnostic tool.

In the last 10 years, PET/CT has revolutionised medical diagnosis[6] and has become the first choice of metabolic imaging modality for applications such as tumour staging, radiotherapy, or surgery planning. However, radiation exposure has always been a concern. In a PET/CT examination with [^{18}F]-fluoro-deoxy-glucose (FDG), the effective dose can range between 8.0 mSv and 26.85 mSv, depending on the scanner and CT protocol. The contribution of the CT component is highly dependent on the CT protocol. It ranges from 32% with a low dose scan to 79% of the total effective dose with a standard dose scan.[7] The CT contribution is also highly dependent on the body part being imaged. In brain studies, CT only contributes 11% of the total effective dose, while in whole-body examinations, CT contributes 69% of the total effective dose.[8] These figures help us understand how important replacing PET/CT with MR-PET could be, whenever the diagnosis is not compromised by this substitution.

Many studies have been presented recently showing that, even if MR-PET produces only equivalent results from the diagnosis point of view when compared with PET/CT, there is a clear benefit from the radiological protection perspective. Schäfer *et al.*[9] performed PET/CT and MR-PET examinations in 18 paediatric patients and assessed lesion detectability with separate readings of both data. They concluded that MR-PET showed rates of lesion detection similar to those obtained with PET/CT (MR-PET detected 61 of the 62 lesions that were detected with PET/CT). The estimated mean effective radiation dose for these patients resulting from the PET/CT examinations was 23 mSv ± 11 with CT accounting for 73% of this value, which corresponds to the dose reduction achievable with the replacement of CT by MR. However, the authors stress that these results were obtained considering a contrast-enhanced diagnostic CT protocol. If the CT was required only for attenuation correction purposes, it would be possible to perform only a low dose CT protocol and the impact of replacing CT with MR would be lower.

Subsequently, the same group presented a new study[10] with younger patients (aged between 1- and 6-years-old). PET/CT and MR-PET detected the same 21 lesions and MR detected one PET-negative lesion that was not detected by CT. Since the CT protocols were weight-adjusted, the contribution to the total effective radiation dose was lower than the one presented in the previous study. However, replacing CT with MR would still result in a reduction of 48% of the effective dose (4.4 mSv from CT in a total of 9.3 mSv).

Cancer staging in pregnant women is commonly done using PET/CT and in these situations, foetal radiation exposure is always a concern. Zanotti-Fregonara *et al.*[11] evaluated foetal radiation dose from ^{18}F-FDG in PET/CT and MR-PET examinations. MR-PET allowed a more accurate region drawing and the delineation of foetal contours, resulting in a lower estimation of the foetal radiation dose from the ^{18}F-FDG. The sample size of the study is very small, but the authors refer a possible reduction of 6–14 mGy in the foetal dose, resulting only from replacing CT by MR, based on previous studies.[12]

As mentioned before, dose reduction resulting from a change to MR-PET can also be critical to patients with conditions that require periodic medical imaging examinations, resulting in high levels of cumulative doses. One of those conditions is lymphoma. Despite being a study with only nine patients, Platzek *et al.*[13] showed that MR-PET could be an important imaging tool for therapy response evaluation in these patients. The authors stressed the possibility of reducing 75% the cumulative effective dose which, based on the study of Nievelstein *et al.*,[14] can reach 97 mSv in 2.5 years in patients with non-Hodgkin's lymphoma, who are regularly monitored with PET/CT.

Another field that has observed an increased interest in PET/CT imaging is the diagnosis of cardiovascular diseases.[15,16] However, its clinical application is still hindered by the radiation dose from PET and CT. MR-PET can, therefore, increase the usage of PET in this domain due to the reduction in the total effective dose. Robson *et al.*[17] presented an MR-PET protocol with ^{18}F-NaF, with an average effective dose of 6.6 mSv and a protocol with ^{18}F-FDG, with an average effective dose of 7.8 mSv. These results represent a reduction of the effective dose by approximately 50% when compared to recent studies presented with PET/CT.[16,18] The authors also consider the possibility of using this dose reduction to open up the possibility of multi-time point studies that would allow the evaluation of disease progression or the monitoring of therapeutic intervention.

Radiation dose resulting from radiopharmaceuticals injected in an MR-PET examination is quantified in the same way as quantified in any other PET exam. Dose evaluation in PET procedures is usually performed by non-patient specific estimates calculated using the medical internal radiation dose (MIRD) method.[19] This method relies on the principle that the total absorbed dose in a certain target organ (D_t) can be calculated as the sum of the contributions from all source organs, which are calculated as the product of time integrated activity for the source organ ($\tilde{A}_s$) and an S_{ts} factor corresponding to the absorbed dose in a target organ per unit cumulated activity in the source organ (eqn (13.1))

$$D_t = \sum_s \tilde{A}_s S_{ts} \tag{13.1}$$

$\tilde{A}_s$ values may be obtained by biokinetic models developed by the ICRP, while S factors are usually calculated by Monte Carlo simulations.

As we have seen in the examples presented above, the comparison of the relative risk of non-uniform exposures is done not with absorbed dose but with effective dose, which can be estimated directly from injected activity following the method proposed by ICRP.[20]

Radiation protection is an issue in an MR-PET facility, as it is in a PET alone or PET/CT facility. From a radiation protection point-of-view, the walls, doors and windows must also be shielded with lead in an MR-PET room.[21] Occupational exposure limits, dose limits to the public and duties of personal monitoring does not differ from a stand-alone PET or PET/CT unit.

13.5 MR Safety

Although MRI is not based on the use of any ionising radiation, as is the case with other modalities such as X-ray projections, CT, SPECT, or PET, it carries its own significant safety considerations. These safety considerations relate to the use of a strong static magnetic field, switched magnetic-field gradients and radiofrequency fields for the imaging process. Each element has distinctive safety-relevant considerations, safety-relevant side effects and potential risks to consider. Every MR site has technological and procedural approaches in place to mitigate any risk to subjects, operators, and the public. In general, the system manufacturers and the MR sites implement the current version of the IEC 60601-2-33 [22] and national safety regulations. However, depending on the examination, one must also consider other additional risks arising from the use of contrast agents or PET tracers (Section 13.4).

13.5.1 Static Magnetic Field

13.5.1.1 Projectile Effect and Magnetic Objects

It is of paramount importance that anybody entering the magnet room, be that patients, operators, or service and maintenance staff, is aware that, once charged, a superconducting magnet is permanently active and will be surrounded by an inhomogeneous magnetic field of significant extent. This will exert a force on any para- or ferro-magnetic object brought into its vicinity. Any object attracted by the magnet can relocate and might experience a dramatic acceleration towards the magnet, commonly referred to as the projectile effect.[23] Such an object may inflict severe injuries to any person in its path and cause significant damage in the scanner room. Consequently, utmost care needs to be taken that any person entering the magnet room is clear of para- and ferro-magnetic objects, which might become attracted to the magnet. Care must also be taken if devices and appliances are brought near to the magnet. Devices tested as MR-safe can be safely used in the magnet room, taking into consideration the boundary conditions given in the system manual.

Modern magnets are often equipped with active shielding, which reduces the stray field of the magnet *via* additional coils. The intention of this feature is to reduce the space and costs requirements for the installation of the MR scanner at the cost of increasing the field gradient present outside the magnetic bore. From a safety point-of-view, this implies that the force experienced by an attractable object is particularly high and varies rapidly with spatial location, making the risk of losing grip on such an object very high.

13.5.1.2 Cryogens and Quenching

Most modern MR scanners use superconducting magnets to generate their static field. A significant amount of energy is stored in the perpetual current flowing in the coil, submerged in a cryogen vessel. The most common cryogens are liquid helium and, for very old magnets, nitrogen. Any contact with the cryogen can cause frostbite as well as asphyxiation and suffocation if inhaled.

It can happen that the superconductivity of the magnet breaks down either spontaneously, or manually induced by the operator. This process is commonly referred to as a quench. It causes the energy stored in the magnet to be rapidly released to the cryogen, resulting in a sudden rise in temperature of the liquid and subsequent evaporation. During this transformation the cryogen expands by approximately 700-fold at 0 °C, thus potentially flooding the magnet room, leading to increased air pressure inside the magnet room and suffocation. Under normal circumstances, the escape of the cryogenic gas is not a concern as a quench pipe, which takes the gas out of the magnet into free air outside the building, is a permanently attached to the system. The magnet and quench pipe are separated from each other by a so-called 'burst disk'—a thin rubber membrane—that ruptures in the event of a quench.

13.5.1.3 Physiological Effects

Historically, a multitude of animal studies and experiments on cell cultures have been carried out to evaluate the influence of high magnetic fields on organisms, cell growth and metabolism. None of which showed any deleterious long-term effects. To date, there have been no reports in the literature indicating cumulative biological effects on humans or animals due to the exposure of strong static magnetic fields and to our current knowledge, harmful long-term effects are highly unlikely. However, physiological and sensory short-term effects due to the static magnetic field may occasionally occur, although these are considered harmless and diminish shortly after exposure.[22]

13.5.2 Radiofrequency Fields

13.5.2.1 RF Heating and Specific Absorption Rate

The use of radiofrequency (RF) fields can cause heating.[24] Temperature increases in a subject caused by RF exposure can lead to local or systemic overheating and consequently to tissue damage. All commercially available systems monitor the amount of RF energy delivered to the subject and prevent the excessive deposition of RF energy. A common measure of RF heating induced by the imaging process is the so-called specific absorption rate (SAR), defined as RF power absorbed per unit mass of the subject. The limitation of SAR aims to ensure that the temperature inside the subject is kept below a safe threshold (usually 1 °C in skin temperature)—limits are provided by the International Electrotechnical Commission (IEC).[22] SAR is dependent on the subject and on the homogeneity of the RF field inside the subject. When a higher amount of locally concentrated RF energy is deposited, it is known as local SAR. In general SAR increases primarily with increasing magnetic field, which is coupled with increasing RF; sequences using many RF pulses, high flip angles and fast repetition rates also have high SAR as, for example, inversion recovery, spin echo and some steady state sequences. If combined with parallel transmission capabilities, ultra-high field systems, particularly with increased excitation frequency, require a more sophisticated treatment of the RF energy deposition by means of careful experimentation and numeric modelling to prevent the formation of local hotspots and to prevent injury to the subject.[25–28]

Extra care must be taken if conductive material is present inside or close to the subject as this material can act as an antenna and cause increased RF energy absorption.[29] Examples of these are: retainer wires, body art, permanent makeup containing conductive materials and the badly placed wires of devices located in the proximity of the subject.

13.5.3 Gradient Fields

13.5.3.1 Peripheral Nerve Stimulation

The switching of magnetic-field gradients can cause electrical potentials and induce electrical currents inside the subject.[30] These manifest in heating as well as signal alterations in the nervous system, *i.e.* peripheral nerve stimulation. In mild cases, this effect can be identified as a tolerable tingling sensation and possibly involuntary muscle twitching. In principle, extremely fast switching gradient fields can induce cramps or cardiac malfunction and pose the potential risk of severe injury or death. Commercial MR scanners are equipped with precautionary measures to limit gradient maximum amplitude and slew rates. In particular, sequences that employ fast switching of strong, monopolar gradients, such as in diffusion measurements, for example, are prone to causing peripheral nerve stimulation. The limits of slew rates for gradient fields are provided by the IEC.[22]

13.5.3.2 Acoustic Noise

In addition to the electrical potential and electrical current associated with gradient switching, the switching of the gradients also causes the entire assembly in the static field to vibrate, causing an effect similar to that of a speaker system. This noise can exceed the threshold found acceptable by the IEC (99 dB)[22] making appropriate hearing protection for the subject/patient a requirement. Often this is done in a two-stage system, with the subject being required to use earplugs and earmuffs.

13.5.3.3 Contraindications and Implanted Medical Devices

Considering the aforementioned risk factors of MR scanners, the utmost care needs to be taken when dealing with a subject's medical condition, implanted devices and prosthetics. Before admitting any subject to the scanner area, careful consideration of the subject's medical history and the possible presence of potentially unsafe implanted devices is crucial.[31] Typically, clips, stents, cardiac pacemakers and neurostimulating devices require particular attention and information on safety and effects of the MR scanner on such devices should be obtained from the manufacturer. The manufacturer will state exactly the conditions under which the device is considered safe.

Acknowledgement

We would like to thank Mrs. Schachoff (TUM, Munich) for her help describing the clinical workflow.

References

1. R. Boellaard, R. Delgado-Bolton, W. J. Oyen, F. Giammarile, K. Tatsch, W. Eschner, F. J. Verzijlbergen, S. F. Barrington, L. C. Pike and W. A. Weber, *Eur. J. Nucl. Med. Mol. Imaging*, 2015, **42**, 328.
2. Y. Noureddine, A. K. Bitz, M. E. Ladd, M. Thürling, S. C. Ladd, G. Schaefers and O. Kraff, *MAGMA*, 2015, **28**, 577.
3. J. Rauschenberg, A. M. Nagel, S. C. Ladd, J. M. Theysohn, M. E. Ladd, H. E. Möller, R. Trampel, R. Turner, R. Pohmann and K. Scheffler, *Invest. Radiol.*, 2014, **49**, 249.
4. I. Uwano, T. Metoki, F. Sendai, R. Yoshida, K. Kudo, F. Yamashita, S. Higuchi, K. Ito, T. Harada and J. Goodwin, *Magn. Reson. Med. Sci.*, 2015, **14**, 35.
5. M. Cornock, *Maturitas*, 2018, **111**, A1.
6. D. W. Townsend, *J. Nucl. Med.*, 2008, **49**, 938.
7. B. Adeleye and N. Chetty, *Radioprotection*, 2018, **53**, 107.
8. J. M. Martí-Climent, E. Prieto, V. Morán, L. Sancho, M. Rodríguez-Fraile, J. Arbizu, M. J. García-Velloso and J. A. Richter, *EJNMMI Res.*, 2017, **7**, 37.

9. J. F. Schäfer, S. Gatidis, H. Schmidt, B. Gückel, I. Bezrukov, C. A. Pfannenberg, M. Reimold, M. Ebinger, J. Fuchs and C. D. Claussen, *Radiology*, 2014, **273**, 220.
10. S. Gatidis, H. Schmidt, B. Gücke, I. Bezrukov, G. Seitz, M. Ebinger, M. Reimold, C. A. Pfannenberg, K. Nikolaou and N. F. Schwenzer, *Invest. Radiol.*, 2016, **51**, 7.
11. P. Zanotti-Fregonara, R. Laforest and J. W. Wallis, *J. Nucl. Med.*, 2015, **56**, 1218.
12. A. M. Takalkar, A. Khandelwal, S. Lokitz, D. L. Lilien and M. G. Stabin, *J. Nucl. Med.*, 2011, **52**, 1035.
13. I. Platzek, B. Beuthien-Baumann, J. Langner, M. Popp, G. Schramm, R. Ordemann, M. Laniado, J. Kotzerke and J. van den Hoff, *Magn. Reson. Mater. Phys., Biol. Med.*, 2013, **26**, 49.
14. R. Nievelstein, H. Q. van Ufford, T. Kwee, M. Bierings, I. Ludwig, F. Beek, J. de Klerk, W. T. M. Mali, P. de Bruin and J. Geleijns, *Eur. Radiol.*, 2012, **22**, 1946.
15. A. Irkle, A. T. Vesey, D. Y. Lewis, J. N. Skepper, J. L. Bird, M. R. Dweck, F. R. Joshi, F. A. Gallagher, E. A. Warburton and M. R. Bennett, *Nat. Commun.*, 2015, **6**, 7495.
16. N. V. Joshi, A. T. Vesey, M. C. Williams, A. S. Shah, P. A. Calvert, F. H. Craighead, S. E. Yeoh, W. Wallace, D. Salter and A. M. Fletcher, *Lancet*, 2014, **383**, 705.
17. P. M. Robson, M. R. Dweck, M. G. Trivieri, R. Abgral, N. A. Karakatsanis, J. Contreras, U. Gidwani, J. P. Narula, V. Fuster and J. C. Kovacic, *JACC: Cardiovasc. Imaging*, 2017, 2154.
18. M. Rubeaux, N. V. Joshi, M. R. Dweck, A. Fletcher, M. Motwani, L. E. Thomson, G. Germano, D. Dey, D. Li and D. S. Berman, *J. Nucl. Med.*, 2016, **57**, 54.
19. R. Loevinger, T. F. Budinger and E. E. Watson, *MIRD Primer for Absorbed Dose Calculations*, Society of Nuclear Medicine, 1988.
20. J. Valentin, *Ann. ICRP*, 1998, **28**, 1.
21. N. Parikh, K. P. Friedman, S. N. Shah and H. Chandarana, *Abdom. Organ Imaging*, 2015, **40**, 1366.
22. International Electrotechnical Commission, *IEC 60601-2-33*, 1995 revised 2010.
23. Medical Devices Agency, Device Bulletin MDA SB 9803, 1998.
24. F. G. Shellock, *J. Magn. Reson. Imaging*, 2000, **12**, 30.
25. S. Wolf, D. Diehl, M. Gebhardt, J. Mallow and O. Speck, *Magn. Reson. Med.*, 2013, **69**, 1157.
26. X. Zhang, P. F. Van de Moortele, J. Liu, S. Schmitter and B. He, *Appl. Phys. Lett.*, 2014, **105**, 244101.
27. D. Shrivastava, T. Hanson, J. Kulesa, L. DelaBarre, P. Iaizzo and J. T. Vaughan, *Magn. Reson. Med.*, 2009, **62**, 888.
28. E. Neufeld, M. C. Gosselin, M. Murbach, A. Christ, E. Cabot and N. Kuster, *Phys. Med. Biol.*, 2011, **56**, 4649.

29. M. F. Dempsey, B. Condon and D. M. Hadley, *J. Magn. Reson. Imaging*, 2001, **13**, 627.
30. C. Ham, J. Engels, G. Van de Wiel and A. Machielsen, *J. Magn. Reson. Imaging*, 1997, **7**, 933.
31. F. G. Shellock, *Magnetic Resonance Procedures: Health Effects and Safety*, CRC Press, 2000.

CHAPTER 14

Parametric Imaging

L. CALDEIRA*[a], H. HERZOG[a], H. A. FERREIRA[b] AND N. J. SHAH[a,c,d]

[a]Forschungszentrum Jülich, Institute of Neuroscience and Medicine – 4, Wilhelm-Johnen-Straße 1, 52428 Jülich, Germany; [b]Institute of Biophysics and Biomedical Engineering, Faculty of Sciences of the University of Lisbon, Campo Grande, 1749-016 Lisboa, Portugal; [c]JARA - BRAIN - Translational Medicine, Aachen, Germany; [d]Department of Neurology, RWTH Aachen University, Aachen, Germany
*E-mail: l.caldeira@fz-juelich.de

14.1 Introduction

Positron emission tomography (PET) images of the whole body, or of a specific organ of interest, primarily show the distribution of the radiolabelled molecules. As the uptake of these molecules is related to certain physiological and metabolic functions, PET directly delivers a qualitative image of this function. For example, the cerebral uptake of [^{18}F]-fluoro-deoxyglucose (^{18}F-FDG) corresponds to the cerebral consumption of glucose. In this way, PET images are able to show an increase in the auditory cortex following acoustic stimulation[1] or a diminished consumption of glucose in the temporo-parietal cortex, which is caused by Alzheimer's disease and leads to typical neuronal damage in this region.[2]

In a similar way, the tissue contrasts seen in native magnetic resonance (MR) images acquired with T_1- or T_2-weighted sequences, for example, are

New Developments in NMR No. 19
Hybrid MR-PET Imaging: Systems, Methods and Applications
Edited by N. Jon Shah
© The Royal Society of Chemistry 2019
Published by the Royal Society of Chemistry, www.rsc.org

able to indicate changes typically associated with disease. Thus, magnetic resonance imaging (MRI) is usually employed to diagnose brain tumours[3] and is also used to detect the neuronal loss which leads to cortical atrophy and is associated with Alzheimer's disease.[4]

In addition to this, the information assessed by PET and MRI can be exploited beyond the primary native images. For example, it is possible to extract additional, more detailed, information relating to physiological and metabolic functions, and corresponding outcome parameters that are representative for these functions. These parameters can be presented on a voxel-by-voxel basis in so-called parametric images.

This chapter presents a comparative overview of parametric imaging in both PET and MRI. Furthermore, the novel use of MR-PET for multiparametric imaging, combining the advantages of the two modalities, is also discussed. Finally, the topic of connectivity is described with regard to the inter-relationship of parameters observed in different brain regions.

14.2 Parametric Imaging in PET

At its most simple, the basic concept of PET imaging is to obtain a static image of an injected tracer over a specific interval of time. Nevertheless, even with such static imaging, useful parametric information can be obtained, and as described in Chapter 8, a static image of cerebral FDG uptake can be converted using eqn (8.13) into a quantified parametric image of the rate of glucose consumption.[5]

However, although static images can provide useful information, in most cases, parametric images are based on the evaluation of dynamic PET imaging, which contains temporal data about the uptake and outwash of a radiotracer. In the early days of dynamic PET imaging, parameters were extracted from time–activity curves (TACs) defined over regions-of-interest (ROIs), so that the data relating to the voxels inside the ROI were averaged. With improvements in PET image quality and the development of appropriate algorithms, it has now become possible to obtain parametric images describing metabolic functions in a voxel-wise manner (see Chapter 8, Section 8.6).

14.2.1 Dynamic PET

In dynamic PET, a PET scan acquires data with high temporal precision (list-mode acquisition) and the data is then subsequently framed, that is to say, data acquired in a certain time interval are grouped together. In addition to the prompts, *i.e.* all measured data, it is also necessary to estimate corrections, such as random and scatter coincidence corrections, for each time frame. Random correction using the delayed time window (and variance reduction) can be applied to short frames. However, it can be challenging to estimate scatter in short frames. Attenuation correction is usually considered to be the same for all time frames, and only if there are large variations in anatomy or large movements during the acquisition time does this have to be taken into consideration. The reconstruction of the dynamic PET data is performed independently for each time frame. Once the dynamic PET images are reconstructed, the activity of each voxel can be obtained from frame to frame, resulting in a TAC of each voxel.

14.2.2 PET Parameters

In the PET literature, the most reported parameters relate to the local cerebral rate of glucose consumption (LCMRglc), the binding potential (BP) of neuroreceptors and the cerebral blood flow (CBF). In order to obtain physiological parameters, it is necessary to apply kinetic models to the activity distribution over time. Kinetic models usually require two different TACs: (1) The input function (IF), which is the time course of the radiotracer delivered to the tissue of interest (sometimes referred to as blood curve); and (2) a TAC of each voxel, which is the variation in radiotracer within the tissue over time.

However, the estimation of kinetic model parameters, which are used to calculate physiological parameters, can present difficulties due to the statistical noise contained in both IF (see next section) and TAC. One way to circumvent this problem is to use simplified or linearised kinetic models that are usually more stable for parametric imaging. The different kinetic model options are described in detail in Chapter 8.

14.2.3 Challenges

14.2.3.1 Reconstruction

In order to obtain a good time resolution in the TAC, the dynamic PET data have to be framed in short-time intervals. However, this results in PET images with high noise levels. This is a consequence of the limited statistics as well as the reconstruction methods, which usually do not sufficiently regularise the noise, applied to obtain these images. Commonly, the single frames of dynamic PET data are reconstructed independently of each other. Alternatives to such standard reconstruction methods may be used to reduce the level of noise and improve the quantification accuracy of dynamic PET data.[6] Most alternative approaches assume that the reconstruction of individual dynamic frames is not independent. These alternative methods have a benefit in terms of noise and bias[6] and parametric image reconstruction (PIR) is currently the most widely accepted method.[7] Although the PIR method is promising, it is difficult to extend from linear kinetic models to complex kinetic models. The PIR method combines kinetic modelling and emission image reconstruction into a single formula. However, when using complex kinetic models, the kinetic parameters are non-linearly involved in the reconstruction formulation.

14.2.3.2 *Input Function*

The IF is the time course of the radioactivity delivered *via* the arterial vessels to the organ. Cannulation of the radial or brachial artery is the most reliable method, but this is an invasive and laborious procedure, discouraging clinical research in brain PET studies. As an alternative, an image-derived IF (IDIF) can be used to obtain the IF using the images obtained with PET. However, this is still a topic of active research. More details on this can be found in Chapter 11, Section 11.4.

14.3 Parametric Imaging in MRI

In addition to the structural images obtained using MRI, different parametric maps, such as cerebral blood flow (CBF) or permeability (parameter used for brain tumours), can also be realised. In MRI, parametric maps can be obtained with techniques such as arterial spin labelling (ASL), dynamic contrast enhanced (DCE) MRI, and dynamic susceptibility contrast (DSC) MRI. These are so-called dynamic techniques. However, there is also interest in obtaining tissue parameters such as T_1, T_2 and T_2^* relaxation parameters (see Chapter 3) as well as advanced parameters *e.g.* from diffusion imaging, water and fat mapping (from T_1/T_2^* multi-echo sequences), iron quantification (from susceptibility-weighted imaging) and velocity-encoded mapping (from phase-contrast sequences), among others.

14.3.1 Static Parameters

Static parameters are parameters that do not change with time, that is to say, parameters that are unchanged during an interval of time. This encompasses tissue parameters such as relaxation times T_1, T_2, T_2^* (and their relaxivity counterparts R_1, R_2 and R_2^*), water content and magnetisation transfer (MT); and also advanced parameters such as diffusion parameters, chemical exchange saturation transfer (CEST), pH and susceptibility.

Tissue parameters are intrinsic to each tissue and specific sequences are used to estimate these values. However, the mapping of such parameters is not usually done routinely, as they can require long acquisition times and processing times. Changes in T_1, T_2 and T_2^* are associated with several diseases (*e.g.* multiple sclerosis and hepatic encephalopathy); water content can provide information on oedema or treatment response; MT gives information about bound protons, which are able to provide more interesting information than free protons, as reduced values are found in pathologies such as demyelination.

Diffusion constants in biological tissues can be measured by using the appropriate sequences (see Chapter 3). The measured diffusion constants are usually represented by the apparent diffusion coefficient (ADC). An area of acute infarction is seen as being bright on a diffusion-weighted image (reduced mobility of the water molecules) and will appear dark on the corresponding ADC map (smaller diffusion constant). It is possible to obtain more detailed information from diffusion tensor imaging (DTI). Parameters such as mean diffusivity (MD) and fractional anisotropy (FA) can also be obtained, as well as axial (AD) and radial (RD) diffusivities. Other diffusion-related parameters can be obtained depending on the diffusion model used, such as mean (MK), axial (AK) and radial (RK) kurtosis derived from diffusion kurtosis imaging (DKI)[8,9] or from tractography reconstructions such as mean fibre length. More information can be found in Chapter 3, Sections 3.4.3 and 3.4.4.

In contrast to diffusion, by using the appropriate sequences, spectroscopy can provide metabolite information (*e.g.* creatinine), CEST indicates proton exchange between water and other compounds and is still being explored in pathologies, *e.g.* brain tumours,[10] pH maps can also be obtained from CEST measurements, as well as ^{31}P spectroscopy, and susceptibility imaging gives local magnetic field variations maps. Research by the MRI community is continuing to find more endogenous contrasts and more useful parameters that can be beneficial in a clinical setting.

14.3.2 Dynamic Parameters

There are several approaches to the measurement of dynamic parameters with MRI. The first method is with the intravascular application of an exogenous contrast agent, usually a gadolinium-based contrast agent: DSC MR perfusion and DCE MR perfusion. CBF can be measured using the MR technique ASL, which employs saturation pulses to magnetically label arterial blood as an endogenous contrast. For more information on these sequences, please refer to Chapter 3. A detailed comparison of quantitative parametric mapping of CBF performed in healthy volunteers using a pseudo-continuous ASL (pcASL) sequence, in comparison to the corresponding PET procedure using ^{15}O-labelled water, has been reported by Zhang *et al.*, showing that MRI-derived CBF was very close to that from PET (Figure 14.1).[11]

In MRI, parameters are considered quantitative if they have a physiological meaning and semi-quantitative if they do not. Semi-quantitative perfusion parameters include the time-to-peak (TTP) signal enhancement, measured from the moment the first change is observed; the signal change over time (enhancement slope, wash-in or wash-out); or the area under the curve (AUC). These parameters can provide interesting information but are limited because results vary with the pulse sequence used and are also affected by other scan characteristics. Hence, these parameters have to be calibrated after each change in the experimental setup, and results are difficult to compare among different study centres.

It is possible to obtain quantitative parameters following the application of quantitative methods to dynamic data. From these data, parametric maps of relative cerebral blood volume (rCBV) and flow (rCBF) and mean transient time (MTT) can be derived (see Section 3.2.2.1). In neuro-oncology, CBV is the most robust and widely used parameter.[12]

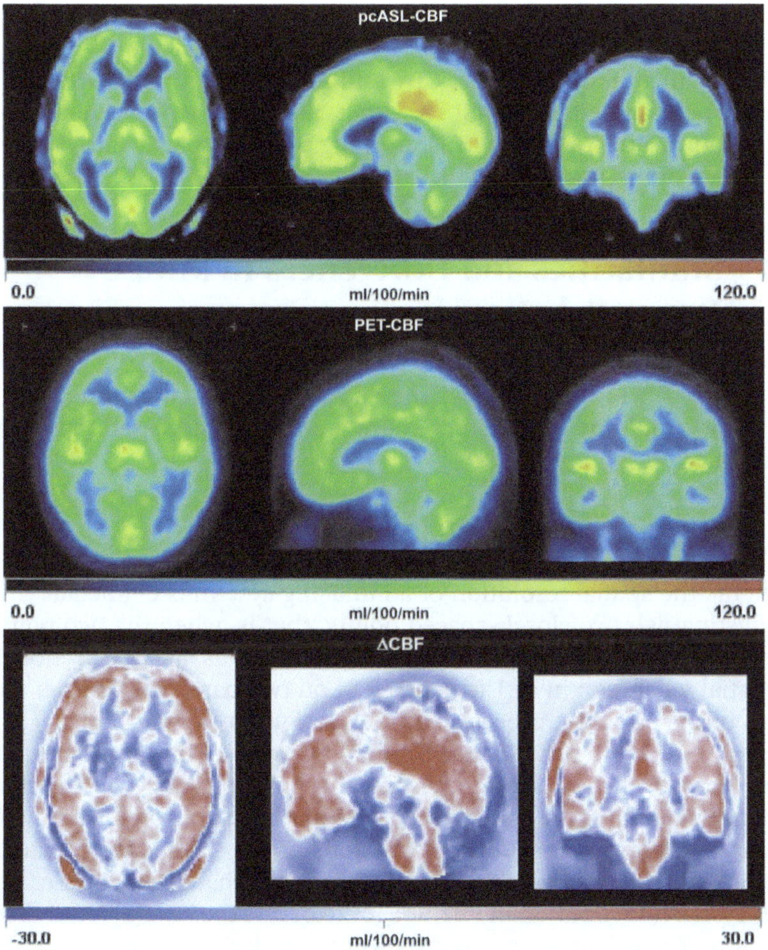

Figure 14.1 Averaged quantitative CBF maps obtained from 10 healthy volunteers with pcASL and ^{15}O-water-PET and difference CBF (ΔCBF) map after subtracting PET-CBF from pcASL-CBF. The unit of the data is mL 100 g^{-1} min^{-1}. In spite of the general similarity between pcASL-CBF and PET-CBF, the difference image clearly exhibits that pcASL-CBF is higher in cortical regions and lower in white matter compared to PET-CBF. Detailed information on this study has been provided by Zhang *et al.*[11]

Most often, DCE MR perfusion imaging is based on a two-compartmental (plasma space and extravascular-extracellular space) pharmacokinetic model. With pharmacokinetic modelling of DCE MR perfusion data, several parameters are commonly derived: the transfer constant (k^{trans}), the fractional volume of the extravascular-extracellular space (v_e), the rate constant (k_{ep}, where $k_{ep} = k^{\text{trans}}/v_e$), and the fractional volume of the plasma space (v_p) (see Section 3.2.2.2).[13,14] The most frequently used parameter in DCE MR perfusion is k^{trans}, which can have different interpretations depending on blood flow and permeability.[33]

With technical modifications, CBV and mean transit time (MTT) can theoretically be obtained using ASL; however, these methods are not yet widely used.[15–19] In brain tumour perfusion imaging, CBF, particularly from ASL, has been an emerging focus.

14.4 Similarities and Differences in MR-PET Parametric Imaging

Following simultaneous acquisition with MR-PET, one particular question presents itself: 'How do I analyse the data?' Before analysis, there are a few issues that should be considered. These are briefly addressed in the following sections.

14.4.1 Spatial Resolution

In general, the spatial resolution of an imaging system is defined as the ability to distinguish two closely-spaced points, that is, when the two points are closer than the spatial resolution value they are seen as one (see Section 5.7.1). The spatial resolution in PET is dependent on two parts: the scanner and the physics of the emission process. Scanner design has a major impact on spatial resolution, but the physics of PET imaging imposes limiting factors. These factors include photon non-collinearity and positron range. In the development of PET imaging, the spatial resolution has been improved with increased detector granularity, *i.e.* smaller dimensions of the scintillator crystals. As a rule of thumb, the resolution of a scanner is approximately the same as the size of the crystal. It is common in PET images to define the voxel size as half of the crystal size. Typical image matrix dimensions are 128×128 or 256×256 pixels, with the latter being frequently selected for high-resolution brain imaging.

Although in MR imaging, the spatial resolution also depends on scanner attributes such as gradient strength and the strength of the magnet, it is mainly dependent on the sequence parameters chosen to acquire the image. Although it is common to define the spatial resolution as the size of the voxel, in reality the true spatial resolution might differ; in simple sequences such as FLASH or a spin echo, the spatial resolution is equal to the voxel size but in demanding sequences, such as EPI or spiral scanning, the point spread function is convolved with T_2^* decay, leading to a poorer spatial resolution.

Because MR and PET have different spatial resolutions and different voxel sizes, registration algorithms are often used to interpolate (sample) one image to the other, and sometimes, an alignment might also be necessary. If MR images are resampled to fit the PET grid with its (usually) larger voxel size, a loss of resolution has to be accepted. However, if PET images are adapted to the finer grid of the MR images, files with dynamic PET can become quite large. In this case, appropriate computer hardware must be available to handle such files. For correcting partial volume effects, co-registration is assumed. A critical review on possible problems when co-registering the bimodal images obtained with MR-PET has been published by Pietrzyk and Herzog.[20]

14.4.2 Temporal Resolution

In principle, simultaneous MR-PET measurement allows time-dependent physiological processes from the dynamic tracer distribution, detected with PET, to be linked with those from MRI, operated with appropriate functional MR sequences. Except for the purpose of methodological comparisons, PET and MR do not necessarily have to measure the same parameter. Instead, the acquisitions may be focused on the simultaneous measurement of complementary information. This allows, for example, the PET measurement of neuroreceptor distribution to be related to the MR-based measurement of CBF by using an MR sequence such as ASL to rule out possible flow-influenced changes in neuro-receptor uptake. The challenge is to adapt the respective acquisition times required by both systems.

Furthermore, synchronising PET and MR data acquisitions enables sophisticated procedures to be used to track any unintentional motion of the patient, as reported for the application of brain imaging by Catana *et al.* and Ullisch *et al.*[21,22] To maintain the advantages of simultaneous MR-PET examinations, such movements have to be recorded and corrected because blurred PET images and motion-affected recording of tracer uptake may hamper a meaningful evaluation.

14.5 Analysis of Parametric Imaging

It is important to distinguish between individual analysis and group analysis. For individual analysis, subjects are investigated in one or more states separately, *e.g.* a brain tumour subject is compared for different parameters in the tumour region and in the contralateral region (one individual and one state) or the tumour-to-brain ratio is compared before and after treatment (one individual and two states).

In a group analysis, several patients in one or more states are measured. The analysis can be paired, that is, each individual is compared with himself (or a matching individual) in a different state, or it can be made on solely on groups. In this latter case, the group data usually have to be normalised and also often smoothed.

14.5.1 Region-of-interest and Voxel Comparison

The first, and simpler, analysis that can be done with parametric images is based on localisation; regions-of-interest (ROIs) and/or voxels can be compared for average values and standard deviations. This usually provides clarification if there are differences between modalities, between states and between groups. However, it does not show that the differences are significant. For that, a statistical test has to be performed. Another interesting analysis is to evaluate a region with its histogram, texture or centre-of-mass. Alternatively, a plot of the two images can also be drawn and a correlation (or similarity metric) can be derived.

14.5.2 Statistical Mapping

In a simple way, statistical mapping is a voxel-by-voxel calculation of a statistical test. The statistical test answers a simple premise of rejecting or accepting the null hypothesis. This can be applied to the parametric images, but also can be applied to a time series. Statistical mapping is commonly used in fMRI time series (see Chapter 3).

14.5.3 Connectivity

Connectivity is a recent approach for studying the structural[23,24] and functional organisation of the brain[25,26] to provide an understanding of neurological and psychiatric illnesses as disconnection diseases.[27,28]

When using DTI data, tractography is computed and the number of fibre tracts between different brain regions is counted, translating into the so-called structural connectivity. This information is typically represented in the form of an adjacency or connectivity matrix, in which each axis represents different brain regions and each matrix element represents the number of fibres connecting each pair of brain regions.

The neuronal activation delivered with fMRI in a specific brain region, or the corresponding parametric information of, *e.g.* cerebral oxygen and glucose consumption and neuroreceptor binding obtained with dynamic PET can be statistically correlated to every other region to translate into dynamic communication of brain regions, the so-called functional connectivity. For this purpose, the simultaneity of integrated MR-PET is a most important advantage. Additionally, the same time series data can be used to compute the directionality of information flow or causality between brain regions, which is given the name of effective connectivity.[29,30]

The approach to the study of brain connectivity can be ROI- or voxel-based, depending on the desired spatial resolution and research questions at stake, which can then lead to the use of more specific methodologies and software (frequently resorting to widely used neuroimaging software such as Freesurfer, FSL or SPM).[30] Typically, the approach involves the use of specific pre-processing steps, such as image registration within the same modality (*e.g.* volumetric T_1-weighted data and DTI) or multimodal (*e.g.* MRI and PET data), and registration to a template (*e.g.* Montreal Neurological Institute—MNI—template). After registration, different atlases are often used to derive relations or connections between brain regions or voxels, which are dependent on the imaging technique (see Figure 14.2).[30]

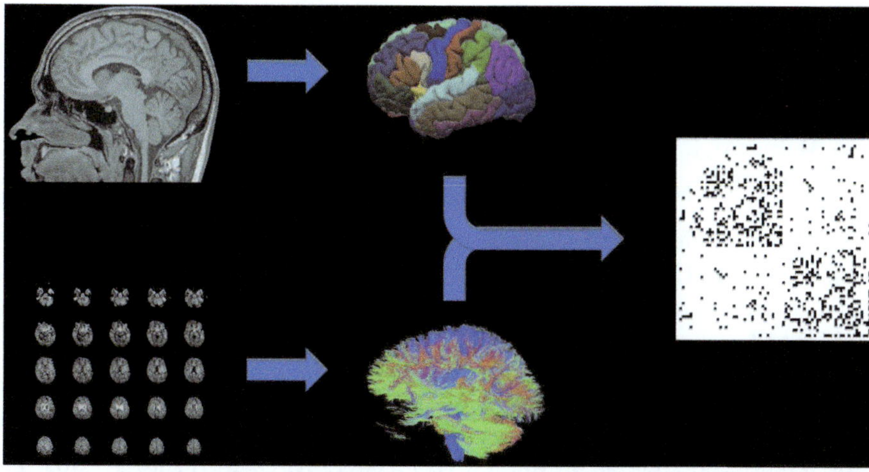

Figure 14.2 Structural connectivity processing. Top row: volumetric T_1-weighted data is segmented into grey and white matter and cerebral spinal fluid. It is then parcellated into different regions according to brain atlases. Bottom row: diffusion tensor imaging data is processed and white matter tracts are computed. Joining arrow: the parcellated brain and tracts are registered to one another and the number of tracts between brain region pairs is determined, from which a connectivity or adjacency matrix is computed (utmost right image) showing in black the brain regions structurally connected *via* tracts.

Connectivity image processing and the computation of network metrics is complex and time intensive, particularly in a multimodal setting such as MR-PET. In order to facilitate research in the topic, the Multimodal Imaging Brain Connectivity Analysis (MIBCA) toolbox was developed.[30] MIBCA automates all the analysis in one batch, computes imaging and network metrics, computes group statistical tests, and provides visualisation routines. Furthermore, it has been used to demonstrate applicability to both Alzheimer's disease and brain tumours, enabling a more comprehensive and innovative view into such illnesses.[31,32]

References

1. M. Reivich, R. Gur and A. Alavi, *Hum. Neurobiol.*, 1983, **2**, 25.
2. M. J. de Leon, S. H. Ferris, A. E. George, D. R. Christman, J. S. Fowler, C. Gentes, B. Reisberg, B. Gee, M. Emmerich, Y. Yonekura, J. Brodie, I. I. Kricheff and A. P. Wolf, *Am. J. Neuroradiol.*, 1983, **4**, 568.
3. P. Y. Wen, D. R. Macdonald, D. A. Reardon, T. F. Cloughesy, A. G. Sorensen, E. Galanis, J. Degroot, W. Wick, M. R. Gilbert, A. B. Lassman, C. Tsien, T. Mikkelsen, E. T. Wong, M. C. Chamberlain, R. Stupp, K. R. Lamborn, M. A. Vogelbaum, M. J. van den Bent and S. M. Chang, *J. Clin. Oncol.*, 2010, **28**, 1963.

4. G. B. Frisoni, N. C. Fox, C. R. Jack, P. Scheltens and P. M. Thompson, *Nat. Rev. Neurol.*, 2010, **6**, 67.
5. M. E. Phelps, S. C. Huang, E. J. Hoffman, C. Selin, L. Sokoloff and D. E. Kuhl, *Ann. Neurol.*, 1979, **6**, 371.
6. A. Rahmim, J. Tang and H. Zaidi, *Med. Phys.*, 2009, **36**, 3654.
7. C. Tsoumpas, F. E. Turkheimer and K. Thielemans, *Med. Phys.*, 2008, **35**, 3963.
8. R. N. Henriques, M. M. Correia, R. G. Nunes and H. A. Ferreira, *Neuroimage*, 2015, **111**, 85–99.
9. R. N. Henriques, H. A. Ferreira and M. M. Correia, *Magma*, 2015, **28**, 511–512.
10. N. A. da Silva, P. Lohmann, J. Fairney, A. W. Magill, A. M. Oros-Peusquens, C. H. Choi, R. Stirnberg, G. Stoffels, N. Galldiks, X. Golay, K. J. Langen and N. J. Shah, *Eur. J. Nucl. Med. Mol. Imaging*, 2018, **45**, 1031.
11. K. Zhang, H. Herzog, J. Mauler, C. Filss, T. W. Okell, E. Rota Kops, L. Tellmann, T. Fischer, B. Brocke, W. Sturm, H. H. Coenen and N. J. Shah, *J. Cereb. Blood Flow Metab.*, 2014, **34**, 1373.
12. S. Cha, E. A. Knopp, G. Johnson, S. G. Wetzel, A. W. Litt and D. Zagzag, *Radiology*, 2002, **223**, 11.
13. M. J. Paldino and D. P. Barboriak, *Magn. Reson. Imaging Clin. North Am.*, 2009, **17**, 277.
14. P. S. Tofts, G. Brix, D. L. Buckley, J. L. Evelhoch, E. Henderson, M. V. Knopp, H. B. Larsson, T. Y. Lee, N. A. Mayr, G. J. Parker and R. E. Port, *J. Magn. Reson. Imaging*, 1999, **10**, 223.
15. T. Kim and S. G. Kim, *Magn. Reson. Med.*, 2005, **54**, 333.
16. T. Kim and S. G. Kim, *Open Neuroimaging J.*, 2011, **5**, 136.
17. E. T. Petersen, T. Lim and X. Golay, *Magn. Reson. Med.*, 2006, **55**, 219.
18. D. L. Thomas, M. F. Lythgoe, F. Calamante, D. G. Gadian and R. J. Ordidge, *Magn. Reson. Med.*, 2001, **45**, 853.
19. J. Wang, D. C. Alsop, H. K. Song, J. A. Maldjian, K. Tang, A. E. Salvucci and J. A. Detre, *Magn. Reson. Med.*, 2003, **50**, 599.
20. U. Pietrzyk and H. Herzog, *Magma*, 2013, **26**, 137.
21. C. Catana, T. Benner, A. van der Kouwe, L. Byars, M. Hamm, D. B. Chonde, C. J. Michel, G. El Fakhri, M. Schmand and A. G. Sorensen, *J. Nucl. Med.*, 2011, **52**, 154.
22. M. G. Ullisch, J. J. Scheins, C. Weirich, E. Rota Kops, A. Celik, L. Tellmann, T. Stöcker, H. Herzog and N. J. Shah, *PLoS One*, 2012, **7**, e48149.
23. P. Hagmann, J. P. Thiran, L. Jonasson, P. Vandergheynst, S. Clarke, P. Maeder and R. Meuli, *Neuroimage*, 2003, **19**, 545.
24. O. Sporns, G. Tononi and R. Kötter, *PLoS Comput. Biol.*, 2005, **1**, e42.
25. M. D. Greicius, B. Krasnow, A. L. Reiss and V. Menon, *Proc. Natl. Acad. Sci.*, 2003, **100**, 253.
26. B. Horwitz, *Neuroimage*, 2003, **19**, 66.
27. X. Delbeuck, M. Van der Linden and F. Collette, *Neuropsychol. Rev.*, 2003, **13**, 79.

28. M. Liang, Y. Zhou, T. Jiang, Z. Liu, L. Tian, H. Liu and Y. Hao, *Neuroreport*, 2006, **17**, 209.
29. K. J. Friston, *Brain Connect.*, 2011, **1**, 13.
30. A. S. Ribeiro, L. M. Lacerda and H. A. Ferreira, *PeerJ*, 2015, **3**, e1078.
31. A. S. Ribeiro, L. M. Lacerda, N. A. da Silva and H. A. Ferreira, *IEEE Trans. Nucl. Sci.*, 2015, **62**, 604.
32. A. C. Mendes, A. M. Oros-Peusquens, A. S. Ribeiro, K. J. Langen, C. Weiß Lucas, N. J. Shah and H. A. Ferreira, *Proc. Int. Soc. Magn. Reson. Med.*, 2016, **24**, e0956.
33. S. P. Sourbron and D. L. Buckley, *Magn. Reson. Med.*, 2011, **66**, 735.

CHAPTER 15

Technical and Methodological Aspects of Whole-body MR-PET

J. CABELLO*, F. SCHILLING AND S. NEKOLLA

Klinikum rechts der Isar, Technische Universität München, Nuklearmedizinische Klinik und Poliklinik, Ismaninger Straße 22, Munich, 81675, Germany
*E-mail: jorge.cabello@tum.de

15.1 Introduction

There are important applications in oncology, and potentially paediatric imaging and pregnant women, where whole-body magnetic resonance-positron emission tomography (MR-PET) could provide significant benefits over PET-computed tomography (CT) and individual PET and magnetic resonance imaging (MRI) scans. Compared to PET-CT, MR-PET exposes the patient to less radiation (up to 70% of the received dose in a PET-CT scan is due to the CT) and provides high soft-tissue contrast and multi-parametric information that can be exploited in different ways. Simultaneous MR-PET makes it possible to perform measurements in one single session, whereas usually, using individual scanners, two examinations need to be coordinated, which has an additional impact on the overall logistics and patient comfort.[1] Quasi-perfect co-registration between anatomical and functional information is always a benefit of simultaneous MR-PET compared to sequential MR-PET, for which re-sampling and registration between the PET and MRI images

New Developments in NMR No. 19
Hybrid MR-PET Imaging: Systems, Methods and Applications
Edited by N. Jon Shah
© The Royal Society of Chemistry 2019
Published by the Royal Society of Chemistry, www.rsc.org

may potentially introduce errors. In the case of sensitive patients, such as in paediatric imaging, women during or after pregnancy, or for patients of reproductive potential, the avoidance of the CT radiation exposure is a major benefit of MR-PET.[2]

From a technical perspective, motion represents a major problem in whole-body PET that affects image quality and quantification.[3] Simultaneous MRI can provide accurate motion measurements that can be modelled and integrated in the PET reconstruction process, producing potentially motion-free PET images. In addition, PET provides quantitative functional information as long as certain data corrections are accurately applied, such as attenuation and scatter of γ-photons in the patient's tissue. In contrast to CT, which provides electron density maps and in this way information on tissue attenuation, MRI delivers data on proton density of imaged tissue. Different strategies to convert proton density into electron density have been explored for the specific case of brain PET.[4] Approaches in whole-body MR-PET have focused on including the major bone structures in the body, such as spine and pelvis, and in extracting patient-specific bone and lung tissue density.[5,6] The combination of PET and MRI requires important considerations regarding both instrumentation, such as novel RF coil designs, and strategies related to software, as for attenuation and motion correction.

15.2 MR-PET System Design

The MRI part of a whole-body MR-PET scanner consists of three major components: the magnet, the gradient coils and the RF coils for transmission and reception of the magnetic resonance signals. Clinical scanners usually use a superconducting magnet in the range of 1.5–3 T and maintain a B_0 field homogeneity below 4 ppm within a spherical region of 50 cm diameter.[7] Linear magnetic field gradients along the x-, y-, and z-direction of the magnet are mounted concentrically within the magnet bore. These allow spatial encoding of magnetic resonance signals and are used for shimming. Every time the bed position is moved in a multi-bed-position MR-PET scan, the shim currents are automatically adjusted to improve the B_0 homogeneity. Typically, nuclear spins are excited using a whole-body transmit volume coil mounted within the magnet bore to create a time-varying magnetic field B_1 (on the order of several μT) on resonance with the Larmor frequency. Typically, such a volume coil provides uniform B_1 across the entire homogeneous region of the magnet.

For a PET system to function in combination with an MRI system, the components need to be MR-compatible, including the scintillator crystal, photodetector, front-end electronics and data-acquisition system. In the case of photodetectors, for instance, avalanche photodiodes (APD) in the case of the Siemens Biograph mMR (Siemens Healthcare GmbH), and silicon photomultipliers (SiPM) in the case of the GE SIGNA (GE Healthcare), are employed. The SiPM detectors have a gain of 10^6 and a rise-time of 300 ps, compared to 10^2 and 1 ns from the APD detectors, providing the GE SIGNA with time-of-flight (TOF) capabilities (400 ps full-width at half-maximum). Further developments in the acquisition electronics of the GE SIGNA, such as placing the ASIC by the detectors and the capability of recovering scattered events, results in higher interaction position determination accuracy of the γ-photons in the block detector, and increased sensitivity compared to the Siemens Biograph.

To mitigate radiofrequency (RF) interferences, shielding the PET detectors, front-end electronics and digitiser is the most common approach. The possible different shielding materials, thicknesses and layouts with which to efficiently block electromagnetic interferences at different frequencies, while at the same time minimising the eddy currents that can occur on the shielding material, are ongoing matters of research.

The integration of a PET system within a whole-body MRI can lead to several interferences: susceptibility changes, particularly caused by the PET scintillators, electronics or connectors, or eddy currents in shielding materials may lead to B_0 inhomogeneities. Therefore, magnetic materials should be avoided in the PET detectors. Also, unwanted sources of RF at the Larmor frequency that can be caused by the PET electronics or power cables (carrying noise from outside the Faraday cage into the system), which are present during the acquisition of the MR signal, will degrade the MRI images. Careful RF shielding design of the PET system can potentially resolve this issue.

15.2.1 Dedicated RF Coils

The sensitivity of an MRI acquisition depends on the distance between the coil, the region being examined and the coil geometry. The closer the coil can be brought to the measured object and the better the coil dimensions match the region-of-interest, the higher the sensitivity. Therefore, dedicated radiofrequency coils for MR-PET have been developed to aid a variety of clinical questions. These coils differ from MR-only coils in their design in two key elements: first, they are designed to minimise radiofrequency interferences with the PET electronics and second, the amount of material is reduced as much as possible in order to minimise γ-photon attenuation and scatter to avoid inaccurate PET images. Typical numbers for the reduction of PET counts with MR-only coils are in the order of 20%, while dedicated MR-PET coils only reduce the number of PET counts by 10–15%.[8] There are specific coils for breast, carotid artery, lungs, spine, prostate, knee, neonatal, and multi-nuclear applications among others. Receive only coils, such as phased arrays, can also be used to allow acceleration of image acquisition for large volumes.

15.2.2 PET Bore Design

Due to the increased field-of-view (FoV) in whole-body PET, it is necessary to take several differences into account when comparing brain and whole-body MR-PET.

The increased diameter of whole-body PET, compared to brain PET systems, has a negative impact on the intrinsic PET resolution due to the non-negligible acollinearity between the two γ-photons originating from each β+ annihilation. This acollinearity is 0.47° full-width at half-maximum,[9] with an impact in spatial resolution of 2.2×10^{-3} times the scanner diameter.[10] However, a large bore implies that the distance between the scintillation crystals from the PET detectors and the origin of the γ-photons is larger compared to smaller FoV scanners, thereby reducing the angle of interaction with the front face of the scintillator crystals. This effect reduces the positioning error of the end-points for each line-of-response between each pair of γ-photons in coincidence and is known as depth-of-interaction degradation. The result is that spatial resolution is more uniform in the main central region of the PET FoV, and while spatial resolution degradation occurs close to the FoV edges, it is less critical compared to scanners with reduced FoV like brain or breast PET scanners.

15.3 Software Considerations

15.3.1 Attenuation Correction

In MR-PET, attenuation correction is considered to be challenging due to the limited information that MRI can provide about tissue density. In PET-CT, electron density information is extracted from a CT scan and is then translated by a bi/tri-linear scaling into the linear attenuation coefficient (LAC) valid for the 511 keV annihilation photons occurring in PET. In the case of MR, only proton density information is obtained. Furthermore, in the case of MR-PET it is not only the patient who needs to be considered for attenuation correction, but also the coils and any other hardware that is placed between the patient and the PET system.

15.3.1.1 Subject Attenuation Correction

PET attenuation correction in whole-body applications is, in general, not as sensitive to the accurate determination of bone as in neurological studies. A direct approach can be applied by calculating an attenuation map containing soft tissue, fat, and air, using segmentation algorithms applied to data from a two-point Dixon volume-interpolated breath-hold (VIBE) sequence for each bed position.[11] Figure 15.1 shows exemplary water and fat images from a torso using the Dixon VIBE sequence, the resulting MR-based attenuation map, and the corresponding CT-based attenuation map. Using this approach, it was determined that ignoring bone produces a consistent underestimation in SUV measurements. However, the obtained errors were within test-retest intra-patient errors. Maximum SUV changes of 13% in the pelvic area were observed, leading to the conclusion that bone does not have an important impact in all analysed lesions in a variety of regions, with the exception of lesions close to dense bone structures. Including bone in the attenuation

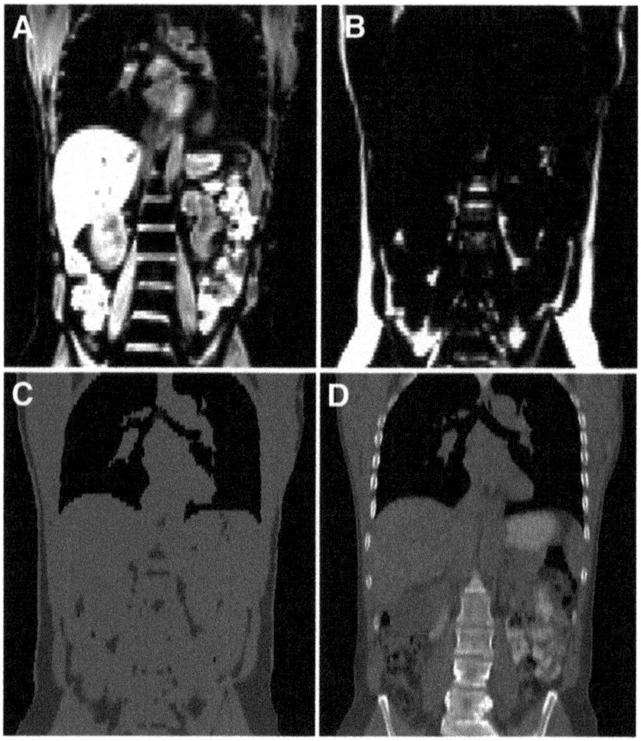

Figure 15.1 Water (A) and fat (B) images from a 2-point Dixon imaging sequence and resulting attenuation map (C). Corresponding CT-based attenuation map (D). Reproduced from ref. 11 with permission, Copyright © SNMMI.

map for whole-body MR-PET is more challenging than in brain imaging. Dedicated MRI sequences to extract bone information, such as ultra-short echo-time and zero echo-time sequences, have been mainly applied to brain imaging but are not yet routinely applied for whole-body imaging. The amount of time required to operate these sequences in whole-body MR-PET is too long to be feasible in a clinical setting. Therefore, to date, approaches to include bone in whole-body MR-PET are based on atlas registration.[12] However, atlas-based methods can potentially fail when patients present a physiognomy that significantly differs from the atlas model.

Assigning a single LAC value to each tissue type may produce inaccuracies, which may not be clinically acceptable in some cases. The lungs represent a special case. Here important differences in lesions in the lungs and nearby structures were found when a single-value non-patient-specific LAC was applied compared to patient-specific LACs, calculated using a Turbo-FLASH sequence (TR: 120 ms; TE: 750 μs).[6]

The effective transverse FoV of the MRI is smaller than the transverse FoV of the PET, due to non-uniformities of the static magnetic field and gradient nonlinearities. Since MRI scans are more time consuming than CT scans, patients are typically positioned with arms down during an MRI scan, as opposed to arms up in CT scans. The difference in FoV dimensions between the PET and MRI causes part of the arms, along the patient's trunk, to be truncated in the attenuation map. Studies have demonstrated an underestimation of radioactive uptake in different organs[13] due to the truncated attenuation map. Approaches to correct for attenuation map truncation have been developed from both the PET and the MRI communities. The PET reconstruction algorithm based on the maximum likelihood *a posteriori* of attenuation and activity (MLAA) was conceptually designed to simultaneously estimate the emission and attenuation map, under the assumption that if there is radiotracer distribution reconstructed in a volume, tissue should be present in the attenuation map.[14] An important limitation of this approach is that tissue regions without radioactivity are not recovered and are not included in the attenuation map. This limitation is especially important in the case of radiotracers with high target binding specificity *i.e.* low unspecific uptake in blood or skeletal muscle. An additional minor limitation is the increased processing time required to reconstruct the attenuation map from the MLAA algorithm. Furthermore, MLAA was extended by including TOF-PET data to eliminate the known cross-talk between emission and transmission data present in non-TOF systems.[15] A purely MR-based approach uses B_0 uniformity optimisation of readout gradients to compensate for distortions and truncations.[16] Figure 15.2 shows an axial slice at the level of the pelvis of the attenuation map obtained from the two-point Dixon VIBE, the corresponding slice with completed attenuation map using MLAA, and the same slice with an extended MRI FoV using the B_0 homogenisation with a gradient enhancement (HUGE) sequence. While the MLAA algorithm completes the missing information from the Dixon VIBE, there is a clear overestimation in

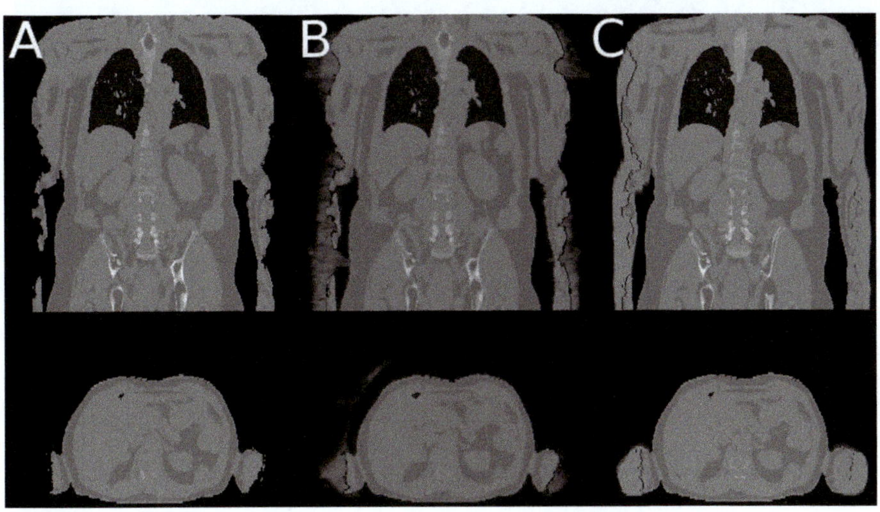

Figure 15.2 Coronal and axial views of an attenuation map calculated with the two-point Dixon sequence (A), MLAA (B) and the HUGE sequence (C).

the attenuation map; in contrast, HUGE produces an accurate attenuation map description. The clinical impact of MLAA and HUGE has been assessed in patients with different body masses and with a variety of lesions distributed across different body parts.[17] Results showed maximal, single voxel quantification differences of 30–50% close to the truncated areas and also important differences in the entire trunk, especially in the thorax. The study by Blumhagen *et al.* concluded that the improvement in image quantification showed by both MLAA and HUGE using [^{18}F]-FDG was comparable.[17]

15.3.1.2 Attenuation Correction of RF Coils

Attenuation and scatter by dedicated coils, surface coils or any other hardware need to be addressed for accurate PET image quantification.[8] The coils embedded in the patient bed and head coils are fixed, hence the position is known. However, the positioning of surface coils differs from patient to patient. In some cases, fiducial markers are used to position the coil with respect to the attenuation map of the subject.[18] As with head coils, a highly-sampled CT of the coil is re-sampled to match the voxel size of the subject's attenuation map. In a study focused on breast cancer, seven patients were scanned with and without a dedicated RF breast coil.[19] The impact of including the attenuation map of the coil was investigated in 14 lesions, resulting in an average underestimation of 21 ± 9% SUV without the breast coil attenuation map. Figure 15.3 illustrates the coronal, sagittal and axial

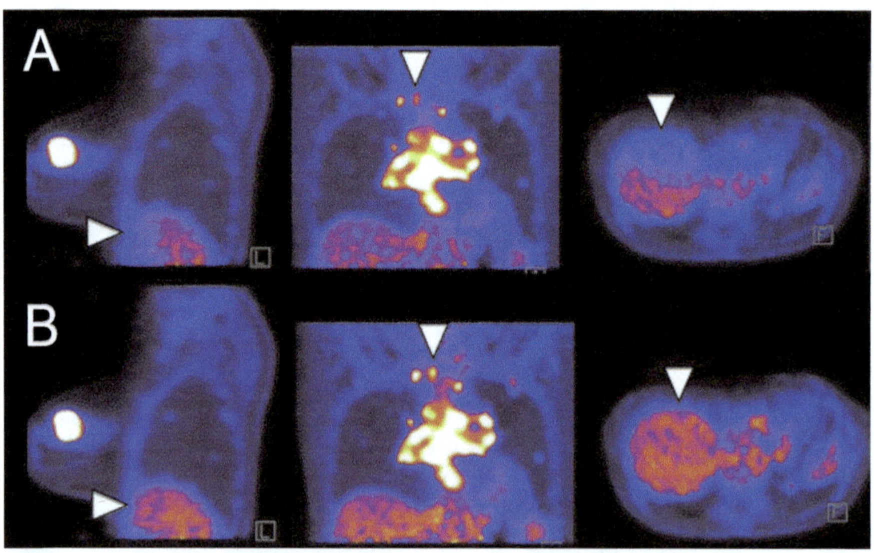

Figure 15.3 Sagittal, coronal and axial PET images reconstructed without (A) and with (B) breast coil attenuation map. Adapted from ref. 19 with permission from Springer Nature, Copyright 2014, European Society of Radiology.

PET slices from the same patient without and with the coil attenuation map, showing non-uniformities in the liver and low contrast in small lesions when the attenuation of the breast coil is not considered. For most flexible coils, like cardiac, carotid or body matrix coils, localisation is currently not implemented clinically. This is because, not only can the coils change position between patients, but also when scanning a single patient, their position can change during the scan due to patient motion. These coils are specially designed for minimum attenuation and scatter, to avoid the need for correction or at least limit the clinically acceptable error. Still, some studies have approached this problem by using MLAA[20] or Microsoft Kinect.[21] Further developments, such as the new blanket-air like MRI coils developed by GE, allow one to accurately adjust to the patient shape, minimizing the attenuation and scatter in the coil.

15.3.2 Scatter Correction

Scatter in PET is one of the major sources of degradation and bias in quantification. Typical scatter fractions in brain PET scanners are ~20–30% (energy window: 420–580 keV),[22] while in whole-body scanners they go up to ~30–40% (435–650 keV).[23,24] Scatter is tightly linked to the amount of tissue located between the radioactivity in the patient and the PET detectors. Therefore, knowing the composition and amount of attenuating media is important, which, as mentioned above, is challenging to overcome in MR-PET. Whole-body scatter calculation is a computation-intensive process. Analytical methods, like the model-based simulated-single-scatter (SSS) method,[25] are widely used due to their relative simplicity and swift computation. However, there are more accurate methods based on Monte Carlo simulations[26] available, although these are not currently feasible for use in clinical routine due to their long computation time. The use of SSS entails a scaling step to match the scatter estimate to the emission data. There are two methods currently available to perform such scaling: relative scaling, where the SSS scatter estimate is fitted to the scatter tails from the region outside the subject outline, and absolute scaling, where the distribution of un-scattered events is additionally calculated so a relation with the scattered events distribution is obtained.

In the case of relative scaling, fitting the scattered estimate distributions can entail problems in situations where the PET FoV limits the extension of the scatter tails (potentially problematic in brain or breast scanners), or in regions with significant lesion/organ-to-background activity concentration where tail fitting can lead to an over-estimation of the scatter distribution and result in regions devoid of activity. The latter scenario is circumvented by appropriately limiting the maximum possible scatter estimate by a threshold. This scatter over-estimation is the main reason for the halo artefacts that are sometimes observed in MR-PET and PET/CT images in areas with high organ-to-background contrast. These typically include the bladder, kidneys or liver and are dependent on the excretion route of the radiotracer.[20]

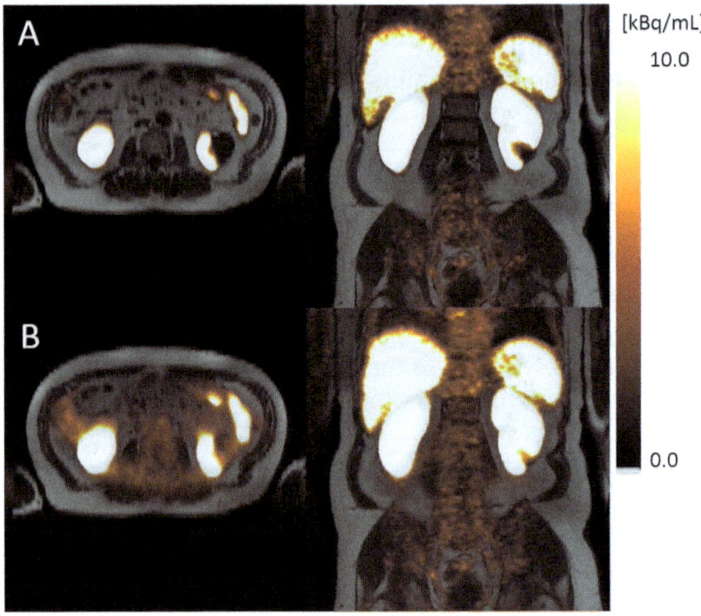

Figure 15.4 Axial and coronal [^{18}F]-PSMA-PET images of a patient corrected for scatter, using relative (A) and absolute (B) scatter scaling, overlaid on the corresponding T1w TSE MRI slice.

Alternatively, absolute scaling is not required to fit the scatter estimate, and hence it is less sensitive to the halo artefacts. Figure 15.4 shows an example of an ^{18}F-PSMA-PET scan in the pelvic area, overlaid on a T_1-weighted Turbo Spin Echo (TSE) scan from the Siemens Biograph mMR, where the PET data were corrected with relative and absolute scatter scaling.

15.3.3 Motion Correction

Motion, both voluntary and involuntary, is one of the main degradation effects in reconstructed PET images and impacts both image quality and quantification. Cardiac and respiratory motion cause the most degradation effects due to motion in PET images, especially in the cranio-caudal direction in the lungs, heart, liver, spleen and pancreas. Motion-free spatial resolution is around 4–5 mm in clinical PET but can be degraded by up to 10 mm due to organ motion. Measuring motion for application to either reconstructed PET images or as additional information in the reconstruction process is an active area of research. Traditionally, patient motion is synchronised with the PET data (gating) through external devices (ECG and respiratory belt) and divided into frames according to different motion phases. Each data frame is then reconstructed, resulting in noisy (due to the limited data statistics) motion-free PET frames.[27] External sensors have been used to extract approximated motion models to apply to the PET data.

However, complicated internal organ motion can hardly be inferred from external motion and as the correct application of these external sensors in clinical routine is technician-dependant, the sensors themselves can often be a source of problems. Although PET data-driven approaches have also been implemented and have successfully corrected for motion, without the need of additional sensors or imaging modalities,[28] these approaches are limited to regions with significant uptake. A more sophisticated approach is to directly measure the internal motion through fast dynamic CT or MRI, and then to correlate it with an external surrogate signal.[29] In both cases, a 4D acquisition results in a series of 3D volumes, from which, through rigid or elastic registration, motion fields are calculated. In the case of CT, there is a significant radiation exposure, which may not be acceptable in some cases. Simultaneous MR-PET represents an ideal tool with which to measure motion, exploiting the high temporal resolution of MRI compared to PET.[30] Additional benefits of MRI compared to CT are: the potentially more accurate motion fields due to the higher soft-tissue contrast; the simultaneity of motion measurement and PET acquisition (at least in principle in whole-body imaging) and avoiding inaccurate motion correction due to changes in the cardiac and respiratory cycles during the PET scan.

In terms of using MRI, there are approaches to extract motion information from the heart and lungs/diaphragm.[31] To correct for respiratory motion, one of the simplest approaches is using 1D diaphragmatic navigators, interleaved at different time points. Self-navigation has been proposed to avoid the need of navigators by measuring the *k*-space spirally or radially. More advanced methods involve the acquisition of 2D and 3D image-based navigators to measure accurate motion models, including rotation and non-rigid deformations. However, these come at the cost of long scan times.

To measure cardiac motion, tagging is a common approach, where an orthogonal 'stripe-pattern' is imprinted on the heart using RF-saturation pulses.[32] Deformations of the imprinted tagging pattern caused by subsequent cardiac motion can be visualised in the reconstructed MR images and allow to track cardiac motion dynamically. The motion model can then be extracted from the tagging patterns through active contours or optical flow methods. However, problems with tagging include the additional scan time and the fact that the patterns fade away due to the longitudinal magnetisation relaxation, although this can partially be addressed by using increased flip angles.[33] Other alternatives are balanced steady-state free precession gradient echo methods, which can acquire either a single slice or multiple 2D slices. These sequences exploit the high T_2–T_1 ratio of blood producing a high intravascular signal; and by using short TR and TE values, multiple echoes can be acquired in one single heartbeat.

Once a motion model for each frame is available, it can then be applied to individually-reconstructed PET frames, to the raw PET data (line-of-responses), or it can be included in the PET reconstruction process by adapting the system matrix according to the motion model.[34,35] Figure 15.5 shows the

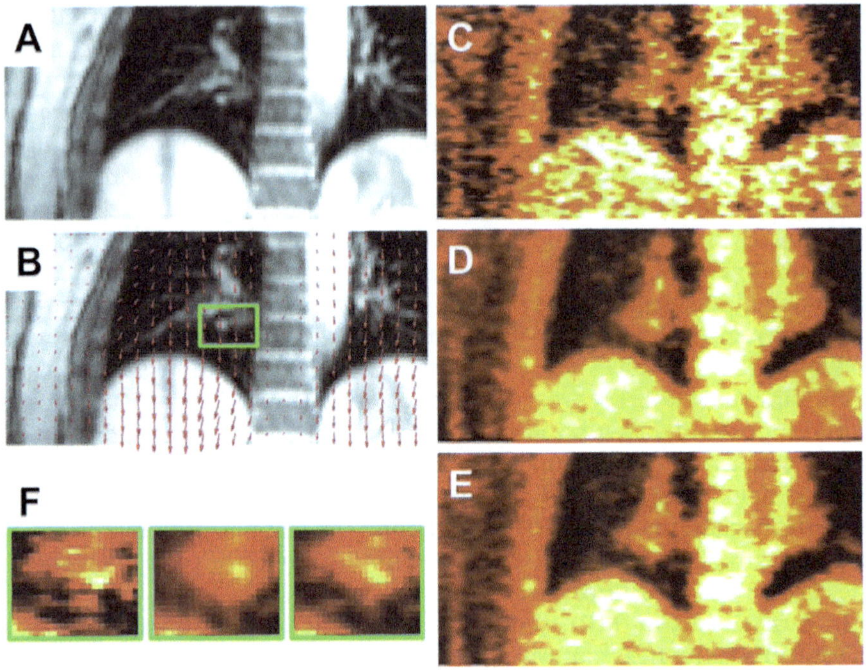

Figure 15.5 Example of PET respiratory motion correction based on MRI data (A) from where the motion is calculated (B). (C) shows the PET data without motion correction, (D) shows the result of using one single gate and (E) shows the result of including the motion model in the image reconstruction. (F) shows a zoom-in from the lungs where different detail recovery is observed from (C), (D) and (E). Adapted from ref. 34 with permission from John Wiley and Sons, Copyright 2016.

comparison of an un-gated, a one-gate and a motion model included in the reconstruction of a PET image measured with simultaneous MRI which was originally degraded by respiratory motion.[33]

From the clinical perspective, continuous motion measurement with MRI is not easily feasible, as it prevents the acquisition of other clinically relevant sequences. To circumvent continuous MRI measurement, approaches to find accurate surrogates for motion are required. Initial approaches established a correlation between an accurate motion model measured with MRI image navigators and an external motion signal, which was later used as a surrogate. This approach was based on the assumption that the initial motion model did not change and the external motion signal remained consistent throughout the entire scan.[36]

15.3.4 Kinetic Modelling

The kinetics of PET radiotracers can be mathematically modelled such that physiological parameters specific to a biological function are extracted, providing valuable information about the function under study (see

Chapter 8). One of the critical parameters required for kinetic modelling is the input function, which represents the temporal course of the radiotracer in the blood. The most accurate method to extract the input function is to take regular blood samples from an artery, which is currently considered as being the standard method. However, arterial sampling requires trained staff and is uncomfortable and painful for patients. Therefore, this approach is typically limited to research settings. Other non-invasive approaches, like mathematical estimation[37] and image-derived input function (IDIF)[38] (see Chapter 11), have been shown to be valid approximations to arterial sampling. The latter is especially interesting due to its relative simplicity. In neuroimaging studies, the carotid artery can be used.[38] However, with a typical diameter of 5 mm and given the spatial resolution of clinical PET scanners (4–6 mm), the contribution of the partial volume effect (PVE) is important. Therefore, the IDIF needs to be corrected for the PVEs.[39] The availability of high resolution and high soft-tissue contrast information makes MRI a valuable tool to identify main blood pools for subsequent partial volume correction.[40] In the case of whole-body studies, access to large blood pools like cardiac structures, aortic segments or femoral arteries makes the impact of the PVE less critical. Since whole-body studies require several consecutive scans (bed positions), as opposed to brain scans, kinetic modelling represents a challenge. However, there are already approaches available to perform short multi-bed-position scans for whole-body PET kinetic analysis.[41]

15.4 Discussion

Recently, the combination of whole-body MR-PET has swiftly developed in several areas. In terms of the software, different approaches for attenuation correction, the main challenge for PET data correction in MR-PET, have been developed, validated, and recently integrated into clinical scanners. Most research studies have focused on brain MR-PET, mainly because of three reasons: (1) accurate quantification in neurological pathologies is of special importance compared to whole-body scans; (2) brain MR studies can be easily registered to CT data; and (3) the covered FoV for any special MR sequence devoted to bone imaging is relatively small. However, several approaches have also been developed for whole-body MR-PET and their feasibility and accuracy compared to CT has been demonstrated. Furthermore, MRI has not only been used in research studies for accurately measuring high soft-tissue contrast anatomical information with PET data, but also for making improvements in PET image quality by correcting for motion or by allowing/facilitating a more accurate measurement of the input function for kinetic studies. Integrating MRI information in the PET reconstruction process, in order to enhance PET spatial resolution and reduce image noise has been already explored for brain MR-PET studies.[42] However, it has not yet been exploited in whole-body imaging. The benefits for MR image quality from PET data have only been exploited in regularised MRI reconstruction brain imaging when under-sampled MR data were acquired.[43] Consequently, it is not clear yet whether similar approaches could be taken for whole-body imaging.

In terms of MRI, RF coils were the component that required most modifications when combined with PET technology, in order to minimise the PET image degradation. Compared to brain studies, the problem of position uncertainty of whole-body flexible RF coils has been ignored until relatively recently and a number of studies have begun work to address the issue.

One of the main drawbacks of whole-body MR-PET is that scan times are usually longer compared to brain MR-PET and PET-CT. Each bed position requires between 3–5 min of PET scan, while several MR sequences run simultaneously with the PET acquisition. To be competitive with PET-CT, MR-PET scans would ideally not last longer than 30 min. However, whole-body MR-PET scans last 60 min (only imaging) on average, where MR imaging is the limiting factor. It should be noted nonetheless that the MRI scans result in large amounts of information. Therefore, the development of accelerated MR sequences is an especially attractive area for whole-body MR-PET.

Even though important advances in whole-body MR-PET were made by a number of groups, most of them still need to be integrated into commercial scanners to be used in clinical routine.

15.5 Conclusions

In this chapter, we have discussed several recent developments in the design of MR-PET systems and in data reconstruction and processing for whole-body MR-PET. With regard to the MR-PET hardware, most efforts have been focused on minimising interferences between the two systems. On the MRI side, RF coils have undergone modifications to minimise attenuation and scatter of the PET signal, since they are located between the PET system and the imaged subject. In the case of PET instrumentation, important advances have been developed from the initial whole-body MR-PET systems in 2010, significantly improving the acquisition electronics and PET detector technology, resulting in improved image quality of the current MR-PET systems based on SiPM technology.[44] These technological advances have also been adopted by other dedicated organ-specific PET systems. Whole-body MR-PET has made particular progress in PET motion correction, especially where cardiac and respiratory motion have a significant impact, for which whole-body MR-PET represents the optimal tool to accurately track the respective motion. Attenuation correction has represented one of the major challenges in whole-body MR-PET. Several strategies for attenuation correction, such as bone inclusion and patient-specific LACs were developed and assessed even in challenging settings like lung imaging, demonstrating that the accuracy of MR-PET attenuation correction nowadays is comparable to that obtained with PET-CT scanners in a wide number of applications. Besides the above-mentioned technical hurdles, one of the main challenges that whole-body

MR-PET still faces is the harmonisation of protocols for different studies among different institutions, while, at the same time, ensuring optimal patient care through personalised diagnosis and treatment.

Acknowledgement

The authors would like to acknowledge the Deutsche Forschungsgemeinschaft (DFG), Großgeräteinitiative for funding the MR-PET facility at the Technische Universität München.

References

1. G. Delso, E. ter Voert, F. de Galiza Barbosa and P. Veit-Haibach, *Semin. Nucl. Med.*, 2015, **45**, 552.
2. S. Purz, O. Sabri, A. Viehweger, H. Barthel, R. Kluge, I. Sorge and F. W. Hirsch, *J. Nucl. Med.*, 2014, **55**, 32S.
3. C. Catana, *Semin. Nucl. Med.*, 2015, **45**, 212.
4. J. Cabello, M. Lukas, E. Rota Kops, A. Ribeiro, N. J. Shah, I. Yakushev, T. Pyka, S. G. Nekolla and S. I. Ziegler, *Eur. J. Nucl. Med. Mol. Imaging*, 2016, **43**, 2190.
5. D. H. Paulus, H. H. Quick, C. Geppert, M. Fenchel, Y. Zhan and G. Hermosillo, *J. Nucl. Med.*, 2015, **56**, 1061.
6. H. R. Marshall, F. S. Prato, L. Deans, J. Théberge, R. T. Thompson and R. Z. Stodilka, *J. Nucl. Med.*, 2012, **53**, 977.
7. S. Vandenberghe and P. K. Marsden, *Phys. Med. Biol.*, 2015, **60**, R115.
8. M. Eldib, J. Bini, D. D. Faul, N. Oesingmann, C. Tsoumpas and Z. A. Fayad, *PET Clin.*, 2016, **11**, 151.
9. P. Colombino, B. Fiscella and L. Trossi, *Nuovo Cim. (1955–1965)*, 1965, **38**, 707.
10. C. S. Levin and E. J. Hoffman, *Phys. Med. Biol.*, 1999, **44**, 781.
11. A. Martinez-Möller, M. Souvatzoglou, G. Delso, R. A. Bundschuh, C. Chefd'hotel, S. I. Ziegler, N. Navab, M. Schwaiger and S. G. Nekolla, *J. Nucl. Med.*, 2009, **50**, 520.
12. D. H. Paulus, H. H. Quick, C. Geppert, M. Fenchel, Y. Zhan, G. Hermosillo, D. Faul, F. Boada, K. P. Friedman and T. Koesters, *J. Nucl. Med.*, 2015, **56**, 1061.
13. G. Delso, A. Martinez-Möller, R. A. Bundschuh, S. G. Nekolla and S. I. Ziegler, *Med. Phys.*, 2010, **37**, 2804.
14. J. Nuyts, P. Dupont, S. Stroobants, R. Benninck, L. Mortelmans and P. Suetens, *IEEE Trans. Med. Imaging*, 1999, **18**, 393.
15. A. Rezaei, M. Defrise, G. Bal, C. Michel, M. Conti, C. Watson and J. Nuyts, *IEEE Trans. Med. Imaging*, 2012, **31**, 2224.
16. J. O. Blumhagen, R. Ladebeck, M. Fenchel and K. Scheffler, *Magn. Reson. Med.*, 2013, **70**, 1047.
17. J. O. Blumhagen, H. Braun, R. Ladebeck, M. Fenchel, D. Faul, K. Scheffler and H. H. Quick, *Med. Phys.*, 2014, **41**, 0223031.

18. M. Eldib, J. Bini, C. Calcagno, P. M. Robson, V. Mani and Z. A. Fayad, *Invest. Radiol.*, 2014, **49**, 63.
19. I. Dregely, T. Lanz, S. Metz, M. F. Mueller, M. Kuschan, M. Nimbalkar, R. A. Bundschuh, S. I. Ziegler, A. Haase, S. G. Nekolla and M. Schwaiger, *Eur. Radiol.*, 2015, **25**, 1154.
20. T. Heußer, C. M. Rank, Y. Berker, M. T. Freitag and M. Kachelrieß, *EJNMMI Phys.*, 2017, **4**, 12.
21. L. J. Frohwein, M. Hess, D. Schlicher, K. Bolwin, F. Büther, X. Jiang and K. P. Schäfers, *Phys. Med Biol.*, 2017, **63**, 0250331.
22. H. Herzog, K. J. Langen, C. Weirich, E. Rota Kops, J. Kaffanke, L. Tellmann, J. Scheins, I. Neuner, G. Stoffels, K. Fischer, L. Caldeira, H. H. Coenen and N. J. Shah, *Nuklearmedizin*, 2011, **50**, 74.
23. G. Delso, S. Fürst, B. Jakoby, R. Ladebeck, C. Ganter, S. G. Nekolla, M. Schwaiger and S. I. Ziegler, *J. Nucl. Med.*, 2011, **52**, 1914.
24. B. W. Jakoby, Y. Bercier, M. Conti, M. E. Casey, B. Bendriem and D. W. Townsend, *Phys. Med. Biol.*, 2011, **56**, 2375.
25. C. C. Watson, D. Newport and M. E. Casey, in *Three-Dimensional Image Reconstruction in Radiology and Nuclear Medicine*, ed. P. Grangeat and J.-L. Amans, Springer, Netherlands, Dordrecht, 1996, p. 255.
26. K. Magota, T. Shiga, Y. Asano, D. Shinyama, J. Ye, A. E. Perkins, P. J. Maniawski, T. Toyonaga, K. Kobayashi, K. Hirata, C. Katoh, N. Hattori and N. Tamaki, *J. Nucl. Med.*, 2017, **58**, 2020.
27. A. Rahmim, O. Rousset and H. Zaidi, *PET Clin.*, 2007, **2**, 251.
28. S. Fürst, R. Grimm, I. Hong, M. Souvatzoglou, M. E. Casey, M. Schwaiger, S. G. Nekolla and S. I. Ziegler, *J. Nucl. Med.*, 2015, **56**, 261.
29. H. Fayad, F. Lamare, T. Merlin and D. Visvikis, *Q. J. Nucl. Med. Mol. Imaging*, 2016, **60**, 12–24.
30. C. Catana, *Semin. Nucl. Med.*, 2015, **45**, 212.
31. J. Ouyang, Q. Li and G. El Fakhri, *Semin. Nucl. Med.*, 2013, **43**, 60.
32. E.-S. H. Ibrahim, *J. Cardiovasc. Magn. Reson.*, 2011, **13**, 36.
33. S. E. Fischer, G. C. McKinnon, S. E. Maier and P. Boesiger, *Magn. Reson. Med.*, 1993, **30**, 191.
34. J. Dutta, C. Huang, Q. Li and G. El Fakhri, *Med. Phys.*, 2015, **42**, 4227.
35. F. Lamare, M. J. L. Carbayo, T. Cresson, G. Kontaxakis, A. Santos, C. C. L. Rest, A. J. Reader and D. Visvikis, *Phys. Med. Biol.*, 2007, **52**, 5187.
36. T. Küstner, M. Schwartz, P. Martirosian, S. Gatidis, F. Seith, C. Gilliam, T. Blu, H. Fayad, D. Visvikis, F. Schick, B. Yang, H. Schmidt and N. F. Schwenzer, *Med. Image Anal.*, 2017, **42**, 129.
37. D. Feng, S.-C. Huang and X. Wang, *Int. J. Bio-Med. Comput.*, 1993, **32**, 95.
38. P. Zanotti-Fregonara, K. Chen, J.-S. Liow, M. Fujita and R. B. Innis, *J. Cereb. Blood Flow Metab.*, 2011, **31**, 1986.
39. M. Soret, S. L. Bacharach and I. Buvat, *J. Nucl. Med.*, 2007, **48**, 932.
40. K. Erlandsson, in *Basic Science of PET Imaging*, ed. M. M. Khalil, Springer International Publishing, Cham, 2017, p. 355.
41. N. Karakatsanis, M. A. Lodge, A. K. Tahari, Y. Zhou, R. L. Wahl and A. Rahmim, *Phys. Med. Biol.*, 2013, **58**, 7391.

42. A. Mehranian, M. A. Belzunce, F. Niccolini, M. Politis, C. Prieto, F. Turkheimer, A. Hammers and A. J. Reader, *Phys. Med. Biol.*, 2017, **62**, 5975.
43. F. Knoll, M. Holler, T. Koesters, R. Otazo, K. Bredies and D. K. Sodickson, *IEEE Trans. Med. Imaging*, 2017, **36**, 1.
44. A. M. Grant, T. W. Deller, M. M. Khalighi, S. H. Maramraju, G. Delso and C. S. Levin, *Med. Phys.*, 2016, **43**, 2334.

Part C

Human MR-PET Applications

CHAPTER 16

Brain

P. LOHMANN[a*], K.-J. LANGEN[a,c], N. GALLDIKS[b,d,e], R. RAJKUMAR[a,f], C. RÉGIO BRAMBILLAA[f] AND I. NEUNER[a,f]

[a]Forschungszentrum Jülich, Institute of Neuroscience and Medicine – 4, Wilhelm-Johnen-Straße 1, 52428 Jülich, Germany; [b]Forschungszentrum Jülich, Institute of Neuroscience and Medicine – 3, Wilhelm-Johnen-Straße 1, 52428 Jülich, Germany; [c]Department of Nuclear Medicine, RWTH Aachen University Hospital, Aachen, Germany; [d]Department of Neurology, University of Cologne, Cologne, Germany; [e]Centre of Integrated Oncology (CIO), Universities of Cologne and Bonn, Germany; [f]Department of Psychiatry, Psychotherapy and Psychosomatics, RWTH Aachen University Hospital, Aachen, Germany
*E-mail: p.lohmann@fz-juelich.de

16.1 Receptor Binding

The study of neurotransmitters using positron emission tomography (PET) is beneficial for the diagnosis of brain disorders and therapy control. Activation studies using a combination of PET and functional magnetic resonance imaging (fMRI) permit the analysis of functional networks in both health and disease and opens the possibility of being able to cross-calibrate different parameters. Furthermore, by using simultaneous MR-PET data acquisition many confounding parameters and changes in physiological parameters that might result from repositioning the patient are reduced.

New Developments in NMR No. 19
Hybrid MR-PET Imaging: Systems, Methods and Applications
Edited by N. Jon Shah
© The Royal Society of Chemistry 2019
Published by the Royal Society of Chemistry, www.rsc.org

Table 16.1 Radiotracers for neuroreceptor studies in PET.

PET radiotracer	Neurotransmitter group
[^{18}F]-flubatine	α4 β2 nicotinic acetylcholine receptors
[^{18}F]-2-fluoro-A85380	nAChR subunits
[^{18}F]-flumazenil	$GABA_A$
[^{11}C]-flumazenil	$GABA_B$
[^{18}F]-ASEM	α7-nAChR
[^{18}F]-fallypride, [^{18}F]-desmethoxyfallypride, [^{11}C]-raclopride, [^{11}C]-NMSP	Dopamine D2/D3 receptors
[^{18}F]-mefway, [^{11}C]-WAY-100635	Serotonin $5HT_{1A}$ receptors
[^{18}F]-nifene	Nicotinic acetylcholine or enzyme substrates
[^{18}F]-FPEB, [^{11}C]-ABP688	Metabotropic glutamate 5 receptors
[^{11}C]-carfentanil, [^{11}C]-diprenorphine	Opiate

MR-PET brain scans open a possibility for research applications, by combining metabolic markers with receptor binding, perfusion, and neurotransmitter release. Tracers for transmitters, receptors, and enzymes further elucidate the involvement of synaptic function in special tasks and uncover changes by diseases and drugs.[1] The highest potential from the simultaneous hybrid approach lies in pharmacological challenges in research studies, *e.g.* with substances like nicotine or ketamine.

A wide range of carbon-11 and fluorine-18 radiotracers have been developed for PET imaging to target specific neuroreceptor subtypes (Table 16.1). However, these radiotracers are mainly used for research purposes in PET centres with a cyclotron and qualified radiochemistry staff and are generally used to test for selectively binding at specific neuroreceptors, which are of interest in psychiatric disorders. PET radiotracers that bind to dopamine, glutamate, serotonin, opioid receptors and other sites have been used in human studies. Research has evaluated the changes in the neuroreceptors of psychiatric patients compared to healthy controls for schizophrenia, depression, substance abuse, mood disorders and other psychiatric conditions.[2–4]

Adding additional information relating to transmitter release and receptor occupancy, as well as on metabolism in the connected areas of functional networks, may be of special interest. Further to information acquired regarding structural connectivity, resting-state fMRI procedures allow the identification of functional connectivity between different brain areas through observation of the interregional correlation of physiological noise on a time scale of seconds.[5] MR-PET may provide a better understanding of the interaction between neurochemistry/transmitter-mediated processes and their functional, metabolic, and hemodynamic consequences both in healthy brains and in patients suffering various disorders.

Another highly interesting scientific application of MR-PET is for the simultaneous measurement of neurochemical changes (PET) and changes in regional brain activity using blood-oxygen-level dependent (BOLD) imaging, including the effects of activation by specific tasks on transmitter release or receptor binding. For example, during the learning of motor skills, the effects of regional neuronal activation might be dynamically monitored by BOLD-fMRI and, simultaneously detected by a PET study in which a dopamine receptor ligand is displaced from the receptor site.[6] The clever design and analysis of data obtained with specific tracers allows for the estimation of changes in receptor binding in response to a stimulus, a concept directly complementary to fMRI activation studies to investigate brain physiology. In our quest for understanding how the brain works, the great challenge and opportunity created by simultaneous MR-PET is to use dynamic measures of brain physiology from MRI and new generations of receptor-specific tracers to perform receptor-specific activation mapping using fMRI-like paradigms.[7]

Typically, the purpose of pharmacological or task challenges in receptor imaging studies with PET is to measure reductions in PET signal that can be attributed to reductions in specific receptor availability. It is vital, however, that these novel-imaging paradigms be expanded upon in order to strengthen our understanding of *in-vivo* ligand–receptor relationships as they apply to clinical disorders such as schizophrenia, addiction, attention deficit hyperactivity disorder (ADHD) and Tourette's syndrome. The ability to measure the stimulant-induced displacement of [^{11}C]-raclopride in the striatum with PET led to a series of human investigations to measure dopamine (DA) release following challenges such as caffeine, smoking (nicotine), alcohol, and Δ9-tetrahydrocannabinol (THC).[8–11] This kind of dopaminergic neuroimaging is particularly important for the investigation of substance abuse: the challenge being to understand the interactions of receptors and enzymes with the substances abused.[12] Based on dopaminergic neuroimaging, PET has revealed that the main effect of the blockade of dopamine transporters by cocaine, amphetamine, or methamphetamine is that dopamine occupancy is increased in the dopamine receptors. This is thought to be due to increased extracellular dopamine and is evidenced by a decline in the binding potential of the dopamine receptor ligand.[12]

Rosa-Neto and colleagues[13] showed the use of binding potentials to estimate the release of a neurotransmitter in patients with ADHD. Patients were treated orally with 20 mg of a cocaine analog and dopamine reuptake inhibitor methylphenidate. This treatment is believed to increase the synaptic concentration of dopamine and hence block the binding of the radioligand [^{11}C]-raclopride to such an extent that the binding is significantly less than in the untreated state of the patients.

One biomarker for the treatment of depression is the electrophysiological measure loudness dependence of auditory evoked potentials (LDAEP). In an exploratory fMRI set-up, PET imaging with [^{11}C]-flumazenil was conducted simultaneously with an EEG recording in a healthy male subject during acoustic stimulation.[14] The aim of this approach was both to investigate the success of combining MR-PET with EEG, and also to investigate the

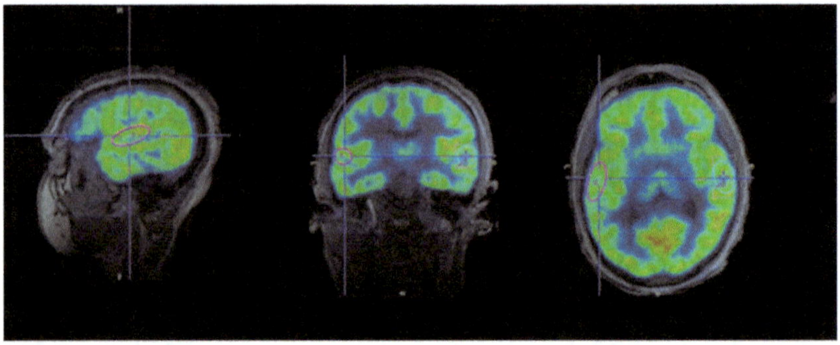

Figure 16.1 PET image in different orientations showing a region of interest (here primary auditory cortex), selected on both the left and right hemispheres of the brain.

role of the GABAergic system in the LDAEP response. The preliminary analysis of the results demonstrated the feasibility of the trimodal approach in humans. PET analysis was performed in the primary auditory cortex (region of interest) and the pons (region for extraction of control data as a region with low GABA receptors). The binding potential during the time frame of auditory stimulation in the auditory cortex was found to be 4.9 ± 0.3 (Figure 16.1).

The availability of metabotropic glutamate receptor 5 (mGluR5) radiotracers such as [^{11}C]-ABP688 and [^{18}F]-FPEB has led to investigations exploring their utility to measure endogenous glutamate levels.[15] Ceccarini and colleagues[16] found that reduced cerebral mGluR5 availability in alcohol-dependent subjects recovers during abstinence and has been linked to reduction in craving behaviours. Thus, higher striatal mGluR5 availability at the start of detoxification may be a predictive marker for long-term relapse.[16]

Neuroreceptor binding can be assessed by evaluating the changes in the distribution volumes or non-displaceable binding potential (BP_{ND}) of the radiotracer. However, confounding factors such as smoking status can influence the BP_{ND} values (Figure 16.2). A study from Akkus *et al.*[17] showed a global reduction in the analysis of receptor binding in studies with [^{11}C]-ABP688.

This non-invasive technique allows the same subject to be measured under different conditions and with different radiotracers, which grants the researcher the ability to measure differences in activity within one subject as well as between groups of subjects. Over the last two decades, methods in PET acquisition have been developed to image endogenous ligands and their relation to cognitive or pharmacological challenges. This significant advance has enabled researchers to explore the neurological mechanisms behind various neuropsychiatric disorders.[18] However, it is important that these novel imaging paradigms be expanded in order to strengthen our understanding of *in-vivo* ligand–receptor relationships and their role in

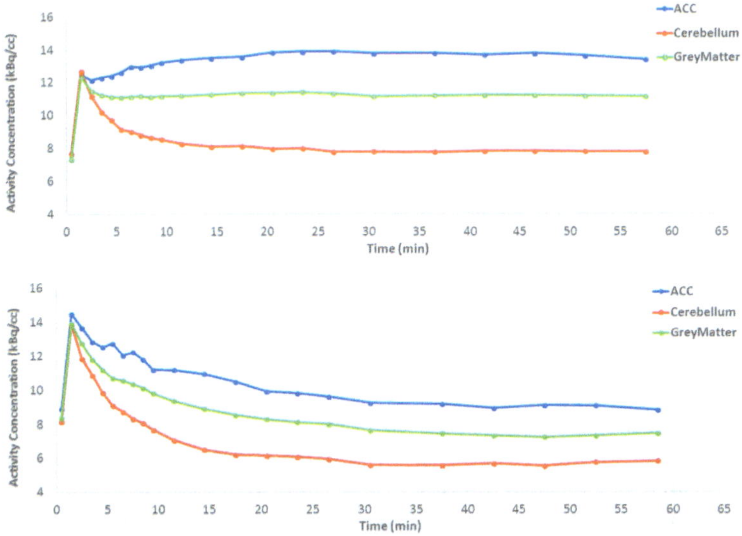

Figure 16.2 Time activity curves in anterior cingulate cortex, cerebellum grey matter and whole brain grey matter based on Hammer's atlas in a healthy non-smoker subject (upper) and a healthy smoker subject (lower). The differences in BP_{ND} showed a downregulation of the glutamatergic (mGluR5) receptors due to smoking status at about 20% with high significance $p < 0.001$.

clinical disorders, as some results appear to be inconsistent. For example, a number of studies reported a nicotine effect on DA release despite differences in the following: the nicotine-free period prior to their PET scans; administration of the nicotine challenge (oral, inhalation, and smoking); image acquisition and analysis methodology (bolus plus constant infusion or bolus administration of radiotracer; equilibrium, graphical or simplified reference tissue image analysis). Furthermore, in some studies, smoking nicotine-containing cigarettes produced modest changes in [^{11}C]-raclopride binding.[19–21]

Moreover, many studies have investigated whether D2 receptor availability is altered in schizophrenia compared to healthy control subjects. There have been more than twenty published studies, but results have been inconsistent. The involvement of glutamatergic mechanisms in schizophrenia has been hypothesised for many years and in 2017 Akkus and colleagues demonstrated that the mGluR5 receptors in the brains of subjects with schizophrenia showed no differences compared to healthy controls.[22]

As a result of recent advances in imaging paradigms, and with the developments in novel MR-PET methods, new possibilities to examine neurochemical abnormalities in various psychiatric and addiction disorders are possible. The future use of these new approaches may amplify our knowledge of neurochemical transmission *in-vivo* and allow the development of better treatments in brain disorders.

16.2 Neurodegeneration

Simultaneous MR-PET allows the correlation of both spatial and temporal signals, creating opportunities that might be impossible to realise when acquisitions are performed sequentially. Considering that MRI represents the first-option diagnostic imaging modality for a large number of clinical indications, the availability of this technology has important applications in translational research in the field of neuroscience. Its significance is additionally underlined by the ability of PET technology to detect radiation from radioactive labelled biomarkers, thus offering sensitivity in the picomolar range.[23] Furthermore, the range of PET specific tracers available for research has grown substantially and some PET radiotracers used for neurodegenerative applications are listed in Table 16.2.

MRI combines good soft-tissue contrast with no additional ionizing radiation and adds further functional data from spectroscopy, functional MRI, and dynamics of contrast media.[24] For clinical and research applications, hybrid MR-PET opens new possibilities to improve our knowledge of the complex functions of the brain and further understanding of disorders of the pathophysiology system. The high diagnostic value of MRI and PET procedures may turn MR-PET into the modality of choice for diagnostic questions directed towards brain pathologies.

Alzheimer's disease is one of the principal health-care challenges in modern society. Furthermore, it is expected that the prevalence of the disease will account for 1/3 of the total population over the age of around 85-years-old.[25] The early diagnosis of Alzheimer's dementia can be made by using MRI to detect hippocampal atrophy and PET to measure glucose consumption and also the accumulation of amyloid and tau. Furthermore, information relating to these parameters can be used for the selection of patients in

Table 16.2 PET radiotracers for neurodegenerative applications.

Targeting	Radiotracer	Features/assessment
Amyloid-β	[^{11}C]-PiB	Tracks amyloid plaques *in-vivo*.
	[^{18}F]-AZD4694	Evaluation of therapeutic purposes and disease modifying therapies.
	[^{18}F]-florbetapir, [^{18}F]-florbetaben and [^{18}F]-flutemetamol	Shows or excludes brain amyloid load in mild cognitive impairment.
Glucose metabolism	[^{18}F]-FDG	Study the characteristics of brain metabolism.
Tau protein	[^{18}F]-THK5351	Early detection of nerve fibre lesions in patients with Alzheimer's disease.
	[^{18}F]-T807, [^{18}F]-AV-1451	Imaging tau protein aggregates.
	[^{18}F]-FDDNP	Marking senile plaques and neuronal fibrous entanglement.
Microglial activation	[^{11}C]-PK11195	Brain inflammation and reactive gliosis.
Dopamine	[^{18}F]-fluorodopa	Dopamine synthesis and vesicular storage.

treatment trials. In addition to Alzheimer's dementia, this imaging modality also enables the differential diagnosis of other degenerative diseases, such as frontotemporal lobar degeneration, Huntington's disease and Parkinson's disease.[26,27]

In a study comparing the use of imaging modalities in patients with dementia, the combination of PET with [^{18}F]-FDG and MRI was found to be superior to the single modality approach and reported an accuracy of 94% for the differentiation of Alzheimer's disease and frontotemporal lobar degeneration.[28] In the investigation of Alzheimer's disease, it also might be clinically relevant to study the state of functional connectivity together with PET. Recent studies have been able to demonstrate a regional interrelation between hypometabolism using [^{18}F]-FDG and disruption of functional connectivity using fMRI in patients at risk of developing Alzheimer's disease.[29] Deep brain stimulation (*e.g.* repetitive transcranial magnetic stimulation or direct current stimulation of selected cortical areas) has been explored for the treatment of Alzheimer's disease.[7] Multiparametric functional imaging could improve our knowledge about the effect of stimulation therapy on functional networks (fMRI) and neuroreceptors (PET). The opportunity to obtain PET and MRI data simultaneously in a single imaging session also represents a major convenience for patients and clinical staff.

Another application of PET with [^{18}F]-FDG imaging is in epileptic patients, as it can be used to determine the regions in the brain which show the epileptogenic foci.[4]

The use of PET imaging to study epilepsy was first introduced in the 1980s following the observation of regional glucose hypometabolism in patients with partial seizures using [^{18}F]-FDG. A scan performed during the ictal period creates better conditions to detect the focus and further clinical relevance can be obtained when fusion imaging with MRI permits a precise anatomical localisation. When performed during a seizure, MR-PET remains the most sensitive, non-invasive tool for imaging temporal lobe seizures, both modalities enabling better anatomical location and detection in between seizures. The radiotracer [^{11}C]-flumazenil has been used to study the role of BZD receptors in epilepsy.[30] These BZD receptors are situated in the same region as the γ-aminobutyric acid (GABA) receptors.[30]

Dopamine is a 'phenethylamine' neurotransmitter and is produced in several areas of the brain, including the substantia nigra.[30,31] The radiotracer [^{18}F]-DOPA is widely used in the study of the dopaminergic system in movement disorders. [^{18}F]-DOPA is taken up by the terminals of dopaminergic neurons and is converted to [^{18}F]-dopamine by the enzyme dopa decarboxylase and other dopamine metabolites. Parkinson's disease occurs as a result of a loss in dopaminergic neurons in the substantia nigra.[32] The greatest loss of neurons is seen in the ventrolateral tier of the pars compacta, with lesser involvement in the dorsomedial tier.[31,32] These areas can be detected by PET using [^{18}F]-DOPA and show a decline in [^{18}F]-DOPA levels in the putamen in the contralateral area of the brain. The use of MR-PET in this area of research could open up a future clinical role not only for an early diagnosis and a prognostic evaluation, but also for a better definition of therapeutic strategies.

16.3 Neuro-oncology

Imaging plays an essential role in the management of brain tumour patients and directly influences treatment planning and ultimately patient survival. The most common and most aggressive primary malignant brain tumour in adults is the glioblastoma (WHO grade IV). Despite treatment (surgery, chemoradiation and adjuvant chemotherapy), prognosis is poor with a median survival of 1.5 years.

Until now, conventional MRI, including T_1- and T_2-weighted sequences, has been the method of choice for brain tumour imaging as it provides information about both the size and the location of a lesion. Furthermore, breakdown of the blood–brain barrier is a typical feature of malignant brain tumours and can be visualised by contrast enhancement following an intravenous injection of the paramagnetic contrast medium Gd-DTPA. This information can then be used for tumour grading or treatment monitoring. However, 9–53% of malignant brain tumours, more precisely astrocytomas, can appear as non-enhancing[33,34] or may extend beyond the area of contrast enhancement. Thus, an early and correct delineation of biologically active tumour tissue and a differentiation from non-specific, treatment-related changes can be challenging with standard MRI alone. Advanced MR methods such as magnetic resonance spectroscopic imaging (MRSI), diffusion-weighted imaging (DWI), perfusion-weighted imaging (PWI) or chemical exchange saturation transfer (CEST) imaging might be useful for brain tumour diagnostics, in addition to conventional MRI, and are currently under clinical investigation (Table 16.3).

Table 16.3 Current significance of amino acid PET and advanced MRI in brain tumour diagnostics.[b]

Clinical question	MRI	Amino acid PET	PWI	DWI	MRSI
Differential diagnosis	++	+	—	+	+
Tumour extent	+	++	+	—	++
Biopsy guidance	+	++	++	—	++
Grading	++	+[a]	++	+	+
Prognosis	+	+	++	++	+
Recurrence	+	++	++	—	—
Therapy monitoring	+	++	+	+	—

[a]Increased accuracy when dynamic [^{18}F]-FET–PET is used. DWI, diffusion-weighted imaging; MRSI, magnetic resonance spectroscopic imaging; PWI, perfusion-weighted imaging. Adapted from ref. 35.

[b]Legend: ++ high diagnostic accuracy, + limited diagnostic accuracy, — not helpful.

PET imaging has already proven its potential for brain tumour diagnostics in numerous studies in recent decades and is helpful to overcome some of the drawbacks of conventional MRI. PET measures the distribution of a radiolabelled biomolecule in the human body with a very high sensitivity and detects neoplastic tissue with a higher specificity than conventional MRI. Nearly any biomolecule can be labelled with a positron-emitting isotope for PET imaging, gaining insight into physiological or pathological processes of any kind. Although more than 70 different PET tracers have been investigated for brain tumour imaging in the last 30 years, radiolabelled amino acids have become the most important group of radiotracers for brain tumour diagnostics.[35]

Among the radiolabelled amino acids, most experience has been gained with [^{11}C]-methyl-L-methionine ([^{11}C]-MET). However, due to the short 20 min half-life of the carbon-11 isotope, the use of [^{11}C]-MET is restricted to centres with on-site radiotracer production. Thus, amino acid tracers labelled with fluorine-18, and with a longer half-life of 110 min, have been developed. The two most prominent examples used in brain tumour imaging are O-(2-[^{18}F]-fluoroethyl)-L-tyrosine ([^{18}F]-FET) and 3,4-dihydroxy-6-[^{18}F]-fluoro-L-phenylalanine ([^{18}F]-DOPA). [^{18}F]-DOPA was originally developed to assess dopamine synthesis in patients with movement disorders but is also used for brain tumour imaging.[36]

Despite the other options available, [^{18}F]-FET is, at present, the best evaluated fluorine-18 labelled amino acid PET tracer for brain tumour diagnostics and was approved by the Swiss Agency for Therapeutic Products as radiopharmaceutical for brain imaging with PET in 2014.[37] [^{18}F]-FET has shown its potential for tumour grading, delineation of the biologically active tumour volume for treatment planning, diagnosis of tumour recurrence, and treatment monitoring in several studies.[38–41]

Additionally, the time–activity curves of [^{18}F]-FET uptake, derived from dynamic PET scans, have been shown to yield additional information useful for the grading of gliomas,[42–44] the prognostication of untreated gliomas[45–47] and the differentiation of recurrence from radiation-induced changes.[48–50] Consequently, the Response Assessment in Neuro-Oncology (RANO) working group has recommended the use of amino acid PET in addition to MRI for brain tumour management.[51]

The introduction of hybrid MR-PET systems offers new possibilities to acquire PET and multiple MR contrasts simultaneously, thereby improving the overall diagnostic accuracy by combining information from both imaging modalities. Hybrid MR-PET systems offer improved patient comfort, reduced scanning times and a lower radiation exposure compared with PET/CT, which makes it an especially interesting option for paediatric patients.

Since no repositioning of the patient is necessary, the precision of the co-registration between PET and MR images is excellent. This is particularly important for the detection of small lesions and for the accurate planning of target sites for biopsy or radiotherapy.

Several studies have already demonstrated the value added by combining MRI and PET in brain tumour diagnostics. Pauleit *et al.* showed that the combination of MRI and [^{18}F]-FET PET in patients with cerebral gliomas improved the identification of tumour tissue; as a reference, areas showing abnormal PET and MRI signals were biopsied. The diagnostic accuracy for the distinction of tumour tissue from peritumoral brain tissue was 68% for MRI alone and increased to 94% when MRI and [^{18}F]-FET PET were combined.[52]

In a first study using a dedicated hybrid brain MR-PET scanner, Boss *et al.* showed good agreement between tumour to background ratios from MR-PET in comparison with those derived from a clinically approved PET-CT in 10 patients with intracranial lesions.[53]

Bisdas *et al.* investigated the combination of magnetic resonance spectroscopy (MRS) and [^{11}C]-MET PET for the classification of inconclusive MRI findings as low-grade or high-grade gliomas and the detection of the most malignant part within brain tumours. They showed good correlation of the area of maximum [^{11}C]-MET uptake with tumour grading, whereas the correlation between areas of high [^{11}C]-MET uptake and increased metabolite concentrations as measured by MRS was found to be poor.[54]

A more recent study from Mauler *et al.* also compared the spatial relationship of glioma volume derived from [^{18}F]-FET PET and MRSI and also showed a poor spatial overlap between the two methods. They concluded that the [^{18}F]-FET uptake and MRSI might represent different properties of glioma metabolism and should be further investigated.[55]

Filss *et al.* compared [^{18}F]-FET PET and PWI in 56 glioma and eight meningioma patients. Again, the spatial congruence was poor and the locations of the local maxima of each modality differed considerably. The authors concluded that the metabolically active tumour tissue as depicted by amino acid PET was not reflected by the relative cerebral blood volume as measured with PWI.[56]

A comparable study conducted by Henriksen *et al.* compared [^{18}F]-FET PET and blood volume obtained from contrast enhanced T_1-weighted MRI with respect to their spatial and quantitative agreement in 32 glioma patients. Although the tumour volumes determined by the two methods were comparable, the spatial congruence was also poor, and the methods yielded different information. The two methods might provide complementary information on brain tumour biology but require further investigation in order to identify a potential clinical value.[57]

Dunet *et al.* combined DWI with dynamic [^{18}F]-FET PET for glioma grading and showed the superiority of the combined approach over the single modality approach.[58]

As mentioned above, at present, the group most likely to benefit from hybrid MR-PET investigations is paediatric patients. A study by Preuss *et al.* performed simultaneous [^{11}C]-MET PET and MRI on a whole-body MR-PET scanner for biopsy planning and neuro-navigation in paediatric brain tumour patients. In all cases, representative tissue samples could be gained using the image data acquired in a single session, whereby additional anaesthesia was avoided, and the radiation exposure compared to PET-CT was reduced.[59]

However, in addition to the considerable cost factor, some further limitations of hybrid MR-PET systems have to be mentioned. The attenuation correction for quantitative PET based on MRI still leaves room for improvement[60] and the clinical value of the simultaneous acquisition of PET and MRI needs to be proven in further studies, as does the benefit of adding more advanced MRI sequences to the measurement protocols in brain tumour diagnostics. Additionally, patients with metal implants or other contraindications against MRI cannot be investigated in a hybrid MR-PET scanner.

The combination of metabolic and morphological imaging in hybrid MR-PET systems offers the potential to gain further insight into the biology of brain tumours and to differentiate different types of tissue. This will lead to improvements in the differential diagnosis of brain tumours, treatment monitoring and the early identification of tumour progression. Furthermore, it will provide a more reliable identification of treatment-related tissue changes.

References

1. P. Qin, X. Wu, N. W. Duncan, W. Bao, W. Tang, Z. Zhang, J. Hu, Y. Jin, X. Wu, L. Gao, L. Lu, Y. Guan, T. Lane, Z. Huang, Y. G. Bodien, J. T. Giacino, Y. Mao and G. Northoff, *Hum. Brain Mapp.*, 2015, **36**, 3867.
2. S. L. Kitson, 5-Hydroxytryptamine (5-HT) receptor ligands, *Curr. Pharm. Des.*, 2007, **13**, 2621.
3. R. V. Parsey and J. Mann, *Semin. Nucl. Med.*, 2003, **33**, 129.
4. S. L. Kitson, V. Cuccurullo, A. Ciarmiello, D. Salvo and L. Mansi, *Curr. Radiopharm.*, 2009, **2**, 224.
5. B. Biswal, F. Z. Yetkin, V. M. Haughton and J. S. Hyde, *Magn. Reson. Med.*, 1995, **34**, 537.
6. R. D. Badgaiyan, A. J. Fischman and N. M. Alpert, *Neuroimage*, 2007, **38**, 549.
7. C. Catana, A. Drzezga, W. D. Heiss and B. R. Rosen, *J. Nucl. Med.*, 2012, **53**, 1916.
8. M. Laruelle and A. Abi-Dargham, *J. Psychopharmacol.*, 1999, **13**, 358.
9. N. D. Volkow, J. S. Fowler and G. J. Wang, *J. Psychopharmacol.*, 1999, **13**, 337.
10. N. B. Urban, L. S. Kegeles, M. Slifstein, X. Xu, D. Martinez, E. Sakr, F. Castillo, T. Moadel, S. S. O'Malley, J. H. Krystal and A. Abi-Dargham, *Biol. Psychiatry*, 2010, **68**, 689.
11. N. Miyake, J. Thompson, M. Skinbjerg and A. Abi-Dargham, *CNS Neurosci. Ther.*, 2011, **17**, 104.
12. D. F. Wong and A. Gjedde, in *Encyclopedia of Neuroscience*, ed. L. R. Squire, Academic Press, Oxford, 2009, vol. 5, pp. 939–952.

13. P. Rosa-Neto, H. C. Lou, P. Cumming, O. Pryds, H. Karrebaek, J. Lunding and A. Gjedde, *Neuroimage*, 2005, **25**, 868.
14. I. Neuner, C. Wyss, R. Rajkumar, J. Mauler, F. Boers and L. Tellmann, Simultaneous trimodal imaging approach using [11C]-Flumazenil for MR-PET-EEG: feasibility of set-up, in PSMR 2016 – 5th Conference on PET/MR and SPECT/MR, 2016 May 23–25, Cologne, Germany, 2016.
15. N. Miyake, J. Thompson, M. Skinbjerg and A. Abi-Dargham, *CNS Neurosci. Ther.*, 2011, **17**, 104.
16. J. Ceccarini, G. Leurquin-Sterk, C. Crunelle, B. De Laat, G. Bormans, H. Peuskens and K. Van Laere, *J. Nucl. Med.*, 2017, **58**(supplement 1), 14.
17. F. Akkus, S. M. Ametamey, V. Treyer, C. Burger, A. Johayem, D. Umbricht, B. Gomez Mancilla, J. Sovago, A. Buck and G. Hasler, *Proc. Natl. Acad. Sci. U. S. A.*, 2013, **110**, 737.
18. L. Deuitch and R. Narendran, in *Imaging of the Human Brain in Health and Disease*, ed. P. Seaman and B. Madras, Elsevier Inc., New York, 2014, vol. 19, pp. 453–484.
19. A. L. Brody, R. E. Olmstead, E. D. London, J. Farahi, J. H. Meyer, P. Grossman, G. S. Lee, J. Huang, E. L. Hahn and M. A. Mandelkern, *Am. J. Psychiatry*, 2004, **161**, 1211.
20. A. L. Brody, M. A. Mandelkern, R. E. Olmstead, D. Scheibal, E. Hahn, S. Shiraga, E. Zamora-Paja, J. Farahi, S. Saxena, E. D. London and J. T. McCracken, *Arch. Gen. Psychiatry*, 2006, **7**, 808.
21. A. L. Brody, M. A. Mandelkern, R. E. Olmstead, Z. Allen-Martinez, D. Scheibal, A. L. Abrams, M. R. Costello, J. Farahi, S. Saxena, J. Monterosso and E. D. London, *Neuropsychopharmacology*, 2009, **34**, 282.
22. F. Akkus, V. Treyer, S. M. Ametamey, A. Johayem, A. Buck and G. Hasler, *Schizophr. Res.*, 2017, **183**, 95.
23. V. C. Spanoudaki and S. I. Ziegler, *Handb. Exp. Pharmacol.*, 2008, **185**, 53.
24. W. D. Heiss, *J. Nucl. Med.*, 2016, **57**, 993.
25. H. W. Querfurth and F. M. Laferla, *N. Engl. J. Med.*, 2010, **362**, 329.
26. A. Drzezga, H. Barthel, S. Minoshima and O. Sabri, *J. Nucl. Med.*, 2014, **55**, 47S.
27. R. Harada, N. Okamura, S. Furumoto, K. Furukawa, A. Ishiki, N. Tomita, T. Tago, K. Hiraoka, S. Watanuki, M. Shidahara, M. Miyake, Y. Ishikawa, R. Matsuda, A. Inami, T. Yoshikawa, Y. Funaki, R. Iwata, M. Tashiro, K. Yanai, H. Arai and Y. Kudo, *J. Nucl. Med.*, 2016, **57**, 208.
28. J. Dukart, K. Mueller, A. Horstmann, H. Barthel, H. E. Möller, A. Villringer, O. Sabri and M. L. Schroeter, *PLoS One*, 2011, **6**, e18111.
29. A. Drzezga, J. A. Becker, K. R. A. Van Dijk, A. Sreenivasan, T. Tanveer, C. Sullivan, A. P. Schultz, J. Sepulcre, D. Putcha, D. Greve, K. A. Johnson and R. A. Sperling, *Brain*, 2011, **134**, 1635.
30. S. Bouvard, F. Costes, F. Bonnefoi, F. Lavenne, F. Mauguiere, J. Delforge and P. Ryvlin, *Brain*, 2005, **128**, 1330.
31. N. D. Volkow, J. S. Fowler, S. J. Gatley, J. Logan, G. J. Wang, Y. S. Ding and S. Dewey, *J. Nucl. Med.*, 1996, **37**, 1242.

32. P. Piccini and A. Whone, *Lancet Neurol.*, 2004, **3**, 284.
33. F. G. Barker 2nd, S. M. Chang, S. L. Huhn, R. L. Davis, P. H. Gutin, M. W. McDermott, C. B. Wilson and M. D. Prados, *Cancer*, 1997, **80**, 936.
34. J. N. Scott, P. M. A. Brasher, R. J. Sevick, N. B. Rewcastle and P. A. Forsyth, *Neurology*, 2002, **59**, 947.
35. K. J. Langen, N. Galldiks, E. Hattingen and N. J. Shah, *Nat. Rev. Neurol.*, 2017, **13**, 279.
36. K. Herrmann, J. Czernin, T. Cloughesy, A. Lai, K. L. Pomykala, M. R. Benz, A. K. Buck, M. E. Phelps and W. Chen, *Neuro-Oncology*, 2014, **16**, 603.
37. Swiss Agency for Therapeutic Products, *J. Swissmedic.*, 2014, **13**, 651.
38. V. Dunet, C. Rossier, A. Buck, R. Stupp and J. O. Prior, *J. Nucl. Med.*, 2012, **53**, 207.
39. N. Galldiks, K. J. Langen, R. Holy, M. Pinkawa, G. Stoffels, K. W. Nolte, H. J. Kaiser, C. P. Filss, G. R. Fink, H. H. Coenen, M. J. Eble and M. D. Piroth, *J. Nucl. Med.*, 2012, **53**, 1048.
40. D. Pauleit, F. Floeth, K. Hamacher, M. J. Riemenschneider, G. Reifenberger, H. W. Muller, K. Zilles, H. H. Coenen and K. J. Langen, *Brain*, 2005, **128**, 678.
41. G. Pöpperl, C. Gotz, W. Rachinger, F. J. Gildehaus, J. C. Tonn and K. Tatsch, *Eur. J. Nucl. Med. Mol. Imaging*, 2004, **31**, 1464.
42. P. Lohmann, H. Herzog, E. Rota Kops, G. Stoffels, N. Judov, C. Filss, N. Galldiks, L. Tellmann, C. Weiss, M. Sabel, H. H. Coenen, N. J. Shah and K. J. Langen, *Eur. Radiol.*, 2015, **25**, 3017.
43. M. L. Calcagni, G. Galli, A. Giordano, S. Taralli, C. Anile, A. Niesen and R. P. Baum, *Clin. Nucl. Med.*, 2011, **36**, 841.
44. G. Pöpperl, F. W. Kreth, J. H. Mehrkens, J. Herms, K. Seelos, W. Koch, F. J. Gildehaus, H. A. Kretzschmar, J. C. Tonn and K. Tatsch, *Eur. J. Nucl. Med. Mol. Imaging*, 2007, **34**, 1933.
45. N. L. Jansen, V. Graute, L. Armbruster, B. Suchorska, J. Lutz, S. Eigenbrod, P. Cumming, P. Bartenstein, J. C. Tonn, F. W. Kreth and C. la Fougère, *Eur. J. Nucl. Med. Mol. Imaging*, 2012, **39**, 1021.
46. N. L. Jansen, B. Suchorska, V. Wenter, S. Eigenbrod, C. Schmid-Tannwald, A. Zwergal, M. Niyazi, M. Drexler, P. Bartenstein, O. Schnell, J. C. Tonn, N. Thon, F. W. Kreth and C. la Fougère, *J. Nucl. Med.*, 2014, **55**, 198.
47. N. L. Jansen, B. Suchorska, V. Wenter, C. Schmid-Tannwald, A. Todica, S. Eigenbrod, M. Niyazi, J. C. Tonn, P. Bartenstein, F. W. Kreth and C. la Fougère, *J. Nucl. Med.*, 2015, **56**, 9.
48. N. Galldiks, G. Stoffels, C. P. Filss, M. D. Piroth, M. Sabel, M. I. Ruge, H. Herzog, N. J. Shah, G. R. Fink, H. H. Coenen and K. J. Langen, *J. Nucl. Med.*, 2012, **53**, 1367.
49. G. Ceccon, P. Lohmann, G. Stoffels, N. Judov, C. P. Filss, M. Rapp, E. Bauer, C. Hamisch, M. I. Ruge, M. Kocher, K. Kuchelmeister, B. Sellhaus, M. Sabel, G. R. Fink, N. J. Shah, K. J. Langen and N. Galldiks, *Neuro-Oncology*, 2017, **19**, 281.
50. N. Galldiks, V. Dunkl, G. Stoffels, M. Hutterer, M. Rapp, M. Sabel, G. Reifenberger, S. Kebir, F. Dorn, T. Blau, U. Herrlinger, P. Hau, M. I. Ruge, M.

Kocher, R. Goldbrunner, G. R. Fink, A. Drzezga, M. Schmidt and K. J. Langen, *Eur. J. Nucl. Med. Mol. Imaging*, 2015, **42**, 685.
51. N. Galldiks, G. Stoffels, C. Filss, M. Rapp, T. Blau, C. Tscherpel, G. Ceccon, V. Dunkl, M. Weinzierl, M. Stoffel, M. Sabel, G. R. Fink, N. J. Shah and K. J. Langen, *Neuro-Oncology*, 2015, **17**, 1293.
52. N. L. Albert, M. Weller, B. Suchorska, N. Galldiks, R. Soffietti, M. M. Kim, C. la Fougère, W. Pope, I. Law, X. Arbizu, M. C. Chamberlain, M. Vogelbaum, B. M. Ellingson and J. C. Tonn, *Neuro-Oncology*, 2016, **18**, 1199.
53. A. Boss, S. Bisdas, A. Kolb, M. Hofmann, U. Ernemann, C. D. Claussen, C. Pfannenberg, B. J. Pichler, M. Reimold and L. Stegger, *J. Nucl. Med.*, 2010, **51**, 1198.
54. S. Bisdas, R. Ritz, B. Bender, C. Braun, C. Pfannenberg, M. Reimold, T. Naegele and U. Ernemann, *Invest. Radiol.*, 2013, **48**, 295.
55. J. Mauler, A. A. Maudsley, K. J. Langen, O. Nikoubashman, G. Stoffels, S. Sheriff, P. Lohmann, C. P. Filss, N. Galldiks, E. Rota Kops and N. J. Shah, *J. Nucl. Med.*, 2018, **59**, 603.
56. C. P. Filss, N. Galldiks, G. Stoffels, M. Sabel, H. J. Wittsack, B. Turowski, G. Antoch, K. Zhang, G. R. Fink, H. H. Coenen, N. J. Shah, H. Herzog and K. J. Langen, *J. Nucl. Med.*, 2014, **55**, 540.
57. O. M. Henriksen, V. A. Larsen, A. Muhic, A. E. Hansen, H. B. Larsson, H. S. Poulsen and I. Law, *Eur. J. Nucl. Med. Mol. Imaging*, 2016, **43**, 103.
58. V. Dunet, P. Maeder, M. Nicod-Lalonde, B. Lhermitte, C. Pollo, J. Bloch, R. Stupp, R. Meuli and J. O. Prior, *Nuklearmedizin*, 2014, **53**, 155.
59. M. Preuss, P. Werner, H. Barthel, U. Nestler, H. Christiansen, F. W. Hirsch, D. Fritzsch, K. T. Hoffmann, M. K. Bernhard and O. Sabri, *Childs Nerv. Syst.*, 2014, **30**, 1399.
60. C. N. Ladefoged, I. Law, U. Anazodo, K. St. Lawrence, D. Izquierdo-Garcia, C. Catana, N. Burgos, M. J. Cardoso, S. Ourselin, B. Hutton, I. Mérida, N. Costes, A. Hammers, D. Benoit, S. Holm, M. Juttukonda, H. An, J. Cabello, M. Lukas, S. Nekolla, S. I. Ziegler, M. Fenchel, B. Jakoby, M. E. Casey, T. Benzinger, L. Højgaard, A. E. Hansen and F. L. Andersen, *Neuroimage*, 2017, **147**, 346.

CHAPTER 17

Clinical Applications of Whole-body MR-PET

K. AUF DER SPRINGE[b], B. NOTO[a], F. BÜTHER[a], M. WILDGRUBER[b] AND L. STEGGER*[a]

[a]University Hospital Münster, Department of Nuclear Medicine, Albert-Schweitzer-Campus 1, 48149 Münster, Germany; [b]University Hospital Münster, Department of Radiology, Albert-Schweitzer-Campus 1, 48149 Münster, Germany

*E-mail: lars.stegger@ukmuenster.de

17.1 General Considerations

The combination of positron emission tomography (PET) and computed tomography (CT) in one device (PET-CT) has proven to be extremely useful and has made PET-CT a very popular tool for whole-body imaging, especially in oncology. Indeed, following its commercial introduction at the beginning of the millennium, combined PET-CT devices have replaced standalone PET for new sales within a relatively short space of time. In comparison to separate PET and CT acquisitions with subsequent side-by-side interpretation of the imaging results, the combined approach offers a much faster completion of the examination by using CT for PET attenuation correction and by obviating the need for a second patient positioning. Additionally, combined PET-CT allows for a much-improved spatial co-registration. However,

New Developments in NMR No. 19
Hybrid MR-PET Imaging: Systems, Methods and Applications
Edited by N. Jon Shah
© The Royal Society of Chemistry 2019
Published by the Royal Society of Chemistry, www.rsc.org

temporal co-registration is still imperfect, since both examinations are performed consecutively and not simultaneously. Consequently, co-registration is still hampered by time-variant physiological processes such as respiration, heart contraction, bowel peristaltic and filling of the urinary bladder. Another crucial advantage of combined PET-CT imaging over separate examinations is improved patient logistics and higher patient comfort. Since only one imaging session is needed to acquire both PET and CT, and since the imaging report covers both CT and PET findings, imaging can give definitive answers to the referring physician more quickly and efficiently.[1] Although PET-CT has proven to be a very powerful tool for whole-body imaging there are still some associated disadvantages, such as the use of ionising radiation and the significantly poorer soft tissue contrast provided by CT in comparison to MRI.

For MR-PET to supplant PET-CT for whole-body applications, it must offer meaningful advantages over PET-CT as well as over stand-alone MRI. First, MRI does not work with ionising radiation, thereby reducing the radiation burden to the patient significantly in comparison to PET-CT. This is especially relevant in the PET imaging of children and adolescents and in patients with frequent follow-up examinations and a high survival rate from their malignant diseases, as well as in patients with benign disorders. For example, Chawla *et al.* investigated the radiation dose for children with malignancies undergoing PET-CT examinations. They calculated the average dose of a PET-CT study to be about 24.9 mSv, with the CT study contributing 20.3 mSv and PET 4.6 mSv.[2] Second, MRI yields different information compared to CT, especially regarding better soft-tissue contrast and multiparametric approaches with functional imaging and tissue characterisation (*e.g.* perfusion, diffusion, spectroscopy). Third, true simultaneous imaging of PET and MRI is possible and may offer the opportunity for MR-based improvement of PET image quality and quantification accuracy. The correction of motion effects due to respiration or heart contraction is possible by measuring the motion of the body with MRI and using this information in the PET reconstruction process. Partial-volume correction, which is routinely used for brain imaging, may also find use in whole-body applications by using the morphoanatomical information provided by MRI.

Despite these clear advantages, there are several disadvantages to be acknowledged for MR-PET in comparison to PET-CT. With current detector technology, PET image acquisition is fast, with imaging durations of 1–3 min per bed-position considered the norm. This matches well with the very short CT image acquisition times. However, due to the poor signal-to-noise ratio (SNR) inherent to MRI and the ever-present desire for high spatial resolution, comprehensive MRI is generally a time-consuming process. Even though advances in fast imaging have accelerated image acquisition considerably, there are still trade-offs between spatial and temporal resolution in MRI that do not apply to CT. While PET is usually applied as a whole-body examination, MRI is strongest when applied for imaging of a circumscribed body region. This is especially relevant for whole-body imaging with MR-PET. An important task for practitioners of MR-PET and for clinical scientists is therefore to devise imaging protocols optimised for the tasks at hand. Protocols may combine fast whole-body imaging with additional in-depth MRI coverage of a relevant body part thereafter. For example, for imaging of patients with neuroendocrine tumours, this may involve further imaging of the liver in addition to whole-body MR-PET.

Furthermore, MRI is generally less well suited to imaging lung tissue and to depicting small pulmonary metastases than CT, even though the capability of MRI for the detection of small pulmonary nodules has increased in recent years. Also, cortical bone does not provide a relevant signal for MRI in most standard sequences due to the absence of water protons. Patients with metal implants of a certain size, location or type and with many implanted devices such as pacemakers or pumps are unsuitable for MR-PET. The narrower patient gantry in MR-PET *vs.* PET-CT, the surface coils around the patient, the significant noise generated by MRI and the generally longer examination durations are further factors that disadvantage MR-PET. Imaging under general anaesthesia is possible with MR-PET, however MR-compatible devices are required, and compared with PET-CT, access to the patient is inferior.

In conclusion, it is highly unlikely that MR-PET will replace PET-CT completely for whole-body applications. Nevertheless, the above advantages have the potential to render MR-PET the method of choice for a variety of indications. In this chapter, examples of promising whole-body clinical applications of MR-PET are presented.

17.2 Applications in Oncology

17.2.1 Prostate Cancer

With nearly 9% of men being diagnosed with prostate cancer by the age of 75, prostate cancer has the highest incidence of all cancers (excluding non-melanoma skin cancers) in developed countries and, indeed, 0.8% of men die from prostate cancer by the age of 75.[3]

Prostate cancer is often diagnosed by screening in a locally confined state and is then actively surveyed or treated locally (surgery, brachytherapy or teletherapy). Most common locations of metastases are the iliac and pelvic lymph nodes and bones of the pelvis and the spine. Imaging plays an important role in the detection and staging of prostate cancer, in guiding treatment and in the localisation of recurrent tumour in cases of increasing PSA levels during follow-up. MR-PET may have several advantages over PET-CT for this important tumour entity. For primary tumours, as MRI can depict the soft-tissue extent and thus the tumour margins better than CT, it may help to identify the tumour by its decreased diffusion due to high cellularity (see example in Figure 17.1).

As the uptake of FDG by prostate cancer cells is low in most cases (it can increase in de-differentiated prostate cancer cells), FDG-PET has limited

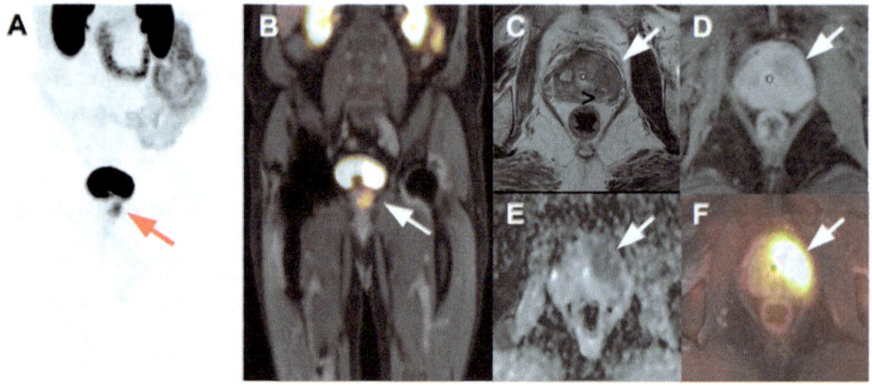

Figure 17.1 ^{68}Ga-PSMA-MR-PET of a patient with prostate cancer. (A) Maximum intensity projection (MIP) of PET. (B) Coronal slice of PET fused with MRI, (C)–(F) T2w, CE T1w, ADC MRI and fusion images dissecting the prostate gland. Arrow pointing to prostate carcinoma.

clinical applications.[4] FDG-PET may help during radiotherapy to identify de-differentiated tumour locations and for the ductal variant of prostate cancer when PSMA expression is low.[5] Therefore, other PET tracers have been evaluated that better capture differentiated prostate cancer. The two most important of these are choline or fluorocholine, depicting membrane synthesis, and more recently, PSMA ligands. Offering a higher contrast and often a better uptake in patients with low values of PSA, PSMA has invigorated the hybrid imaging of prostate cancer with PET. PSMA PET-CT has shown promising results in the depiction of intraprostatic tumour extent in primary staging[6] as well as in detecting local recurrences and metastases in patients with biochemical relapse of prostate cancer.[7]

Regarding MR-PET, one study was able to prove that MR-PET significantly outperformed multiparametric MRI and stand-alone PET in the localisation of the primary tumour.[8] Other studies showed the superiority of MR-PET over PET-CT for the detection of local recurrences.[9]

The first preliminary results also hint at a superiority of PMSA-MR-PET compared to PSMA-PET-CT in the detection of nodal and osseous metastases in the setting of biochemical recurrence of prostate cancer. In two studies comparing PSMA-MR-PET to PSMA-PET-CT, comprising 26 and 21 patients, respectively, a high concordance of PSMA PET positive findings between the two modalities was found.[10,11] However, PET positive osseous metastases that could not be morphologically confirmed in CT were morphologically confirmed using MRI. However, there are some possible limitations to PSMA-MR-PET. Concerning both MR-PET and PET-CT, PSMA-negative metastases have been reported, possibly due to a heterogeneous PSMA-expression of prostate cancer cells; addressing this calls for high image quality from MRI (or CT) (see example in Figure 17.2).

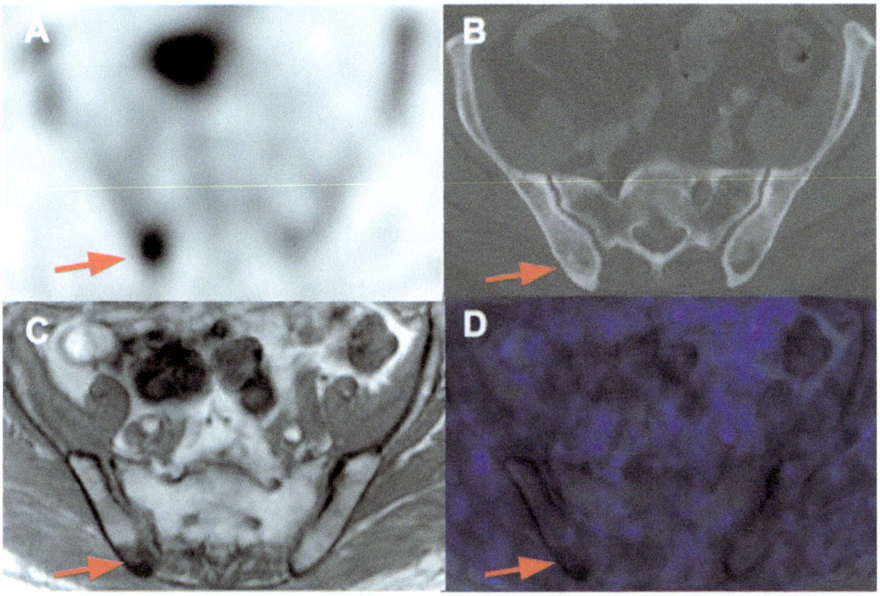

Figure 17.2 ^{99m}Tc-MDP bone SPECT and ^{68}Ga-PSMA-MR-PET of a patient with recurrent prostate cancer. (A) SPECT showing intense focal uptake in the iliac bone. (B) CT of the pelvis with inconclusive findings. (C) T1w MRI shows loss of fatty signal in the lesion. (D) PET fusion images show lack of tracer uptake. Arrows point to tumour lesion.

Halo artefacts, areas of artificially reduced tracer signal surrounding structures with high tracer accumulations (kidneys and bladder), may be more pronounced in MR-PET than in PET-CT (see example in Figure 17.3).[11,12] Although ligands with a different excretion route and newer software for attenuation and scatter correction and subsequently image reconstruction can alleviate this problem.

17.2.2 Liver Malignancies

Primary and secondary liver malignancies are frequently encountered in routine hybrid imaging using PET-CT and MR-PET. Liver metastases resulting from a variety of solid tumours of the gastrointestinal tract, but also other tumour entities, occur far more frequently than primary liver tumours, such as hepatocellular or cholangiocellular carcinomas. MRI can detect small liver lesions very well and usually better than CT;[13] therefore, MR-PET can be expected to be well-suited for imaging of the liver. Small liver lesions may also be missed in the FDG-PET component, due to the physiological tracer uptake in normal liver tissue and consequently weaker lesion to background contrast of malignancies. For the primary characterisation of liver lesions, a liver-specific contrast agent can be used for MR-PET (Figure 17.4). For

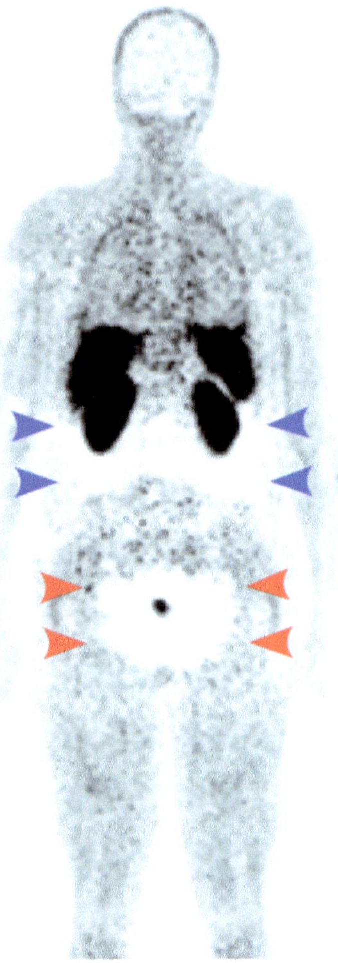

Figure 17.3 ^{68}Ga-PSMA-MR-PET of a patient with prostate cancer. Whole-body MIP of the PET shows relevant halo artefacts around the kidneys (blue arrowheads) and the bladder (red arrowheads).

staging purposes, dynamic imaging of the liver using one of the standard contrast agents is possible.

Several studies have demonstrated that MR-PET is at least as reliable as PET-CT in the depiction of liver malignancies. In a comparison of FDG-MR-PET with a whole-body protocol and PET-CT with a contrast-enhanced protocol in 70 patients with various solid tumours and in 96 liver lesions in a cohort of 36 patients, both methods detected the same number of metastatic lesions (26). However, lesion conspicuity and diagnostic confidence was better in MR-PET, even without specific imaging of the liver (*e.g.* no dynamic contrast enhanced MRI).[14] An example from a patient with medullary thyroid carcinoma and liver metastases is shown in Figure 17.5.

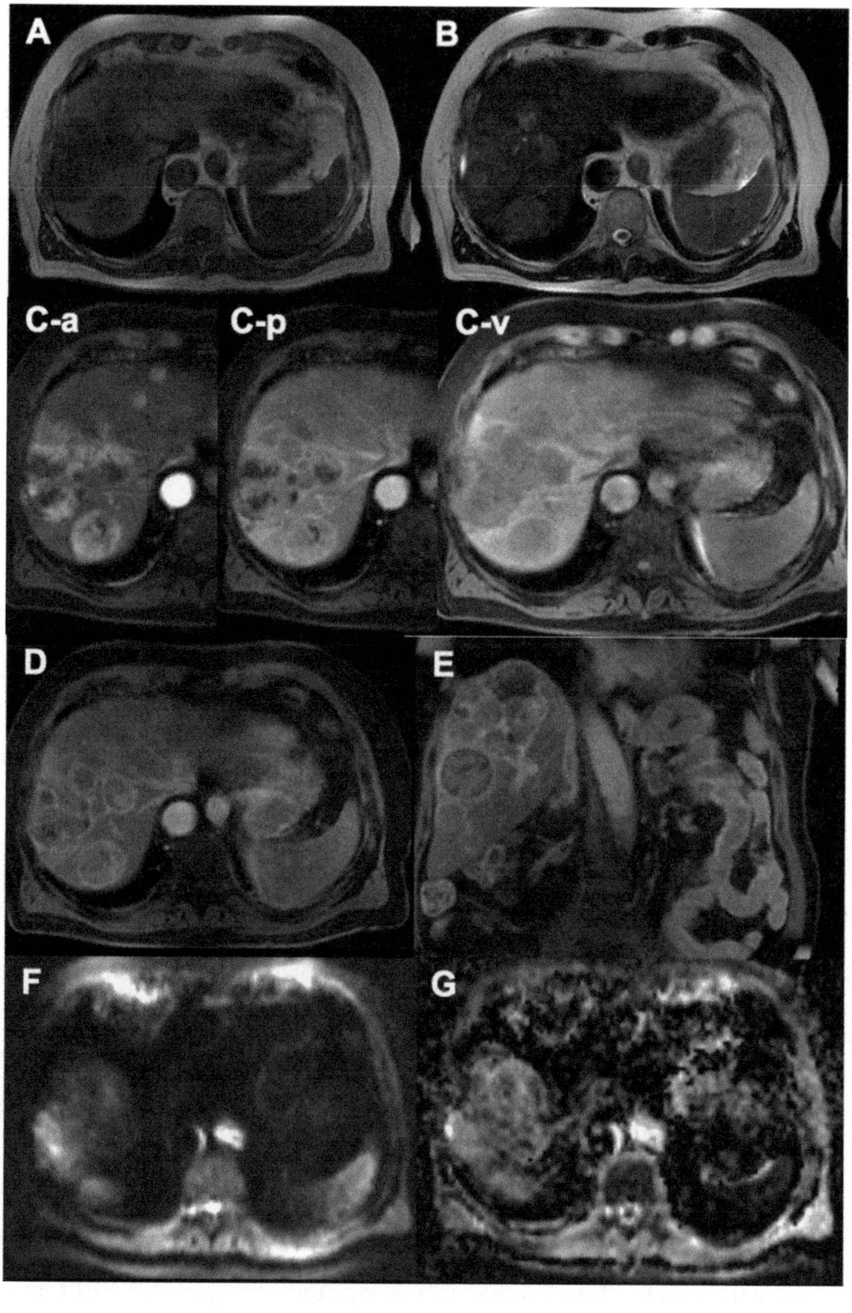

Figure 17.4 MRI evaluation of the liver using a liver-specific contrast agent. (A) T1w MRI. (B) T2w MRI. (C) Arterial, portalvenous and venous phases of contrast-enhancement MRI following contrast injection. Late contrast-enhanced MRI images in two directions (D) and (E). DWI and ADC map (F) and (G).

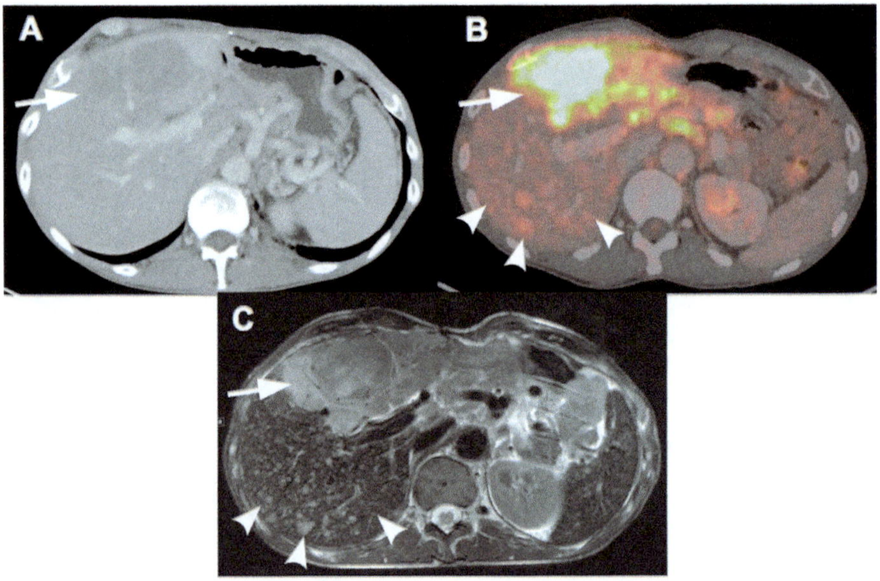

Figure 17.5 [^{18}F]-FDG-MR-PET of a patient with medullary thyroid carcinoma. (A) CT of the liver, depicting the large ventral metastases (arrow) but not the small dorsal metastases. (B) PET fusion images showing marked tracer uptake in the large metastases (arrow) and barely visible increased uptake in several of the smaller metastases (arrowheads) dorsally. (C) T2w MRI depicting the larger (arrow) and disseminated small metastases (arrowheads).

In a study comprising 37 patients, significantly more metastases were detected in a retrospective software fusion of ^{18}F-FDG-PET with an MRI with a dedicated liver protocol and liver-specific contrast agent than with PET-CT (93% *vs.* 76%). The authors concluded that both gadoxetic-acid enhanced MRI alone and software-fused MR-PET were more sensitive in the detection of liver metastases than PET-CT.[15]

A rather specific application of PET is the imaging of the distribution of radioactive particles labelled with the beta emitter yttrium-90 instilled in the liver tumour vasculature during a selective internal radiation therapy procedure (SIRT). Here, particles are administered into the arterial bloodstream of the liver *via* an angiographic catheter. Shunting to the arteries subtending the gastrointestinal organs is the most feared complication in SIRT that should be detected *via* imaging after SIRT. This can be imaged with single-photon emission computed tomography (SPECT) that can detect the *bremsstrahlung* produced when the loaded beta particle is deflected in tissue. A very small positron emission component allows for the alternative detection of yttrium-90 loaded particles with PET-CT and MR-PET (an example is given in Figure 17.6).

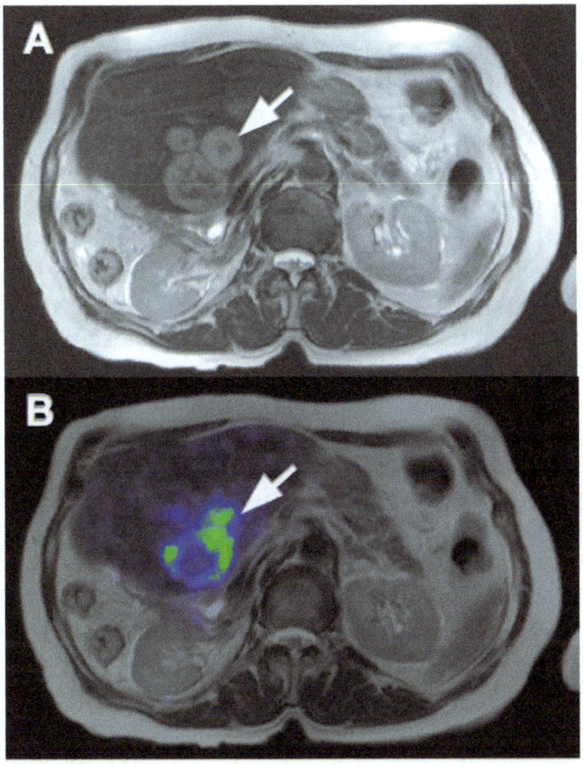

Figure 17.6 ^{90}Y-SIR-Spheres-MR-PET (Sirtex Medical Europe, Germany) of a patient with prostate cancer. (A) T2w MRI and (B) PET fusion image. Arrow points to tumorous uptake.

17.2.3 Neuroendocrine Tumours

Neuroendocrine cells are at the junction of the neuronal and endocrine system and are found throughout the body in different tissues. Tumours arising from these cells, the so-called neuroendocrine tumours, are therefore a heterogeneous group, which includes gastroenteropancreatic tumours (the most common group), catecholamine-secreting tumours (phaeochromocytomas, paragangliomas, neuroblastoma), medullary carcinoma of the thyroid, pituitary tumours, small cell lung cancer and Merkel cell tumours. Although representing a rare tumour entity, the incidence rate of neuroendocrine tumours has risen 6.4-fold from 1973 to 2012, possibly due to detection of early-stage disease and stage migration.[16] The treatment of these tumours is dependent on their grading. G1 and G2 NET with a proliferation rate of less than 20% can be treated with drugs targeting somatostatin receptors (SSTRs) both with or without a radioactive label. For differentiated or de-differentiated G3 neuroendocrine carcinomas, treatment regimens include chemotherapy at an early stage. The presence of SSTRs determines whether a treatment attempt with drugs targeting these receptors can be considered. Detailed treatment recommendations are published by the European Neuroendocrine Tumour Society (ENETS).

For the detection of SSTRs, a variety of radiopharmaceuticals have been developed that allow imaging with conventional nuclear imaging techniques using indium-111 or technetium-99m as radioisotopes attached to the SSTR ligands OCTREOTIDE and its derivatives. Similar radiotracers for PET are also available that afford additional benefits from the better resolution and image contrast of this imaging modality. In recent years, PET-CT with ligands such as DOTATATE, DOTOTOC OR DOTANOC attached to radioactive gallium-68 has replaced conventional scintigraphy more and more in countries with the availability and reimbursement of PET. These PET tracers show a varying avidity to the different subtypes of the SSTRs, with a good avidity to the SSTR-2 subtype.[17] Considering the superior soft tissue contrast of MRI, MR-PET may be a suitable imaging method for this tumour entity (with the possible exception of lung tumours). The first studies comparing the modalities showed that MR-PET was either superior to or equal in performance to PET-CT. A prospective study of 30 patients with histopathologically confirmed NET found that a better lesion conspicuity is obtained for ^{68}Ga-DOTATOC MR-PET when compared to ^{68}Ga-DOTATOC-PET-CT.[18] However, another study found a comparable performance of the two modalities in a study comprising 28 patients.[19] MR-PET targeting SSTRs can be used to characterise known lesions, to perform staging, to assess treatment options and to monitor treatment efficacy (an example is provided in Figure 17.7).

17.2.4 Malignancies in Children and Adolescents

MR-PET can be especially valuable in children and adolescents due to the absence of ionising radiation in MRI. Ionising radiation poses a much higher risk for children and adolescents than for adults: cellular sensitivity to radiation is higher in young people and adverse life-time effects that may be induced by radiation (secondary malignancies, genetic effects

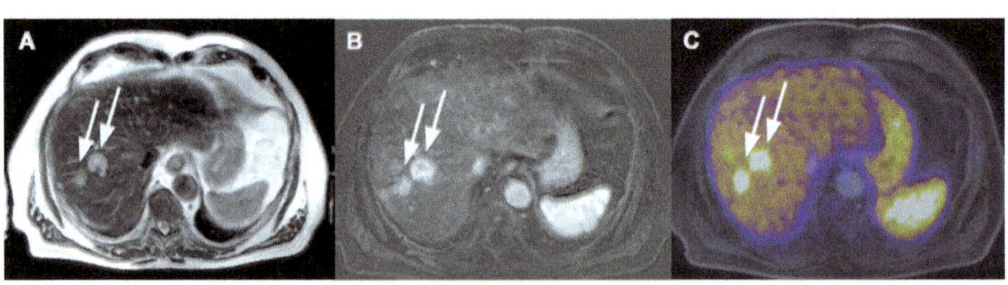

Figure 17.7 ^{68}Ga-Dotatate-MR-PET of a patient with liver metastases from a neuroendocrine tumour. T2w and CE T1w MRI depicting intrahepatic metastases (arrows) (A) and (B). (C) PET fusion images show intense tracer uptake.

in offspring) are more likely to become relevant. Additionally, due to the much-improved treatment regimens, long-term survival has become possible in a large proportion of affected children, especially for lymphoma or leukaemia. Therefore, long-term radiation effects are becoming more relevant. Furthermore, there is the expectation that the clinical performance of MR-PET may be better than PET-CT in a variety of solid tumours and haematological malignancies. An investigation of the radiation burden and the diagnostic performance of MR-PET in adolescents with a variety of malignancies referred for PET-CT imaging brought interesting results.[20] The authors estimate the mean effective dose from PET-CT as 23 ± 11 mSv, where PET contributed 5.6 ± 1.5 mSv and CT accounted for 18.2 ± 10 mSv. MR-PET, therefore, promises to significantly reduce the radiation burden. In addition, they were able to show that the PET component was equally well-suited to depict malignant tracer uptake, despite some observed artefacts caused by attenuation correction. A single pulmonary lesion out of 62 was missed in comparison to PET from PET-CT. The MR component, however, could depict four malignant lesions without discernible tracer uptake, which were invisible in PET-CT (recurrent liver tumour in hepatocellular carcinoma, bone marrow lesions in lymphoma and neuroblastoma, renal infiltration in lymphoma). On the downside, small lung metastases were more visible in CT. These data again show the advantage of MRI for imaging of the liver and the bone marrow. Other groups could also show that MR-PET can replace PET-CT in the management of lymphomas without losing diagnostic efficacy but with significantly reduced radiation exposure.[21,22] An example is provided in Figure 17.8.

The application of MR-PET instead of PET-CT in solid tumours with frequent pulmonary metastases is the subject of continuous debate. MR-PET may have advantages in patients with primary bone tumours, such as osteosarcoma and Ewing sarcoma, for the depiction of soft-tissue components and invasion into adjacent structures and also for the detection of osseous skip metastases. These are typical applications of MRI as the second-line imaging modality of choice. In particular, the detection of bone marrow metastases may be helped significantly by MRI being accompanied by PET (an example is provided in Figure 17.9). However, early detection of pulmonary metastases is equally important. Here CT still has the edge over MRI[23] and to ensure optimal diagnostic performance, MR-PET may have to be complemented by a low-dose CT of the lung.

17.2.5 Other Malignant Tumours

MR-PET has been evaluated, mostly in comparison to PET-CT, in a range of other malignant tumours requiring whole-body imaging for staging, assessment of therapies and during follow-up for recurrence detection. Among these are head-and-neck cancers, gastrointestinal cancers, gynaecological cancers and cancer with an unknown primary (CUP).[24–28] Again, the rationale to replace the standard PET-CT for MR-PET is often centred

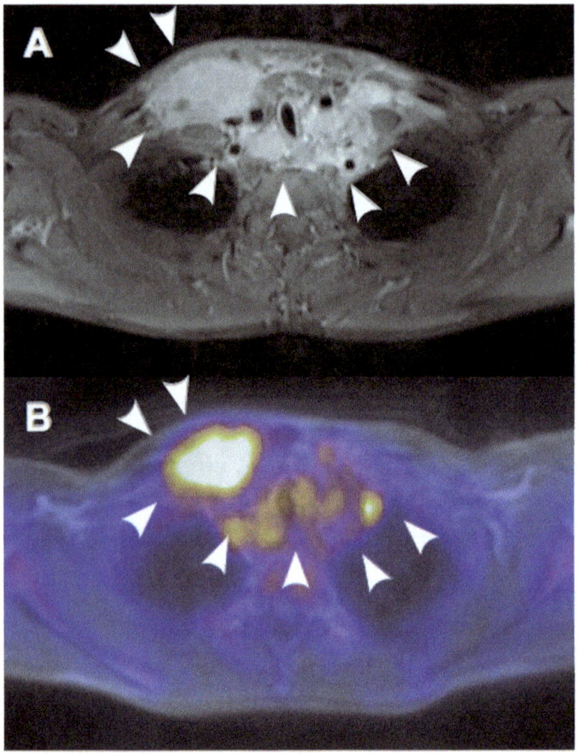

Figure 17.8 [^{18}F]-FDG-MR-PET of a patient with Hodgkin's lymphoma for initial staging. (A) CE T1w MRI depicting lymphoma extent with regard to important structures. (B) PET fusion images depicting mostly intense but varying tracer uptake. Arrowhead enclose lymphoma manifestations.

on the key benefits of MRI *vs.* CT, namely the non-existent ionising radiation, the better soft-tissue contrast and the ability to better assess bone marrow involvement. As an example, for the assessment of patients with differentiated thyroid carcinoma, elevated thyroglobulin and negative iodine scan, it could be demonstrated that FDG-MR-PET is effective. In addition to standard whole-body MR sequences, higher resolution MRI was additionally used for neck imaging. FDG-MR-PET proved equivalent to FDG-PET-CT (with high-dose contrast-enhanced CT) for the neck but inferior for lung nodules.[29] The tracer ^{68}Ga-Dotatate, used with MR-PET, was more sensitive in detecting extrapulmonary tumour manifestations than FDG-PET-CT. Both were markedly superior to whole-body diffusion-weighted MRI.[30] In patients with breast cancer, several authors have used FDG-MR-PET not only for whole-body staging but also for local assessment of the breast. A local breast imaging protocol in a prone position can build on the track-record of MRI for breast imaging, which is far superior to CT; CT has a good sensitivity but lower specificity.[31,32]

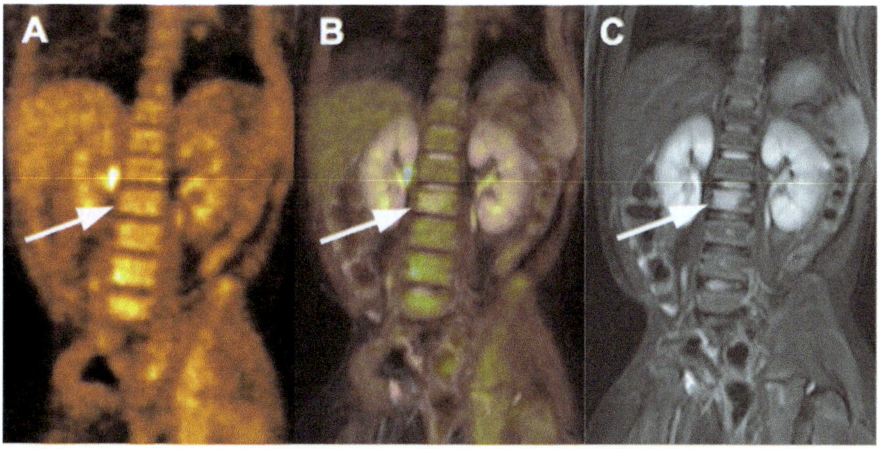

Figure 17.9 [^{18}F]-FDG-MR-PET of a patient with Ewing sarcoma during follow up with several recurrent tumour manifestations in previous years. (A) Coronal slice through the vertebral column depicting not increased tracer uptake in tumour location (arrow). (B) Fused images. (C) CE T1w MRI showing recurrent cancer in the second lumbar vertebra. Images courtesy of C. Franzius, Zentrum für moderne Diagnostik, Bremen, Germany.

17.3 Applications for Benign Diseases

17.3.1 Parathyroid Adenoma

The parathyroid hormone (PTH), produced in the parathyroid glands, plays a major role in calcium and bone metabolism. In a response to low blood calcium levels, PTH is excreted and indirectly stimulates osteoclastic activity, thus elevating blood calcium levels in the short term. The condition of excess production of parathyroid hormone is called hyperparathyroidism and results in a disturbance of calcium, phosphate and bone metabolism. Hyperparathyroidism is usually classified as primary, secondary or tertiary; however, nuclear medicine diagnostics usually focuses on primary hyperparathyroidism. Primary hyperparathyroidism is characterised by PTH over secretion, usually caused by a single parathyroid adenoma and can be cured by surgical resection. It is advantageous to locate the adenoma before surgery. For this purpose, scintigraphy with ^{99m}Tc-Sestamibi, in combination with an ultrasound of the neck, is the most commonly used technique. However, the sensitivity of scintigraphy is only 63–84%.[33] The superior imaging properties of PET can be used for detection of parathyroid adenoma, even in cases with negative scintigraphy.[34] Methionine labelled with carbon-11 is the most thoroughly studied PET tracer used for the detection of parathyroid adenoma and studies have confirmed its efficacy by demonstrating that a significant number of additional adenomas could be located with PET. In a study investigating the additional value of methionine-PET-CT in patients with hyperparathyreoidism, in 11 of 17 (65%) of the patients, a single adenoma was detected by conventional scintigraphy combined with ultrasound.[35] With PET-CT, an adenoma was detected in five out of the six remaining patients; this approach of scintigraphy followed by PET-CT in negative cases identified the adenoma in 94% of the patients. The use of MR-PET instead of PET-CT might be especially relevant for this patient group with benign tumours since it allows a significant reduction in radiation exposure and promises better morphoanatomic imaging in comparison to low-dose, non-contrast-enhanced CT acquired during PET-CT. An example is shown in Figure 17.10.

17.3.2 Imaging of Inflammation

Investigations of inflammatory and infectious diseases are other promising applications of whole-body MR-PET. A range of such conditions are frequently encountered in routine clinical imaging, ranging from systemic inflammatory diseases such as chronic inflammatory bowel diseases, graft-*vs.*-host disease (GvHD) of the intestine, vasculitis, fever of unknown origin

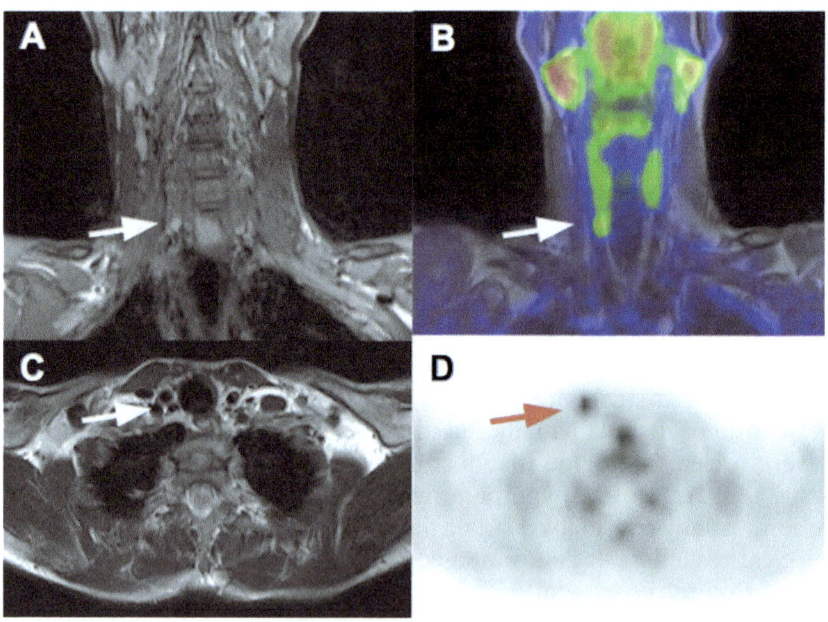

Figure 17.10 ^{11}C-methionine-MR-PET in a patient with suspected parathyroid adenoma but unsuccessful previous scintigraphy for tumour location. (A) Coronal water image calculated from the two-point Dixon images acquired for attenuation correction clearly depicting focal enhanced signal (arrow) co-registered to the tracer uptake. (B) Fusion image depicting focal extrathyroidal uptake (arrow). (C) T2w image at the height of focal tracer uptake without clearly depicting the lesion. (D) Transaxial PET image.

(FUO) to infectious diseases such as osteomyelitis. An example of intestinal GvHD is shown in Figure 17.11. FDG is a tracer that shows uptake in macrophages and inflammatory cells,[36] present in all the above diseases. In addition to its strength in morphological imaging, MRI can give additional information about an inflammatory condition such as perfusion changes, diffusion restriction or depicting oedema in the surrounding fat of infected and inflamed tissue.

Again, the advantages of MRI, the better soft-tissue contrast and the non-existent ionising radiation exposure, might render MR-PET the preferred choice for all the above diseases. Diseases that typically influence the lung structure such as sarcoidosis are, however, not optimal diseases for MR-PET imaging.

17.4 Clinical Applications of Motion Correction

Despite recent advances in PET sensitivity, the duration of PET acquisitions still amounts to several minutes per bed position in most clinical applications. This inherently poses a problem caused by different unavoidable types of patient movement. The most important ones of these include respiratory motion of organs during scans of the thorax and the upper abdomen and cardiac contractile motion during cardiac PET scans. Both types of motion are known to cause image blur and reduces the effective spatial resolution. This potentially leads to a decrease in tracer uptake values and overestimated lesion volumes in PET images. Furthermore, structures close to the border between lung and neighbouring tissue (liver, heart, spleen) can be influenced by erroneous attenuation correction if the mean respiratory phase during the PET acquisition does not match the respiratory phase of the attenuation map.

Arguably the most important method to overcome these problems in clinical PET is the concept of gating the acquired PET data.[37] Here, a signal derived from measuring respiratory/cardiac motion is used to retrospectively sort the PET data into subsets representing a specific motion state, resulting in almost motion-free datasets which can then be reconstructed independently. They are hampered, however, by bad statistics and the

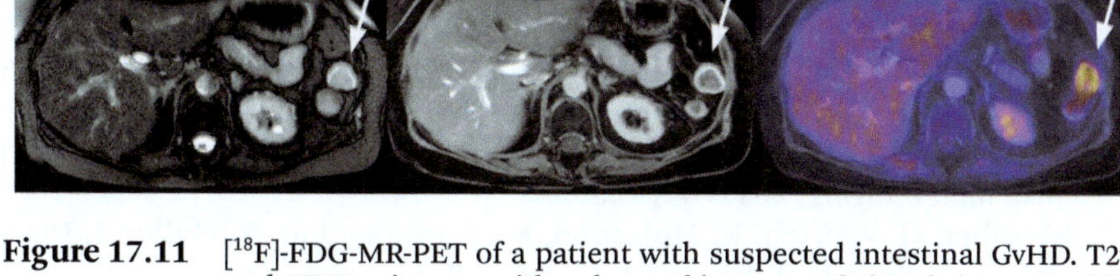

Figure 17.11 [^{18}F]-FDG-MR-PET of a patient with suspected intestinal GvHD. T2w and CE T1w images with enhanced intramural signals in parts of the colon (A) and (B). (C) Fused images showing intense tracer uptake.

subsequent increase in image noise. MR-PET systems are of special interest when looking into motion correction. MR and PET data can be acquired simultaneously and as MRI does not involve ionizing radiation, data can be acquired over extended time periods. This leads to interesting new options for motion correction strategies, *e.g.* the acquisition of respiratory gating signals using MR navigator sequences. This may potentially prove to be superior to surrogate signal generation, as in pressure sensor-based gating, since there may be differences in internal organ motion and external body surface motion.[38]

More advanced motion correction approaches use the calculation of motion vector fields between different respiratory phases/gates and apply these vectors in the PET image reconstruction to calculate a motion-free image of a specific respiratory or cardiac position, where all measured PET events are considered.[39] Fast 3D MR imaging during the PET acquisition, on the other hand, results in complete anatomical motion information well-suited to determine accurate respiratory or cardiac motion vector fields.

Although most of the above methods are now readily available in PET-MRI systems, it remains to be seen how successful motion correction strategies really are in future clinical studies.

Acknowledgements

We acknowledge the contribution of our technologists, namely A. Kanzog and S. Milachowski for image acquisition and the staff of the radiochemistry facilities of the University Hospital Münster for tracer preparation.

References

1. K. Wechalekar, B. Sharma and G. Cook, *Clin. Radiol.*, 2005, **60**, 1143.
2. S. C. Chawla, N. Federman, D. Zhang, K. Nagata, S. Nuthakki, M. McNitt-Gray and M. I. Boechat, *Pediatr. Radiol.*, 2010, **40**, 681.
3. L. A. Torre, F. Bray, R. L. Siegel, J. Ferlay, J. Lortet-Tieulent and A. Jemal, *Ca–Cancer J. Clin.*, 2015, **65**, 87.
4. H. Jadvar, *Semin. Nucl. Med.*, 2016, **46**, 502.
5. L. M. McEwan, D. Wong and J. Yaxley, *J. Med. Imaging Radiat. Oncol.*, 2017, **61**, 503.
6. K. Rahbar, M. Weckesser, S. Huss, A. Semjonow, H. J. Breyholz, A. J. Schrader, M. Schäfers and M. Bögemann, *J. Nucl. Med.*, 2016, **57**, 563.
7. A. Afshar-Oromieh, T. Holland-Letz, F. L. Giesel, C. Kratochwil, W. Mier, S. Haufe, N. Debus, M. Eder, M. Eisenhut, M. Schäfer, O. Neels, M. Hohenfellner, K. Kopka, H. U. Kauczor, J. Debus and U. Haberkorn, *Eur. J. Nucl. Med. Mol. Imaging*, 2017, **44**, 1258.
8. M. Eiber, G. Weirich, K. Holzapfel, M. Souvatzoglou, B. Haller, I. Rauscher, A. J. Beer, H. J. Wester, J. Gschwend, M. Schwaiger and T. Maurer, *Eur. Urol.*, 2016, **70**, 829.

9. M. T. Freitag, J. P. Radtke, A. Afshar-Oromieh, M. C. Roethke, B. A. Hadaschik, M. Gleave, D. Bonekamp, K. Kopka, M. Eder, T. Heusser, M. Kachelriess, K. Wieczorek, C. Sachpekidis, P. Flechsig, F. Giesel, M. Hohenfellner, U. Haberkorn, H. P. Schlemmer and A. Dimitrakopoulou-Strauss, *Eur. J. Nucl. Med. Mol. Imaging*, 2017, **44**, 776.
10. M. T. Freitag, J. P. Radtke, B. A. Hadaschik, A. Kopp-Schneider, M. Eder, K. Kopka, U. Haberkorn, M. Roethke, H. P. Schlemmer and A. Afshar-Oromieh, *Eur. J. Nucl. Med. Mol. Imaging*, 2016, **43**, 70.
11. A. Afshar-Oromieh, U. Haberkorn, H. P. Schlemmer, M. Fenchel, M. Eder, M. Eisenhut, B. A. Hadaschik, A. Kopp-Schneider and M. Röthke, *Eur. J. Nucl. Med. Mol. Imaging*, 2014, **41**, 887.
12. B. Noto, F. Büther, K. Auf derSpringe, N. Avramovic, W. Heindel, M. Schäfers, T. Allkemper and L. Stegger, *EJNMMI Res.*, 2017, **7**, 12.
13. B. R. Kim, J. M. Lee, D. H. Lee, J. H. Yoon, B. Y. Hur, K. S. Suh, N. J. Yi, K. B. Lee and J. K. Han, *Radiology*, 2017, **285**, 134.
14. K. Beiderwellen, B. Gomez, C. Buchbender, V. Hartung, T. D. Poeppel, F. Nensa, H. Kuehl, A. Bockisch and T. C. Lauenstein, *Eur. J. Radiol.*, 2013, **82**, e669.
15. O. F. Donati, T. F. Hany, C. S. Reiner, G. K. von Schulthess, B. Marincek, B. Seifert and D. Weishaupt, *J. Nucl. Med.*, 2010, **51**, 692.
16. A. Dasari, C. Shen, D. Halperin, B. Zhao, S. Zhou, Y. Xu, T. Shih and J. C. Yao, *JAMA Oncol.*, 2017, **3**, 1335.
17. C. B. Johnbeck, U. Knigge and A. Kjær, *Future Oncol.*, 2014, **10**, 2259.
18. L. M. Sawicki, C. Deuschl, K. Beiderwellen, V. Ruhlmann, T. D. Poeppel, P. Heusch, H. Lahner, D. Führer, A. Bockisch, K. Herrmann, M. Forsting, G. Antoch and L. Umutlu, *Eur. Radiol.*, 2017, **27**, 4091–4099.
19. D. Berzaczy, C. Giraudo, A. R. Haug, M. Raderer, D. Senn, G. Karanikas, M. Weber and M. E. Mayerhoefer, *Clin. Nucl. Med.*, 2017, **42**, 669.
20. J. F. Schäfer, S. Gatidis, H. Schmidt, B. Gückel, I. Bezrukov, C. A. Pfannenberg, M. Reimold, M. Ebinger, J. Fuchs, C. D. Claussen and N. F. Schwenzer, *Radiology*, 2014, **273**, 220.
21. M. R. Ponisio, J. McConathy, R. Laforest and G. Khanna, *Pediatr. Radiol.*, 2016, **46**, 1258.
22. A. C. Sher, V. Seghers, M. J. Paldino, C. Dodge, R. Krishnamurthy, R. Krishnamurthy and E. M. Rohren, *Am. J. Roentgenol.*, 2016, **206**, 623.
23. N. F. Schwenzer, C. Schraml, M. Müller, C. Brendle, A. Sauter, W. Spengler, A. C. Pfannenberg, C. D. Claussen and H. Schmidt, *Radiology*, 2012, **264**, 551.
24. M. A. Queiroz and M. W. Huellner, *Semin. Nucl. Med.*, 2015, **45**, 248.
25. C. Brendle, N. F. Schwenzer, H. Rempp, H. Schmidt, C. Pfannenberg, C. la Fougère, K. Nikolaou and C. Schraml, *Eur. J. Nucl. Med. Mol. Imaging*, 2016, **43**, 123.
26. B. Kang, J. M. Lee, Y. S. Song, S. Woo, B. Y. Hur, J. H. Jeon and J. C. Paeng, *Am. J. Roentgenol.*, 2016, **206**, W10.
27. P. Brandmaier, S. Purz, K. Bremicker, M. Höckel, H. Barthel, R. Kluge, T. Kahn, O. Sabri and P. Stumpp, *PLoS One*, 2015, **10**, e0141684.

28. V. Ruhlmann, M. Ruhlmann, A. Bellendorf, J. Grueneisen, L. M. Sawicki, H. Grafe, M. Forsting, A. Bockisch and L. Umutlu, *Eur. J. Radiol.*, 2016, **85**, 1941.
29. A. Vrachimis, M. C. Burg, C. Wenning, T. Allkemper, M. Weckesser, M. Schäfers and L. Stegger, *Eur. J. Nucl. Med. Mol. Imaging*, 2016, **43**, 212.
30. A. Vrachimis, L. Stegger, C. Wenning, B. Noto, M. C. Burg, J. R. Konnert, T. Allkemper, W. Heindel, B. Riemann, M. Schäfers and M. Weckesser, *Eur. J. Nucl. Med. Mol. Imaging*, 2016, **43**, 1765.
31. A. Jena, S. Taneja, A. Singh, P. Negi, R. Sarin, P. K. Das and M. Singhal, *Am. J. Roentgenol.*, 2017, **209**, 662.
32. D. M. Plecha and P. Faulhaber, *Eur. J. Radiol.*, 2017, **94**, A26.
33. W. J. Wei, C. T. Shen, H. J. Song, Z. L. Qiu and Q. Y. Luo, *Hell. J. Nucl. Med.*, 2015, **18**, 127.
34. W. P. Kluijfhout, J. D. Pasternak, F. T. Drake, T. Beninato, J. E. Gosnell, W. T. Shen, Q. Y. Duh, I. E. Allen, M. R. Vriens, B. de Keizer, M. H. Pampaloni and I. Suh, *Langenbecks Arch. Surg.*, 2016, **401**, 925.
35. C. Lenschow, P. Gassmann, C. Wenning, N. Senninger and M. Colombo-Benkmann, *World J. Surg.*, 2015, **39**, 1750.
36. S. Vaidyanathan, C. N. Patel, A. F. Scarsbrook and F. U. Chowdhury, *Clin. Radiol.*, 2015, **70**, 787.
37. M. Dawood, F. Büther, N. Lang, O. Schober and K. P. Schäfers, *Med. Phys.*, 2007, **34**, 3067.
38. P. K. Dasari, M. S. Shazeeb, A. Könik, C. Lindsay, J. M. Mukherjee, K. L. Johnson and M. A. King, *Med. Phys.*, 2014, **41**, 112508.
39. C. Würslin, H. Schmidt, P. Martirosian, C. Brendle, A. Boss, N. F. Schwenzer and L. Stegger, *J. Nucl. Med.*, 2013, **54**, 464.

Part D

Preclinical Applications

CHAPTER 18

Preclinical Hybrid MR-PET Scanner Hardware

C.-H. CHOI*[a], J. FELDER[a], C. W. LERCHE[a] AND N. J. SHAH[a,b,c]

[a]Forschungszentrum Jülich, Neuroscience and Medicine – 4, Wilhelm-Johnen-Straße 1, 52428 Jülich, Germany; [b]JARA – BRAIN – Translational Medicine, Aachen, Germany; [c]Department of Neurology, RWTH Aachen University, Aachen, Germany
*E-mail: c.choi@fz-juelich.de

18.1 Introduction

Currently, the arena of preclinical imaging is undergoing rapid development, with the evolution of hybrid magnetic resonance-positron emission tomography (MR-PET) constituting a significant role. This is by virtue of the fact that hybrid MR-PET has the capability to deliver large amounts of diverse information, offering unique potential for basic biomedical and translational research. The depiction of exquisite anatomical detail and the measurement of brain function are two areas where MRI and PET excel. However, the potential of MR-PET goes far beyond this, given that PET tracers can be used precisely to target complex biochemical pathways or that *in-vivo* pH can be measured by ^{31}P NMR spectroscopy. Furthermore, as measurement time is not so constrained in animal imaging, much can be done to introduce rigour

New Developments in NMR No. 19
Hybrid MR-PET Imaging: Systems, Methods and Applications
Edited by N. Jon Shah
© The Royal Society of Chemistry 2019
Published by the Royal Society of Chemistry, www.rsc.org

with quantitative imaging. The manifold possibilities offered by combining MR and PET become even more potent in simultaneous hybrid MR-PET where the temporal and spatial correlation of the data with high accuracy is a given. Much of the discussion pertaining to clinical multi-modality imaging is also relevant for animal imaging systems, whereby sequences and techniques have to be suitably adjusted to account for the use of anaesthesia and differing physiological characteristics. Nevertheless, the hybrid MR-PET methodology remains very powerful.

In this chapter, the particular challenges relating to the preclinical MR-PET scanner hardware will be described, and the development of hybrid MR-PET systems for imaging small animals will be introduced. Details of individual MRI and PET instrumentation, and human hybrid MR-PET systems have been introduced in the previous chapters and form the basis of the descriptions herein.

18.2 Challenges

Similar to the human hybrid MR-PET systems (see Chapter 10), developing simultaneously operating MR-PET scanners for preclinical uses is challenging and additional technical aspects have to be considered. Both of the separate modalities are compactly and cost-effectively designed and have been optimised for their own functionality. The integration of both modalities into one system capable of simultaneously imaging the same object, therefore, requires a reduction in size and granularity of the PET detectors or a concomitant increase in the size of the magnet bore. At the same time, the sensitivity of the PET or MR image quality should not be degraded. A significant factor in the price of a small-animal MRI system is the magnet, and generally, the price of the magnet scales with radius^3 and $B_0{}^2$, making larger diameters prohibitively expensive at times. Thus, system integration demands the development of a compact PET detector design. The limited space for the PET detectors inside the MRI bore adds another important challenge for the detector technology (in comparison to human-sized installations), and the depth of interaction (DOI) detection technique, for example, becomes very important.

For most animal experiments, the animal is anaesthetised for at least 2 hours and the temperature and respiration are monitored. Consequently, units for measuring these also need to be considered in terms of both MR and PET constraints. These need to be implemented carefully either onto the animal bed, or close enough to the animal in order to avoid any physiological monitoring issues but far enough away to maintain MR-compatibility and not to affect the PET acquisition in terms of attenuation or scattering. In addition, from the viewpoint of preclinical system integration, the followings factors also require consideration: (a) ultra-high magnetic field environment; (b) ultra-high gradient strength; and (c) radiofrequency (RF) interferences.

18.2.1 Ultra-high B_0 Field Strength

It should be emphasised that there are differences in the commonly available B_0 field strength between clinical and preclinical imaging. As mentioned above, ultra-high field (≥7 T) MRI systems tend to be used to scan small animals, since it is relatively easy and economical to construct ultra-high field magnets with smaller magnet bores. Increasing the size of the magnet free bore increases production costs, as the maintenance of field uniformity in the pre-determined diameter of spherical volume is determined by the size of the animal to be examined. The main magnetic field strength of most clinical MRI scanners is 1.5 T or 3 T. However, for small animal systems, the field strength can be three to four times higher than that of the clinical systems, with 7 T and 9.4 T the most commonly used field strengths. Most of the PET scintillation detectors in standalone systems are constructed using an array of scintillator pixels together with position sensitive photomultiplier tubes (PMTs). However, this design does not function at such high magnetic fields and requires either the PMTs to be moved from the high magnetic field regions to low-field regions or the replacement of the PMTs with silicon-based photodetectors that are better able to tolerate magnetic fields.

PET systems placed inside the bore of the magnet can cause problems for MRI, as, for example, they are liable to degrade B_0 homogeneity and to create artefacts due to susceptibility changes. The magnetic susceptibility effect is most intense near the interfaces of media with distinctively different magnetic susceptibilities. The abrupt change of magnetic susceptibility potentially caused by the PET components results in a bound surface current distribution that induces an additional magnetic field perturbing B_0. This results in degradation of the homogeneity of the main magnetic field and requires passive B_0 shimming to be performed again, but with the PET detectors in the bore. Designing the PET detectors using non-magnetic materials is absolutely recommended in order to minimise degradation of the B_0 homogeneity and to avoid susceptibility artefacts. The magnetic susceptibility effect increases with increasing magnetic field strength, leading to a greater impact on image quality in preclinical hybrid scanners.[1]

18.2.2 Ultra-high Gradient Field Strength

Three orthogonal gradient fields are generated by rapidly switching current in the gradient coils and are essential for spatial encoding in MRI. A general aim in imaging is that a 'morphological resolution' should be retained across animals and humans; that is, the number of pixels across a given anatomical structure should at least be maintained when going from humans to animals. This mandates a much higher real spatial resolution in animal imaging. In order to attain this much higher-resolution capability for small animal imaging, much faster (~2700 T m^{-1} s^{-1}) and stronger gradients (~600 mT m^{-1})

are required. Time-varying gradient fields generate physical vibration and possibly induce eddy currents in any conductive components. These include the magnet bore and the PET components, which are RF-shielded and dense gamma shielding materials are used inside the magnet bore.[2,3] Due to the very limited space in the magnet bore of small animal MRI scanners, PET detectors and electronics are likely to be placed very close to the gradient set, and the effect of mechanical vibrations can be much more severe. The induced eddy currents generate magnetic fields opposing the primary gradient fields, which superimpose to the predetermined gradient field. This has the potential to generate serious MR image distortions and artefacts (*e.g.* ghosts) and may also result in damage to the PET components. The induced eddy currents typically flow in circular paths and decay multi-exponentially. Their amplitudes and decay times depend highly on the conductance and geometry of the elements. Removing any loops or areas perpendicular to the changing field and reducing the conductance can minimise the eddy current issue. There have been several methods to minimise eddy currents induced by the gradients while maintaining the RF shielding efficacy for hybrid systems.[4] One method is to reduce the shielding thickness (<skin depth of the conductor at the Larmor frequency), which directly reduces the induced surface current density. Another way is to segment the solid shield using, for example, high value capacitors and this blocks the current loop path on the conductive surface.[5]

18.2.3 RF Interference

Any PET electronics in the magnet room can interfere with the reception of the MR signal. It is observed in many MRI-compatible PET systems that RF transmission during the MRI acquisition can cause a drop in the PET count rates.[6] RF shielding is, however, very effective in this higher frequency range and so, in principle, PET detectors and the associated electronics within the magnet bore are required to be entirely enclosed in an RF shield to prevent RF interference or potential damage. RF leakage from the PET electronics causes an increase in the overall noise level in MRI, which, in turn, decreases the signal-to-noise ratio (SNR) and also has the potential to generate artefacts (*e.g.* zipper) in images. The MR signals generated in imaging or spectroscopy are intrinsically weak and therefore require preamplification and RF shielding to avoid noise contamination. Any unwanted sources of RF present during the measurement destroy the quality of MRI images. Many of the frequencies present in modern digital electronics, *e.g.* clock pulses or harmonics, are indeed in the range close to that of the system's Larmor frequency and care must be taken to minimise interference by shielding the PET components. Unshielded power cables can also easily carry noise into the MRI coil circuitry. The shielding material needs to be carefully chosen and should be transparent for gradient fields to avoid eddy current generation while providing a high RF shielding effectiveness.

18.3 Hybrid Preclinical MR-PET Systems

18.3.1 Achieved by Modification of PET

Most of the developments in the combination of preclinical MRI and PET in research have focused on a compact, MR-compatible, PET detector design located inside the magnet bore operating at ultra-high field. As explained previously, PMTs were commonly used for standalone PET scanners. However, the performance of standard PMTs is severely degraded in the presence of magnetic fields and they do not function properly at ultra-high field.[7] Two approaches have been considered to address this issue. The first retaining the conventional, magnetic field-sensitive PMT-based PET detectors by modification of other aspects of the MR or PET system (see Sections 18.3.1.1 and 18.3.2), or the second is with the use of magnetic field insensitive and reliable MR-compatible solid-state photodetectors, such as avalanche photo-diodes (APDs) (see Section 18.3.1.2)[8–10] and more recently silicon photomultipliers (SiPMs) (see Section 18.3.1.3).[11–15] SiPMs are, in particular, of great interest as an alternative photodetector in the design of viable MR-PET systems capable of simultaneous measurement, since these devices are very compact and rugged. Furthermore, they require a smaller bias voltage of only a few tens of volts and outperform the timing resolutions of PMTs.

18.3.1.1 Small Animal Systems Based on Photomultiplier Tubes (PMTs)

Initial hybrid MR-PET systems were realised using PMTs with scintillators and lightguides, since these contain fewer magnetic components in the imaging FOV. The scintillators were arranged inside the MR magnet, which was connected *via* optical fibres to the PMTs placed outside the magnet's fringe field. The feasibility of this design was examined and reported.[16,17] Since then, as the example shows in Figure 18.1a, several prototype systems were constructed and demonstrated with long optical fibres connecting a detector ring of LSO crystals to PMTs in a low magnetic field region (<10 mT).[8,18,19]

Simultaneous acquisition of ^{18}F-FDG measurements and ^{31}P MR spectroscopy was carried out to investigate cardiac metabolism on a perfused rat heart at 9.4 T.[20] In the system setup, a cylindrical NaI(TI) crystal was placed onto the rat heart in order to collect the scintillation light which was transferred *via* a long lightguide to PMTs located just outside the magnet (~50 mT). The first version of the small animal PET-only scanner[21] was demonstrated utilising short 15 cm optical fibres to transfer scintillation light from the LSO crystal arrays to position sensitive PMTs. The method using optical fibres was further improved with the use of a 90° bend and shorter optical fibre bundles. This enabled an extended PET FOV in an axial direction, within an MRI scanner.[22,23] The modified light guides, however, still resulted in substantially attenuated scintillation light and degradation of crystal

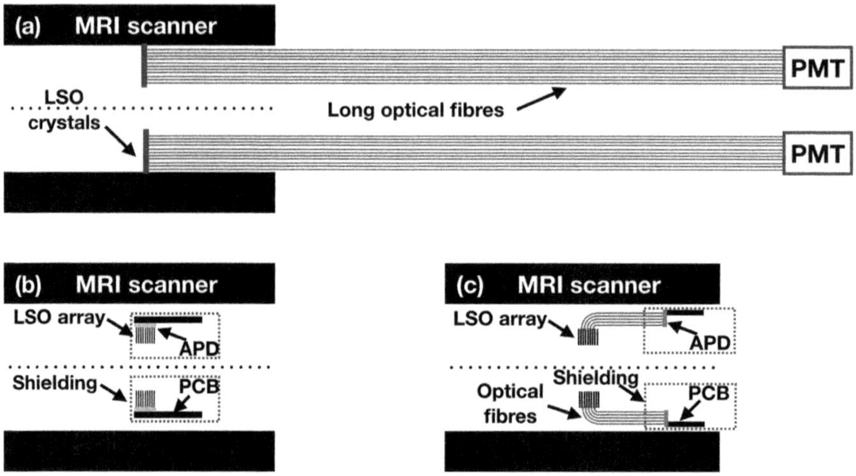

Figure 18.1 Combined MR-PET system achieved with different approaches—long optical fibres used to couple scintillators to PMTs residing outside the magnet (a), APDs coupled directly to scintillator elements (b) and APDs coupled through short optical fibres to scintillators (c). Reproduced from ref. 8 with permission, Copyright © SNMMI.

identification, energy resolution and transaxial PET spatial resolution, compared to those measured by a standalone PET scanner. Nevertheless, this very simple and practical system implementation allows excellent access to the animal and negligible MR and PET interference.

18.3.1.2 Small Animal Systems Based on Avalanche Photodiodes (APDs)

As an alternative to the conventional PMTs, small animal PET scanners based on APDs have been investigated by several laboratories. APDs have the advantage of being compact, and importantly, they are magnetic field insensitive photodetectors that could be used inside an MRI scanner.[7,8,24] Exemplary configurations of the APD detector with short optical fibres and one with direct coupling of APD to scintillators are shown in Figure 18.1b and c.

The operation of APDs is briefly as follows: the operating voltage is usually set below the device breakdown voltage, meaning that the charge signal obtained is proportional to the original number of electron–hole pairs produced and this, in turn, is proportional to the energy deposited in the scintillator by the incident gamma. As the applied voltage increases, so too does the gain; the optimum operating range corresponds to a gain of between 50 to 150 only, whereas the voltages are in the range of a few hundred to a few thousand V. APDs have an advantage in that the quantum efficiency can be as high as 80% (at 420 nm—the peak emission wavelength for LSO). However, a significant shortcoming is that an APD/LSO combination usually only achieves a temporal resolution of a few ns and this is too long to be of use in time-of-flight PET. The comparable value is <600 ps for a PMT/LSO combination.

Following general practice in electronics, low noise readout electronics must be placed as close to the detector as possible, in order to read out the very small signals produced.[25] Another major challenge of using APDs is the fact that the gain is very sensitive to the changes of ambient temperature and the temperature variations can render the normalisation calibration inadequate for the pertaining temperature state of the detectors. This can lead to errors in energy windowing and count rate and substantially degrade the system performance regarding temporal resolutions. Therefore, the strict control of the temperature for APD-based PET scanners is necessary.[26]

The system developed by the UC Davis group[8] was used to detect scintillation light *via* 10-cm long optical fibres connected optically to position-sensitive APDs, which were cooled to −5 °C and operated with a bias voltage of −1.7 kV. By bending the fibres, the PET subsystem could be placed outside the RF excitation volume and the active regions of the gradients, but this limited the axial PET FOV to 12 mm. However, the use of position sensitive APDs in the magnet bore introduced small distortions in the MR images and a certain degree of light loss in PET due to the bent fibres.

Vandenberghe and Marsden evaluated the sensitivity of PET to RF excitation with and without the PET electronics during RF excitation, resulting in a deadtime of 6–28% depending on the MRI sequences. Nevertheless, the PET performance of these prototypes, particularly relating to sensitivity which is typically <1%, was far inferior to that of standalone PET-only small animal systems (typically 4–10%).[25] In order to improve the PET performance, some groups have tried to construct the system without using the optical fibre connections by directly placing the APDs and electronics in the active MR imaging region.[9,27] In this approach, the scintillation light loss due to the optical fibres was removed. However, the interference between the two sub-systems, the APDs and the RF and gradients, increases. This is especially the case when using high-performance MRI pulse sequences demanding intensive gradient-switching and high-intensity RF pulses.

Wehrl *et al.* implemented an APD-based PET insert on a 7 T preclinical MRI scanner and evaluated MR compatibility and performance.[3] Their system comprised a ring of LSO arrays coupled directly to an APD array through a 3 mm thin light guide. The crystals, APD array, and the electronics of each detector module were shielded with 10 µm thick copper to minimise RF interference and to avoid the eddy current problem. The system was operated at room temperature and a small SNR reduction with negligible B_0 and B_1 homogeneity degradation was found, even when a wide range of demanding sequences, including functional MRI and diffusion MRI, and both employing EPI sequences were used.

18.3.1.3 Small Animal Systems Based on Silicon Photomultipliers (SiPMs)

SiPMs are an MR-compatible semiconductor photosensor offering extremely high resolution with high photon count rates. A SiPM-based small animal system was actively developed within the Hyperimage European project. The

PET system for the preclinical use was fully integrated as an insert into a 3 T clinical MRI system.[28,29] This allowed a large transaxial FOV and provided extra capabilities for translational studies. In this system, each detector unit consisted of a 22 × 22 array of lutetium yttrium orthosilicate (LYSO) crystals which were coupled to an 8 × 8 array of 4 × 4 mm^2 analogue SiPMs *via* a 2 mm thick light guide. The group, led by Schulz, took the approach of directly addressing the problems, such as potential cross-talk, associated with a large number of readout channels from the SiPM array. All analogue signals coming from the SiPM elements were digitised immediately adjacent to the sensor and within the magnet. This was done using a custom-designed application specific integrated circuit (ASIC). This ASIC outputs digital energy, timing and channel information. The detector units, comprising the scintillator, SiPM array and ASIC, were enclosed in an RF-shielded module, employing liquid cooling to 22 °C. The MRI compatibility of each component was carefully tested, and diligent consideration was given to the selection and production of non-ferromagnetic components, the differential signalling for all analogue signals, the circuit layout to avoid eddy currents and heating effects, and the minimisation of RF from power supplies. A small decrease in the MRI SNR was observed during PET system operation. Importantly, only very minimal effects on B_0 and gradient homogeneity or RF-related effects were observed. Likewise, despite the use of a continuous RF shield around the PET modules, no influence of gradient switching on the PET was measured.[25]

Figure 18.2 shows another example of a compact SiPM based PET scanner inserted inside the bore of 7 T preclinical MRI scanner.[14] The PET insert used here consists of 16 detector modules including 4 detector blocks each. Each block comprises a 9 × 9 array of LYSO crystals and a SiPM array. The LYSO crystals and SiPM arrays were optically coupled by using a grease without inserting any light guides between them. Temperature and bias voltages were interactively monitored using a real-time gain compensation device. This was able to maintain the quality of performance, such as stable photopeak positions, constant energy resolutions and prompt counting rates. The PET insert was shielded by a 1 mm thick carbon fibre tube to minimise the eddy current interference, but at the same time to sufficiently block unwanted radiofrequency signals.

A second-generation of the MR-compatible, SiPM based PET system has been developed and includes digital technology.[12,30,31] The digital SiPM (dSiPM) provides timing, energy and spatial resolutions comparable to analogue SiPM, but it does not require a readout ASIC, which makes the digital SiPM-based insert a promising candidate for integration into an MRI scanner.[31] The fact that no dedicated readout ASIC is required means that the PET detector size can be even more compact when it comes to the integration with MRI. Furthermore, the calibration and handling of the signals from a number of digital channels become more reliable. As shown in Figure 18.3, the PET system is shielded by a carbon fibre enclosure, which provides effective RF shielding whilst having a negligible effect

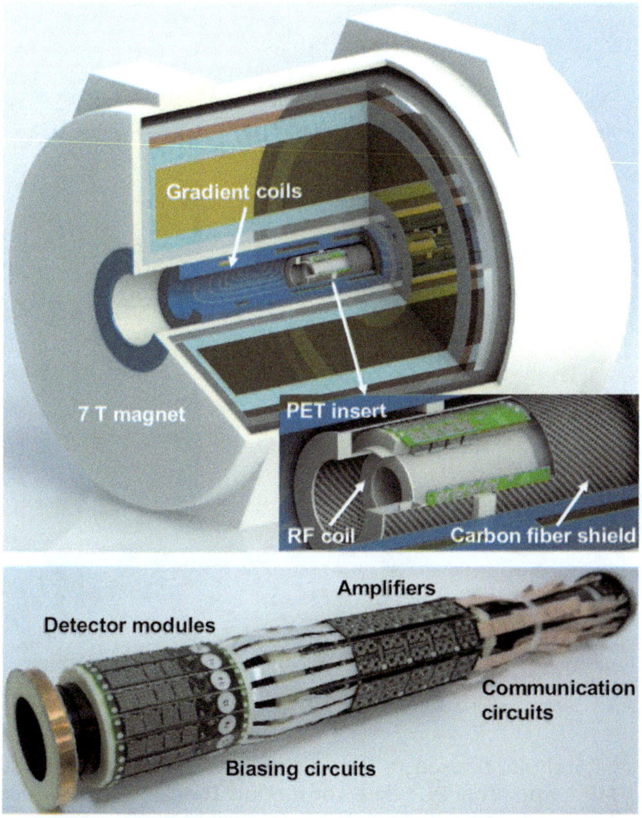

Figure 18.2 SiPM-based PET insert on a 7 T small animal MRI system. Reprinted from ref. 14 with permission, Copyright © SNMMI.

on the eddy current signature. The dissipation of heat from the detector modules was achieved by liquid cooling. The system also includes several advanced functions to synchronise the data acquisition between PET and MRI. For example, MRI sequence data can be fed directly into the PET data stream.

A number of different groups have also designed and constructed small animal PET insert systems based on the SiPM technology. Although the fundamental detector arrangements were similar, different approaches have been taken regarding signal transfer and the shielding configurations. Goertzen *et al.* developed a resistor-based multiplexer to readout the SiPM array. This was done to avoid power dissipation and complex cooling mechanisms, in which the signals and power were transferred *via* high definition media interface (HDMI) cables, providing excellent noise immunity.[15] Hong *et al.*, for example, placed short optical fibre bundles between the crystal arrays and sensors. This arrangement resulted in gaps between the shielded sensor modules, allowing existing MR coils to be used.[32] Kang *et al.* used 3 m

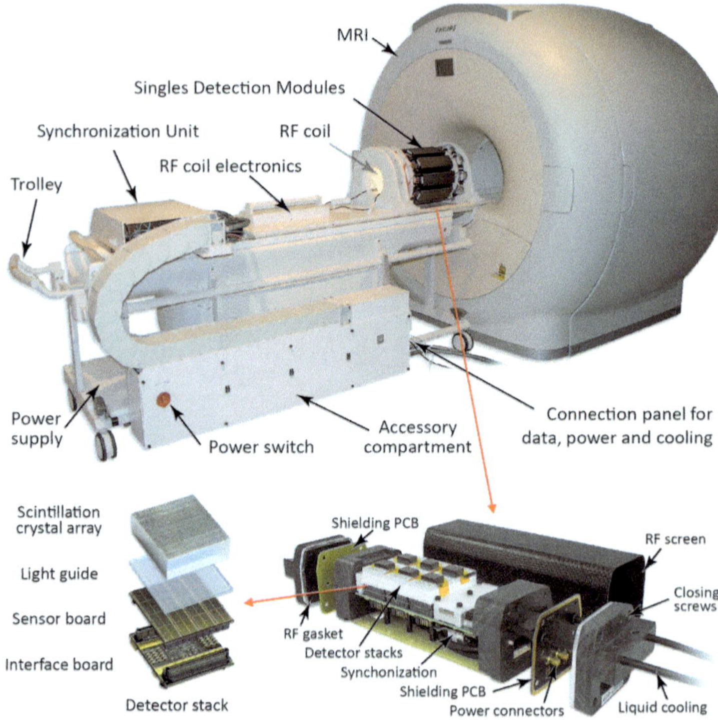

Figure 18.3 Digital SiPM-based PET insert on a 3 T clinical system (top) and an MR-compatible PET detector module (bottom). Reproduced from ref. 12 with permission from IEEE, Copyright 2015.

long flexible flat cables to transfer output signals to a preamplifier located remotely. This was advantageous since the PET subsystems can be located outside of the 5 Gauss line and only non-magnetic PET components, such as scintillation crystals and SiPM arrays need be placed inside the bore of the magnet.[33]

18.3.2 Achieved by Modification of MRI

Another approach to the design of MR-PET systems is to modify the MRI system, allowing a conventional PET detector, such as PMTs, to be used. Although this approach is feasible, it is more complicated and requires more engineering efforts.

18.3.2.1 Split Magnet System

In order that it could be combined with a PET system, an MRI scanner using a split in the 1 T superconducting magnet and gradients was developed by the Cambridge group.[34] The principle of their approach was to provide a substantial gap in the axial direction in the magnet structure, that is, a split magnet, to accommodate a multi-ring PET detector array. The gap in the magnet allows multiple optical fibre bundles to couple the scintillating crystal blocks to conventional PMTs positioned in a region of low magnetic field (~30 mT) just at the outer circumference of the magnet. Another advantage of this arrangement is that by moving the PET electronics out of the region of the RF and gradient coils, any potential artefacts resulted by electromagnetic interference between the PET and MR systems are averted. In order to create a sufficiently low field for the PMTs, the magnet was self-shielded on either side of the gap, and mild steel shields were also inserted. With further mu-metal shielding, the PMTs experienced a field of just 1 mT and overall there was only a small reduction in PET performance compared with the standalone system. The split magnet is an innovative approach, but this configuration does, inevitably, compromise the performance of gradients (because they too are split) and the B_0 homogeneity (because the main solenoid of the magnet is split). The long length (>1 m) of optical fibre bundles, which transfer scintillation light from the scintillator arrays to PMTs also resulted in the loss of scintillation light.

18.3.2.2 Fast-field Cycling System

The fast field-cycling technique using a resistive magnet allows the main magnetic field to be dynamically switched between levels during an MR acquisition.[35] The magnetisation is prepared at the polarisation field (high field) to ensure a sufficient SNR and detection takes place at low, but stable, field. This can produce image quality comparable to that of clinical superconducting systems and can be used to generate new types of MRI contrast.[36] Since the field-cycling magnet is capable of switching the B_0 field rapidly down close to zero, conventional but robust PET technology, such as using PMTs can be combined with this MRI system for hybrid MR-PET operation with very little interference between the modalities. A prototype system has been constructed by the Western Ontario group in Canada.[37,38] Their system consisted of four PMTs optically coupled to a bismuth germanate scintillator block and enclosed within a mu-metal container. The detector block was placed in a gap in the magnet (polarising field of 0.3 T and detection field of 94 mT), similar to the split magnet arrangement mentioned previously.

18.3.3 Achieved Without Modifying Both Systems

Combining individual MR and PET systems with a sequential acquisition can be the most straightforward way. The major advantages of this approach are that it saves enormously on development time for the modification of the separate systems, any existing standalone scanners can potentially be combined, and this approach guarantees the best quality of performance for each system. The sequential measurements simply require a detachable bed which fits both MRI and PET systems. The bed with the animal can be transferred from the PET to MRI, or *vice versa*. Alternatively, an in-line configuration with an automated bed that traverses through the two-scanner assembly sequentially can also be employed. The disadvantage of utilising sequential MRI and PET systems is the long acquisition times required. Furthermore, due to physiological activity and motion during the scan and transition, in the case of using the common bed, co-registration of mismatch may be required. However, these factors might be less important when imaging anaesthetised animals. Importantly, for a number of different types of studies, such as pharmacological, differences in parameters such as blood flow can introduce unwanted confounds which are not present in simultaneously acquired data.

18.4 Commercially Available Preclinical MR-PET Systems

Over recent years, small animal MR-PET systems, capable of sequential or simultaneous operation, have become commercially available. The specifications of a number of such systems, *e.g.* resolution, detector or crystal being used, are listed in Table 18.1, modified from ref. 39. Most of the preclinical MR-PET systems use SiPM photodetectors together with a LYSO crystal.

Table 18.1 Overview of the current commercially available preclinical MR-PET system products. Reproduced from ref. 39 with permission from Elsevier, Copyright 2018.[a]

Vender	Aspect/ Brightonix	Bruker	Cubresa	Inviscan	Mediso	MR solutions
PET type	Insert	Insert & in-line	Insert	Insert	In-line	Insert & in-line
FOV/mm^2 transaxial × axial	50 × 55	80 × 148 (3-ring)	60 × 68	58 × 78	80 × 100 (2-ring)	80/120 × 150 (3-ring)
Resolution/ mm	0.8	0.7	0.8	<1.0	0.9	0.7
Sensitivity (%)	3.4	12	>5.5	>4	7	12
Energy resolution (%)	15	17	<12.5	14	14	15
Detector	SiPM	SiPM	SiPM	SiPM	SiPM	SiPM
Crystal	LSO	LYSO	LYSO	LYSO	LYSO	LYSO
Magnet type	Permanent	Supercon.	(b)	(b)	Supercon.	Supercon.
B_0 strength	1 T & (b)	3 T & (b)	(b)	(b)	1 T (3 T)	7 T & (b)
Grad. strength/ mT m^{-1}	150	450 (900)	—	—	450	Up to 1000
References	40, (a)	41, (a)	(a)	(a)	42	(a)

[a]This table is modified from ref. 39 and updated by personal communication: (a) with the venders; (b) PET insert that can be adapted into high-field MRI systems. Note that the measurement method may differ among the venders.

References

1. Q. Yang, M. Smith and J. Wang, in *Magnetic Susceptibility Effects in High Field MRI*, ed. R.-M. Robitaille and L. Berliner, Springer, Boston, 2006, pp. 249–284.
2. D. Strul, D. Cash, S. F. Keevil, P. Halsted, S. C. R. Williams and P. K. Marsden, *IEEE Trans. Nucl. Sci.*, 2003, **50**, 609.
3. H. F. Wehrl, M. S. Judenhofer, A. Thielscher, P. Martirosian, F. Schick and B. J. Pichler, *Magn. Reson. Med.*, 2011, **65**, 269.
4. M. Alecci and P. Jezzard, *Magn. Reson. Med.*, 2002, **48**, 404.
5. B. J. Lee, R. D. Watkins, C. M. Chang and C. S. Levin, *Magn. Reson. Med.*, 2018, **79**, 1745.
6. N. Omidvari, G. Topping, J. Cabello, S. Paul, M. Schwaiger and S. Ziegler, *Phys. Med. Biol.*, 2018, **63**, 095002.
7. B. J. Pichler, M. S. Judenhofer, C. Catana, J. H. Walton, M. Kneiling, R. E. Nutt, S. B. Siegel, C. D. Claussen and S. R. Cherry, *J. Nucl. Med.*, 2006, **47**, 639.
8. C. Catana, Y. Wu, M. S. Judenhofer, J. Qi, B. J. Pichler and S. R. Cherry, *J. Nucl. Med.*, 2006, **47**, 1968.
9. M. S. Judenhofer, C. Catana, B. K. Swann, S. B. Siegel, W. I. Jung, R. E. Nutt, S. R. Cherry, C. D. Claussen and B. J. Pichler, *Radiology*, 2007, **244**, 807.
10. J. M. Vrigneaud, P. Walker, B. Barbier, A. Camacho, A. Oudot, B. Collin and F. Brunotte, *Biomed. Phys. Eng. Express*, 2017, **3**, 035006.
11. E. Roncali and S. R. Cherry, *Ann. Biomed. Eng.*, 2011, **39**, 1358.
12. B. Weissler, P. Gebhardt, P. Dueppenbecker, J. Wehner, D. Schug, C. W. Lerche, B. Goldschmidt, A. Salomon, I. Verel, E. Heijman, M. Perkuhn, D. Heberling, R. M. Kiessling, F. Kiessling and V. Schulz, *IEEE Trans. Med. Imaging*, 2015, **34**, 2258.
13. J. E. Mackewn, C. W. Lerche, B. Weissler, K. Sunassee, R. T. M. de Rosales, A. Phinikaridou, A. Salomon, R. Ayres, C. Tsoumpas, G. M. Soultanidis, P. Gebhardt, T. Schaeffter, P. K. Marsden and V. Schulz, *IEEE Trans. Nucl. Sci.*, 2015, **62**, 784.
14. G. B. Ko, H. S. Yoon, K. Y. Kim, M. S. Lee, B. Y. Yang, J. M. Jeong, D. S. Lee, I. C. Song, S. K. Kim, D. Kim and J. S. Lee, *J. Nucl. Med.*, 2016, **57**, 1309.
15. A. L. Goertzen, G. Stortz, J. D. Thiessen, D. Bishop, M. S. Khan, P. Kozlowski, F. Retiere, G. Schellenberg, E. Shams, V. Sossi and C. J. Thompson, *IEEE Trans. Nucl. Sci.*, 2016, **63**, 2424.
16. B. Hammer, *Magn. Reson. Imaging*, 1990, **9**, 4.
17. N. L. Christensen, B. E. Hammer, B. G. Heil and K. Fetterly, *Phys. Med. Biol.*, 1995, **40**, 691.
18. Y. Shao, *Phys. Med. Biol.*, 1997, **42**, 1965.
19. J. E. Mackewn, P. Halsted, G. Charles-Edwards, R. Page, J. J. Totman, K. Sunassee, D. Strul, W. A. Hallett, M. Jauregui-Osoro, P. Liepins, S. C. R. Williams, T. Schaeffter, S. F. Keevil and P. K. Marsden, *IEEE Trans. Nucl. Sci.*, 2010, **57**, 1052.

20. M. Buchanan, P. K. Marsden, C. H. Mielke and P. B. Garlick, *IEEE Trans. Nucl. Sci.*, 1996, **43**, 2044.
21. S. R. Cherry, Y. Shao, R. W. Silverman, K. Meadors, S. Siegel, A. Chatziioannou, J. W. Young, W. Jones, J. C. Moyers, D. Newport, A. Boutefnouchet, T. H. Farquhar, M. Andreaco, M. J. Paulus, D. M. Binkley, R. Nutt and M. E. Phelps, *IEEE Trans. Nucl. Sci.*, 1997, **44**, 1161.
22. S. Yamamoto, M. Imizumi, Y. Kanai, M. Tatsumi, M. Aoki, E. Sugiuama, M. Kawakami, E. Shimosegawa and J. Hatazawa, *Ann. Nucl. Med.*, 2010, **24**, 89.
23. R. R. Raylman, S. Majewski, S. K. Lemieux, S. S. Velan, B. Kross, V. Popov, M. F. Smith, A. G. Weisenberger, C. Zorn and G. D. Marano, *Phys. Med. Biol.*, 2006, **51**, 6371.
24. M. S. Judenhofer, H. F. Wehrl, D. F. Newport, C. Catana, S. B. Siegel, M. Becker, A. Thielscher, M. Kneilling, M. P. Lichy, M. Eichner, K. Klingel, G. Reischl, S. Widmaier, M. Röcken, E. E. Nutt, H. J. Machulla, K. Uludag, S. R. Cherry, C. D. Claussen and B. J. Pichler, *Nat. Med.*, 2008, **14**, 459.
25. S. Vandenberghe and P. K. Marsden, *Phys. Med. Biol.*, 2015, **60**, R115.
26. V. Keereman, R. Van Holen, S. Vandenberghe and C. Vanhove, *Med. Phys.*, 2013, **40**, 092506.
27. R. Grazioso, N. Zhang, J. Corbeil, M. Schmand, R. Ladebeck, M. Vester, G. Schnur, W. Renz and H. Fischer, *Nucl. Instrum. Methods Phys. Res. A*, 2006, **569**, 301.
28. V. Schulz, T. Solf, B. Weissler, P. Gebhardt, P. Fischer, M. Ritzert, V. Mlotok, C. Piemonte, N. Zorzi, M. Melchiorri, S. Vandenberghe, V. Keereman, D. Wirtz, S. Renisch, T. Schaeffter and P. Marsden, *IEEE Nucl. Sci. Symp. Med. Imaging Conf. Rec.*, 2009, 2577.
29. B. Weissler, P. Gebhardt, C. W. Lerche, J. Wehner, T. Solf, B. Goldschmidt, J. E. Mackewn, P. K. Marsden, F. Kiessling, M. Perkuhn, D. Heberling and V. Schulz, *Phys. Med. Biol.*, 2014, **59**, 5119.
30. P. M. Düppenbecker, B. Weissler, P. Gebhardt, D. Schug, J. Wehner, P. K. Marsden and V. Schulz, *Biomed. Phys. Eng. Express*, 2016, **2**, 015010.
31. J. Wehner, B. Weissler, P. Dueppenbecker, P. Gebhardt, B. Goldschmidt, D. Schug, F. Kiessling and V. Schulz, *Phys. Med. Biol.*, 2015, **60**, 2231.
32. K. J. Hong, Y. Choi, J. H. Jung, J. Kang, W. Hu, H. K. Lim, Y. Huh, S. Kim, J. W. Jung, M. S. Song and H. W. Park, *Med. Phys.*, 2013, **40**, 042503.
33. H. G. Kang, S. J. Hong, G. B. Ko, H. S. Yoon, I. C. Song, J. T. Rhee and J. S. Lee, *J. Instrum.*, 2015, **10**, 12008.
34. R. C. Hawkes, T. D. Fryer, S. Siegel, R. E. Ansorge and T. A. Carpenter, *Technol. Cancer Res. Treat.*, 2010, **9**, 53.
35. D. J. Lurie, S. Aime, S. Baroni, N. A. Booth, L. M. Broche, C. H. Choi, G. R. Davies, S. Ismail, D. O. Hogain and K. J. Pine, *C. R. Phys.*, 2010, **11**, 136.
36. J. K. Alford, B. K. Rutt, T. J. Scholl, W. B. Handler and B. A. Chronik, *Magn. Reson. Med.*, 2009, **61**, 796.
37. K. M. Gilbert, T. J. Scholl, W. B. Handler, J. K. Alford and B. A. Chronik, *Magn. Reson. Med.*, 2009, **62**, 1017.

38. G. A. Bindseil, K. M. Gilbert, T. J. Scholl, W. B. Handler and B. A. Chronik, *Magn. Reson. Med.*, 2011, **66**, 301.
39. J. G. Mannheim, A. M. Schmid, J. Schwenck, P. Katiyar, K. Herfert, B. J. Pichler and J. A. Disselhorst, *Semin. Nucl. Med.*, 2018, **48**, 332.
40. https://www.aspectimaging.com/wp-content/uploads/2018/02/DOC10001259-Rev.-2-PET-MRI-Brochure.pdf(last accessed July 2018).
41. https://www.bruker.com/fileadmin/user_upload/8-PDF-Docs/PreclinicalImaging/Brochures/PET_MR3T_Brochure_T156188.pdf (last accessed July 2018).
42. http://www.mediso.com/uploaded/NPMPCB_0817_web.pdf(last accessed July 2018).

CHAPTER 19

Preclinical Applications of MR-PET

A. WILLUWEIT*, P. LOHMANN AND C. STEGMAYR

Forschunsgzentrum Jülich, Institute of Neuroscience and Medicine – 4, Wilhelm-Johnen-Straße, Jülich, 52425, Germany
*E-mail: a.willuweit@fz-juelich.de

19.1 Introduction

The latest improvements made in the development of combined positron emission tomography (PET) and magnetic resonance imaging (MRI) scanners for small animals offer various possibilities for preclinical studies. In addition to proof-of-concept and cross-validation of new diagnostic tools for translation into the clinic, MR-PET can also be implemented in preclinical treatment studies for the development of new therapeutic agents and strategies. In addition, validation of new animal models benefits from multimodal monitoring of disease progression in individual animals. Further preclinical applications for combined MR-PET include the analyses of disease mechanisms and drug actions in basic science. The simultaneous use of PET and MRI offers advantages for animal welfare as it reduces imaging times and anaesthetic stress to the animals. Furthermore, throughput is increased and optimal co-registration in a spatio-temporal manner is possible. It is hoped that the current disadvantages associated with combined scanners

New Developments in NMR No. 19
Hybrid MR-PET Imaging: Systems, Methods and Applications
Edited by N. Jon Shah
© The Royal Society of Chemistry 2019
Published by the Royal Society of Chemistry, www.rsc.org

(in comparison to standalone machines), relating to lower resolution and difficulties in attenuation correction, will be solved in the future.

19.2 Imaging of Tumours

Tumour models are often used to assess the feasibility of newly developed simultaneous MR-PET systems, as tumours can be produced easily and reliably in rats or mice and they are big enough to be detected with PET. Furthermore, and in contrast to phantoms, tissue can be processed after the scans and additional experimental techniques, such as immunohistochemistry and autoradiography, can be applied and co-registered with the PET and MR images. Ng *et al.*[1] tested a PET insert using a mouse tumour model. Colorectal adenocarcinoma cells were implanted in the mouse's flank and assessed using both autoradiography and a commercial small animal PET scanner to show quantitative accuracy and spatial concordance. The advantages of processing tools in relation to the specific location of the tumour in the body have been investigated in other tumour model studies. These have focused on scanner technique, optimised image processing (such as spatial and temporal image registration), attenuation correction and robust co-registration.[2,3]

While the majority of hardware-related tumour studies cite the performance of the PET detector as the focus of interest, with MR taken as a reference, contrasting this and using the MR performance as the focus also offers interesting applications. For example, the use of X-nuclei MR imaging in rat 9 L brain tumour models was tested by taking the clinically established PET tracer O-(2-[^{18}F]-fluoroethyl)-L-tyrosine (^{18}F-FET) as a reference, as interest in the use of ^{18}F-FET PET for brain tumour diagnostics is increasing. Furthermore, ^{18}F-FET PET is often used in addition to MRI to define the tumour extent, to monitor treatment and to differentiate treatment-related effects from tumour progression. Using ^{18}F-FET PET as a reference image, the possible benefit of sodium (^{23}Na) imaging and phosphorus (^{31}P) spectroscopy (MRS) for brain tumour diagnostics was tested. Structural T_2-weighted proton (^{1}H) and X-nuclei imaging was performed on a 9.4 T animal scanner with switchable coils.[4] Following this, without moving the anaesthetised rat from the bed, the bed was attached to a preclinical PET. Co-registered images of the tumour in the anterior striatum and ^{31}P MRS of the tumour region are shown in Figure 19.1.

A potential clinical application of these very preliminary sodium MR images and of the differences in the ^{31}P MRS compared to a non-tumour-bearing rat remains to be elucidated, but since the determination of tumour grade and treatment response with conventional MRI is difficult, X-nuclei imaging might be a useful approach to gain information beyond conventional structural imaging.

The studies conducted at our institute, and other proof-of-principle studies with tumour models aim to visualise tumours containing malignant tissue, necrosis and oedema that might not be separable using

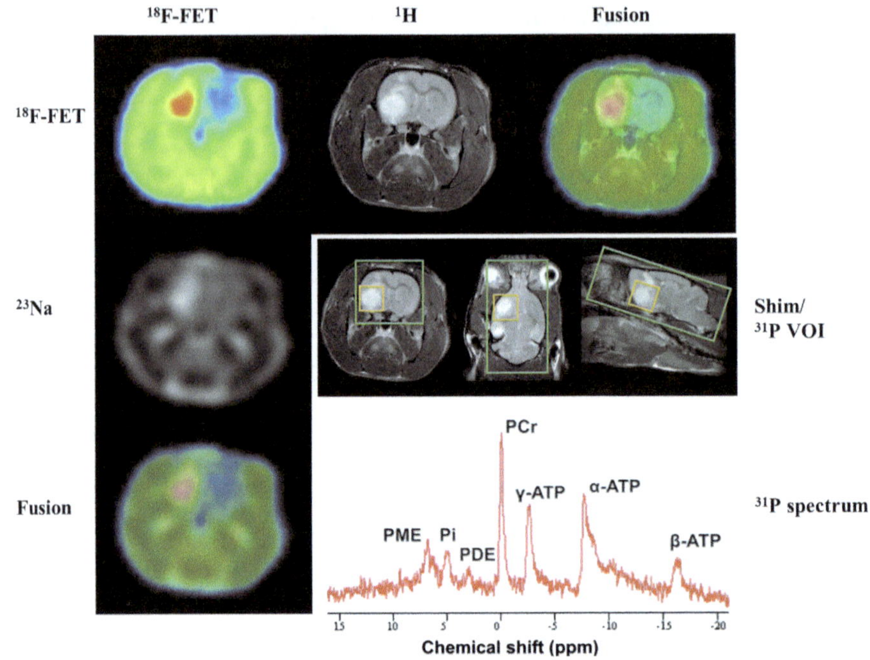

Figure 19.1 ^{18}F-FET PET with co-registered ^{23}Na-MRI and ^{31}P-MRS in a tumour-bearing rat. The 9 L tumour is clearly visible in PET and MRI. The MRS shows the chemical shift of adenosine triphosphate (ATP), inorganic phosphate (Pi), phosphocreatine (PCr), phosphodiester (PDE), phosphomonoester (PME) in the tumour volume-of-interest (VOI).

conventional MRI. Detecting tumour heterogeneity, and thus molecular and cellular differences, in small structures using PET with MRI as an anatomical reference is a benchmark for hybrid instrumentation and makes therapy response assessment studies with animal models eminently feasible.[5] Within those studies, the combination of either new or established tracers with co-registered conventional or advanced MRI aims to detect earlier and/or more reliable treatment-related changes, which is crucial for switching therapy in case of insufficient tumour response. Schmitz and colleagues[6] used a mouse model of breast cancer to demonstrate the benefits of the two established imaging methods [^{18}F]-fluorodeoxyglucose (^{18}F-FDG) PET in combination with diffusion-weighted imaging (DWI) to detect tumour heterogeneity and confirmed the translational potential in five breast cancer patients.

Rommel *et al.*[7] used advanced MR methods to test whether quantitative indexes of ^{1}H MRS and [^{18}F]-fluoromethylcholine (FCH) evolve in a similar way following external radiation therapy. Choline concentration in ^{1}H MRS and FCH uptake seem to address the same anabolic cascade, hence was a correlation expected in choline parameters. MR-PET was performed on days 3, 9 and 16 after therapy and tissue slices were stained post mortem. Surprisingly, an opposite trend between choline parameters from both modalities was observed at early time points after treatment. Despite the fact that the volumes-of-interest (VOIs) of PET and MRS were not congruent, and thus a rigorous comparison was not possible, this study demonstrated that PET and MRS data can show different information, even when a similar metabolic parameter is addressed. The authors outline the sensitivity of PET for detecting early changes after therapy and similar results were achieved in our group when ^{18}F-FET PET images of rat brain tumours were compared with chemical exchange saturation transfer (CEST) images of amino acids (unpublished data). Although PET and MR images shared similarities due to amino acid imaging, the two modalities provided different information relating to the tumour and the surrounding tissue.

Other examples of preclinical studies confirm the high sensitivity with which PET can detect treatment response earlier than MR imaging. Heijmen *et al.*[8] used a mouse model of colorectal cancer to show that ^{18}F-FDG PET may be sensitive enough to assess the efficacy of bevacizumab, an anti-angiogenic antibody, also known as Avastin, beyond progression. While T_2*-weighted imaging, DWI and dynamic contrast-enhanced MRI found no differences between bevacizumab-treated and the untreated group, ^{18}F-FDG uptake showed a decrease in the treatment group and histochemistry confirmed lower vascular density and lower cell proliferation. In another mouse model of colorectal cancer, only combined MR-PET parameters in combination with histopathology resulted in significant differentiation of treatment groups.[9]

As typical MR contrast agents cannot cross the intact blood–brain barrier, MR-PET data are also of high value in brain tumour therapy response studies, using, for example, bevacizumab, especially when treatment affects the impaired tumour–blood barrier. Nedergard *et al.*[10] demonstrated the feasibility of ^{18}F-FET for anti-angiogenic response evaluation and superiority to T_2-weighted MRI.

19.3 Imaging the Cardiovascular System

Applications for combined PET and MRI in the field of cardiovascular diseases are driven by the need for better co-registration in order to localise the PET signal to the exact localisation with high spatial resolution. Here, co-registration to an MR image is superior to CT co-registration because of its better soft tissue contrast, the lack of ionising radiation and the more tolerable contrast agents. In addition, for imaging in the abdomen, movement, such as ventilation and especially heartbeat, is problematic. Respiratory-gated and/or cardiac-triggered acquisition and software corrections are needed for both MRI and PET analysis.[11] Beyond providing anatomical information with high resolution, new MRI techniques offer functional information, which may complement that obtained from PET and *vice versa*. The combination of PET and MRI measurements in cardiovascular preclinical studies often serves as cross-validation and proof-of-concept for the implementation of new techniques and their translation into clinical applications.

For the assessment of the global left ventricular function of the heart, studies were performed in a model of myocardial infarction in a mouse. ^{18}F-FDG PET on myocardial viability was compared with information obtained from volumetric measurements using cine sequences in MRI with sequential or simultaneous MR-PET. The results of the studies found the information yielded to be comparable and with good correlation.[12,13] This combination was further expanded by additionally integrating sequences to evaluate the muscular performance of the left ventricle and late gadolinium enhancement imaging to measure the infarcted region.[14] Likewise, the use of [^{68}Ga]-albumin as a blood pool tracer in dilative and ischemic cardiomyopathy showed a high correlation in measuring left ventricular function compared to cine-MRI, which was used as the reference standard.[15] A very recent study of heart failure under pressure overload (which is a model of non-ischemic cardiomyopathy) compared non-diabetic mice with diabetic mice and quantified cardiac metabolic, structural and functional changes using a combination of several MRI methods and PET. In a longitudinal design, glucose utilisation was followed with ^{18}F-FDG PET, cardiac function with cine-MRI and lipid content and cardiac energy status using ^{1}H MRS and ^{31}P MRS, respectively. All the implemented methods were complementary and provided insight into the development and progression of heart failure in non-diabetic and diabetic mice in this multi-modal study.[16]

For PET imaging of atherosclerotic plaques, co-registration with MRI is also of importance, as the spatial resolution of PET is not sufficient to detect single vessels. Here, simultaneous measurements are advantageous because of organ movement within the abdomen. In a study using atherosclerotic mini pigs, inflammatory cells infiltrating the atherosclerotic plaque could be visualised with ^{18}F-FDG PET, whereas multi-sequence MRI was used for optimal anatomical imaging of the aorta.[17] The detection of inflammatory, phagocytic cells is also possible with the use of MRI and with the application of ultra-small superparamagnetic iron-oxide (USPIO) particles, which are 'swallowed' by macrophages *in vivo*. Although ^{18}F-FDG PET has been shown to be more sensitive, both methods showed good correlation in detecting a statin treatment response in a rabbit model of atherosclerosis.[18]

19.4 Imaging the Central Nervous System (CNS)

With a few exceptions, PET imaging of the brain also requires co-registration with a high-resolution modality, like MRI, in order to locate the metabolic or molecular images precisely within a specific brain region. This is of importance for both the characterisation of new mouse models and new PET tracers but does not particularly require simultaneous acquisition,[19,20] since methods for rigid-body co-registration are mature. Although co-registration with MRI has many advantages over CT imaging, the latter is routinely used for skull attenuation correction, which has been proven to be somewhat difficult to do with MRI.[21] Recently, a study showed the feasibility of ^{31}P zero echo time MRI of hydroxyapatite for bone attenuation correction in rat brain imaging.[22] Using a mouse model of Alzheimer's disease, Maier *et al.*[23] demonstrated that, in addition to high-resolution co-registration, MRI can add functional information regarding regional cerebral blood flow by using arterial spin labelling approaches in relation to amyloid PET.

Moreover, both PET and MRI can detect neuroinflammation, which occurs not only in many neurodegenerative diseases but also in stroke. While the PET tracer [^{11}C]-PK11195 can be used to image mainly activated microglia cells, phagocytes can be detected using MRI after the application of USPIO particles, which, in a study conducted by Walter *et al.*, gave a complementary picture in a rat model of permanent middle cerebral artery occlusion.[24] Furthermore, changes in DWI combined with altered glucose consumption, as measured by ^{18}F-FDG PET, have been followed in individual animals before, during and after hypoxia-ischemia, and gave new insights into the course of the disease. Here, simultaneous acquisition was essential because of the rapid and variable nature of the disease.[25] In addition to these applications, multiparametric imaging has also been successfully used to validate a new mouse model for haemorrhagic stroke by applying both perfusion and diffusion-weighted MRI in combination with PET with ^{18}F-FDG and the hypoxia marker 1-(5-deoxy-5-[^{18}F]-fluoro-α-D-arabinofuranosyl)-2-nitroimidazole (^{18}F-FAZA).[26]

Functional MRI (fMRI) based on blood oxygen level-dependent (BOLD) contrast offers the possibility to study brain activity and networks and is therefore of the utmost interest for basic and clinical research of the CNS. In separate studies, Wehrl *et al.* researched brain function in rats in response to a whisker stimulation of the barrel field cortex. The stimulation was performed during BOLD-fMRI and compared with glucose consumption using ^{18}F-FDG PET or [^{15}O]-H_2O PET as a marker for cerebral blood flow.[27,28] Both studies found spatial and quantitative discrepancies between the PET and MRI methods. This finding underlines the different physiological mechanisms behind each measurement and points to the fact that the combination of PET and MRI is most powerful when applied in a complementary manner. This is also highlighted by a simultaneous [^{11}C]-raclopride PET-fMRI study on dynamic dopamine receptor desensitisation and internalisation that enabled characterisation of agonist-induced neuroreceptor adaptations.[29]

19.5 Dual-probes in Preclinical MR-PET

The introduction of hybrid MR-PET scanners opened new possibilities for the development and application of contrast agents in preclinical and clinical research. Nowadays, contrast agents for each modality can be used in combination to form dual probes.

There are two different approaches for creating dual-probes for MR-PET. The easiest way is to mix two imaging agents from each modality that are already in use for single modality imaging. For example, ^{18}F-FDG PET is commonly used in combination with gadolinium (Gd)-based contrast agents in MRI to assess both tumour metabolism and vascularisation at the same time. This approach simply combines the information from the two different modalities for better tumour characterisation.

The second approach, that will also be the main focus in the following, is to create dual-probes that are detected by both modalities and synergistically combine the advantages of PET and MRI. For example, conventional MRI using Gd-based contrast agents has a very low sensitivity (millimolar range) compared to PET (nanomolar range or lower). However, PET has a low spatial resolution and provides only very limited anatomical information, whereas MRI offers a superior soft tissue contrast that allows for a more precise allocation of the PET signal within the tissue. Combining MRI and PET probes overcomes these limitations. Adding a PET radioisotope to an MRI probe allows the detection of areas where the signal from the contrast-enhancement is below the detection limit of MRI due to the high sensitivity of PET. Furthermore, using MRI to provide an accurate quantification of contrast-enhancement is challenging, whereas PET allows for a quantification of the amount of radiotracer in the tissue of interest with high precision. Thus, an MR-PET dual-probe can precisely quantify the amount of tracer uptake and enables the assessment of biodistribution. In the following, some preclinical applications of dual-probes in MR-PET are briefly introduced.

19.5.1 Sentinel Lymph Node Imaging

Several types of cancers, such as melanoma or breast cancer, spread the disease *via* the lymphatic system. The lymph nodes, which are in connection with the primary tumours and cause the disease to spread, are called sentinel lymph nodes (SLNs). Therefore, the early identification and surgical removal of the SLNs is a common treatment to prevent the disease from spreading. Today, SLNs are identified by the injection of a mixture of a radioisotope (technetium-99m) and a blue dye into the primary tumour. The mixture is slowly taken up by the SLNs and allows their detection either intra-operatively by using a hand-held gamma-probe or pre-operatively using SPECT. In addition, the blue dye helps the surgeon intra-operatively to visually evaluate the amount of SLN resection required. However, in spite of its high sensitivity, very small cancerous SLNs might not be detected due to the low spatial resolution of SPECT, which is in the centimetre range.

Several approaches for SLN detection using dual-probes can be found in the literature. Employing a mouse model, De Rosales *et al.* successfully used copper-64 (^{64}Cu) labelled superparamagnetic iron oxide nanoparticles (SPIONs) to visualise popliteal lymph nodes following a footpad injection of [^{64}Cu]-(dtcbp)2-Edorem using preclinical MR-PET.[30] Other groups have used germanium-69[31] or gallium-68[32] labelled SPIONs for SLN detection. Thorek *et al.* combined the PET radioisotope zirconium-89 with a clinically available formulation of iron oxide nanoparticles for the investigation of axillary drainage in naive C57BL/6 mice in comparison with Hi-Myc mice that express human c-Myc in the prostate, leading to tumour formation. The use of [^{89}Zr]-ferumoxytol enabled the successful identification of lymphatic drainage in this preclinical study and might have the translational potential to improve the preoperative planning for lymph node resection and tumour staging.[33]

19.5.2 Conclusions

Stroke and embolisms are caused by the formation of thrombi in the vasculature. A major component of blood clots is fibrin, which has been used as a target for the detection of thrombi.

MR-PET has high potential for thrombus detection as MRI allows a precise identification of vessel walls with a high spatial resolution. However, the low sensitivity of MRI alone allows for the detection of large thrombi only. For early staging of the disease and treatment planning, the assessment of newly developed small thrombi would be of great potential. Here, the high sensitivity of PET comes into play. Uppal *et al.* developed a dual-probe for thrombus detection in a rat model of carotid artery thrombosis by partially exchanging Gd for the radioisotope ^{64}Cu in the commonly used fibrin-targeted MR probe EP-2104R. While PET was more sensitive, and therefore suitable also for detection of newly formed clots, MRI helped to image the identified regions-of-interest identified by PET with high spatial resolution.[34] Further efforts by the same group identified improved ^{64}Cu-PET tracers binding to fibrin for the detection of thrombi in rat models of arterial and deep venous thrombosis and pulmonary embolism.[35–37]

19.5.3 Determination of the Tissue pH

The pH value of tissue is an attractive potential biomarker of cancer.[38] One interesting property of MR contrast agents is their ability to respond to external factors such as pH by changes in relaxivity or 'signal intensity'.[39] However, a precise quantification of the concentration of contrast agent in the tissue of interest is necessary for a reliable determination of the pH value. As previously highlighted, this is especially challenging in areas with low contrast agent uptake. However, to overcome this issue, Frullano *et al.* found that the sensitivity of the pH-responsive probe could be improved by using a radionuclide. By introducing a pH-responsive MR contrast agent labelled with fluorine-18, improved quantification and detectability was achieved.[40] Another approach to measuring pH in cancerous tissue using MR-PET dual probes was performed by Longo *et al.* using a murine mouse model of breast cancer. Following an injection of the pH-responsive contrast agent, iopamidol, extracellular pH was measured using CEST MRI in combination with

^{18}F-FDG PET. The results showed an inverse correlation between extracellular pH and glucose uptake and demonstrated how the combination of ^{18}F-FDG uptake and extracellular pH can be assessed with this dual-probe to improve characterisation of the tumour microenvironment.[41]

This section only provides a very condensed overview of preclinical applications of MR-PET dual-probes. A more detailed description of dual-probes is beyond the scope of this section but can be found in a review by Lahooti *et al.*[42]

19.6 Conclusions

The preclinical application of MR-PET is both valuable for the validation of new clinical diagnostic tools and also for animal studies in general, as it provides the opportunity to obtain numerous *in vivo* data at once in a longitudinal and non-invasive manner. This would otherwise only be accessible by sacrificing the animal. Furthermore, a number of MR-PET methods offer unique functional information from within the living organism, which cannot be measured *ex vivo*. In addition to the clinical benefits of MR-PET, the simultaneous acquisition of PET and MRI contributes to animal welfare, as multiple data can be recorded in a shorter time. However, despite these advantages, whether simultaneous measurements are of major advantage or whether sequential acquisition can provide equivalent information is dependent on the application and the animal model.

References

1. T. S. Ng, J. R. Bading, R. Park, H. Sohi, D. Procissi, D. Colcher, P. S. Conti, S. R. Cherry, A. A. Raubitschek and R. E. Jacobs, *J. Nucl. Med.*, 2012, **53**, 1102.
2. T. S. Ng, D. Procissi, Y. Wu and R. E. Jacobs, *Med. Phys.*, 2010, **37**, 1995.
3. B. Weissler, P. Gebhardt, P. M. Dueppenbecker, J. Wehner, D. Schug, C. W. Lerche, B. Goldschmidt, A. Salomon, I. Verel, E. Heijman, M. Perkuhn, D. Heberling, R. M. Botnar, F. Kiessling and V. Schulz, *IEEE Trans. Med. Imaging*, 2015, **34**, 2258.
4. C. H. Choi, Y. Ha, P. Veeraiah, J. Felder, K. Mollenhoff and N. J. Shah, *J. Magn. Reson.*, 2016, **273**, 28.
5. C. Catana, D. Procissi, Y. Wu, M. S. Judenhofer, J. Qi, B. J. Pichler, R. E. Jacobs and S. R. Cherry, *Proc. Natl. Acad. Sci. U. S. A.*, 2008, **105**, 3705.
6. J. Schmitz, J. Schwab, J. Schwenck, Q. Chen, L. Quintanilla-Martinez, M. Hahn, B. Wietek, N. Schwenzer, A. Staebler, U. Kohlhofer, O. H. Aina, N. E. Hubbard, G. Reischl, A. D. Borowsky, S. Brucker, K. Nikolaou, C. la Fougere, R. D. Cardiff, B. J. Pichler and A. M. Schmid, *Cancer Res.*, 2016, **76**, 5512.
7. D. Rommel, J. Abarca-Quinones, A. Bol, F. Peeters, R. Lhommel, M. Lonneux, D. Labar, V. Gregoire and T. Duprez, *Nucl. Med. Biol.*, 2010, **37**, 645.

8. L. Heijmen, C. J. Punt, E. G. Ter Voert, L. F. de Geus-Oei, A. Heerschap, J. Bussink, C. G. Sweep, V. Zerbi, W. J. Oyen, P. N. Span, O. Boerman and H. W. van Laarhoven, *Invest. New Drugs*, 2013, **31**, 881.
9. V. S. Honndorf, H. Schmidt, S. Wiehr, H. F. Wehrl, L. Quintanilla-Martinez, A. Stahlschmidt, H. Barjat, S. A. Emmas and B. J. Pichler, *Mol. Imaging Biol.*, 2016, **18**, 249.
10. M. K. Nedergaard, S. R. Michaelsen, T. Urup, H. Broholm, H. El Ali, H. S. Poulsen, M. T. Stockhausen, A. Kjaer and U. Lassen, *PLoS One*, 2015, **10**, e0115315.
11. Y. Yang, S. Rendig, S. Siegel, D. F. Newport and S. R. Cherry, *Phys. Med. Biol.*, 2005, **50**, 2979–2989.
12. K. Buscher, M. S. Judenhofer, M. T. Kuhlmann, S. Hermann, H. F. Wehrl, K. P. Schafers, M. Schafers, B. J. Pichler and L. Stegger, *J. Nucl. Med.*, 2010, **51**, 1277.
13. L. Stegger, E. Heijman, K. P. Schafers, K. Nicolay, M. A. Schafers and G. J. Strijkers, *J. Nucl. Med.*, 2009, **50**, 132.
14. G. Buonincontri, C. Methner, T. Krieg, R. C. Hawkes, T. A. Carpenter and S. J. Sawiak, *Nucl. Instrum. Methods Phys. Res. A*, 2014, **734**, 152.
15. A. Todica, S. Brunner, G. Boning, S. Lehner, S. G. Nekolla, M. Wildgruber, C. Ubleis, C. Wangler, M. Sauter, K. Klingel, P. Cumming, P. Bartenstein, R. Schirrmacher, W. M. Franz and M. Hacker, *Mol. Imaging Biol.*, 2013, **15**, 441.
16. D. Abdurrachim, M. Nabben, V. Hoerr, M. T. Kuhlmann, P. Bovenkamp, J. Ciapaite, I. M. E. Geraets, W. Coumans, J. Luiken, J. F. C. Glatz, M. Schafers, K. Nicolay, C. Faber, S. Hermann and J. J. Prompers, *Cardiovasc. Res.*, 2017, **113**, 1148.
17. S. F. Pedersen, T. P. Ludvigsen, H. H. Johannesen, J. Lofgren, R. S. Ripa, A. E. Hansen, A. J. Ettrup, B. O. Christoffersen, H. D. Pedersen, L. H. Olsen, L. Hojgaard and A. Kjaer, *Am. J. Nucl. Med. Mol. Imaging*, 2014, **4**, 448.
18. A. Millon, S. D. Dickson, A. Klink, D. Izquierdo-Garcia, J. Bini, E. Lancelot, S. Ballet, P. Robert, J. Mateo de Castro, C. Corot and Z. A. Fayad, *Atherosclerosis*, 2013, **228**, 339.
19. A. Manook, B. H. Yousefi, A. Willuweit, S. Platzer, S. Reder, A. Voss, M. Huisman, M. Settles, F. Neff, J. Velden, M. Schoor, H. von der Kammer, H. J. Wester, M. Schwaiger, G. Henriksen and A. Drzezga, *PLoS One*, 2012, 7, e31310.
20. S. Wagner, R. Teodoro, W. Deuther-Conrad, M. Kranz, M. Scheunemann, S. Fischer, B. Wenzel, U. Egerland, N. Hoefgen, J. Steinbach and P. Brust, *J. Labelled Compd. Radiopharm.*, 2017, **60**, 36.
21. J. Bini, P. M. Robson, C. Calgano, M. Eldib and Z. A. Fayad, *EJNMMI Phys.*, 2014, **1**, A70.
22. V. Lebon, S. Jan, Y. Fontyn, B. Tiret, G. Pottier, E. Jaumain and J. Valette, *EJNMMI Phys.*, 2017, **4**, 16.
23. F. C. Maier, M. D. Keller, D. Bukala, B. Bender, J. G. Mannheim, I. M. Brereton, G. J. Galloway and B. J. Pichler, *J. Nucl. Med.*, 2015, **56**, 1593.

24. H. L. Walter, M. Walberer, M. A. Rueger, H. Backes, D. Wiedermann, M. Hoehn, B. Neumaier, R. Graf, G. R. Fink and M. Schroeter, *Neuroscience*, 2015, **292**, 71.
25. Y. Ouyang, M. S. Judenhofer, J. H. Walton, J. Marik, S. P. Williams and S. R. Cherry, *J. Visualized Exp.*, 2015, **103**, e52728.
26. S. Castaneda Vega, C. Weinl, C. Calaminus, L. Wang, M. Harant, W. Ehrlichmann, D. Thiele, U. Kohlhofer, G. Reischl, J. M. Hempel, U. Ernemann, L. Quintanilla Martinez, A. Nordheim and B. J. Pichler, *Neuroimage*, 2017, **155**, 245.
27. H. F. Wehrl, M. Hossain, K. Lankes, C. C. Liu, I. Bezrukov, P. Martirosian, F. Schick, G. Reischl and B. J. Pichler, *Nat. Med.*, 2013, **19**, 1184.
28. H. F. Wehrl, P. Martirosian, F. Schick, G. Reischl and B. J. Pichler, *Neuroimage*, 2014, **89**, 271–279.
29. C. Y. Sander, J. M. Hooker, C. Catana, B. R. Rosen and J. B. Mandeville, *Neuropsychopharmacology*, 2016, **41**, 1427.
30. R. Torres Martin de Rosales, R. Tavare, R. L. Paul, M. Jauregui-Osoro, A. Protti, A. Glaria, G. Varma, I. Szanda and P. J. Blower, *Angew. Chem., Int. Ed. Engl.*, 2011, **50**, 5509.
31. R. Chakravarty, H. F. Valdovinos, F. Chen, C. M. Lewis, P. A. Ellison, H. Luo, M. E. Meyerand, R. J. Nickles and W. Cai, *Adv. Mater.*, 2014, **26**, 5119.
32. R. Madru, T. A. Tran, J. Axelsson, C. Ingvar, A. Bibic, F. Stahlberg, L. Knutsson and S. E. Strand, *Am. J. Nucl. Med. Mol. Imaging*, 2013, **4**, 60.
33. D. L. Thorek, D. Ulmert, N. F. Diop, M. E. Lupu, M. G. Doran, R. Huang, D. S. Abou, S. M. Larson and J. Grimm, *Nat. Commun.*, 2014, **5**, 3097.
34. R. Uppal, C. Catana, I. Ay, T. Benner, A. G. Sorensen and P. Caravan, *Radiology*, 2011, **258**, 812.
35. F. Blasi, B. L. Oliveira, T. A. Rietz, N. J. Rotile, H. Day, P. C. Naha, D. P. Cormode, D. Izquierdo-Garcia, C. Catana and P. Caravan, *J. Nucl. Med.*, 2015, **56**, 1088.
36. F. Blasi, B. L. Oliveira, T. A. Rietz, N. J. Rotile, P. C. Naha, D. P. Cormode, D. Izquierdo-Garcia, C. Catana and P. Caravan, *Arterioscler. Thromb. Vasc. Biol.*, 2015, **35**, 2114.
37. K. L. Ciesienski, Y. Yang, I. Ay, D. B. Chonde, G. S. Loving, T. A. Rietz, C. Catana and P. Caravan, *Mol. Pharm.*, 2013, **10**, 1100.
38. P. Swietach, R. D. Vaughan-Jones, A. L. Harris and A. Hulikova, *Philos. Trans. R. Soc. Lond. B. Biol. Sci.*, 2014, **369**, 20130099.
39. C. Tu, E. A. Osborne and A. Y. Louie, *Ann. Biomed. Eng.*, 2011, **39**, 1335.
40. L. Frullano, C. Catana, T. Benner, A. D. Sherry and P. Caravan, *Angew. Chem., Int. Ed. Engl.*, 2010, **49**, 2382.
41. D. L. Longo, A. Bartoli, L. Consolino, P. Bardini, F. Arena, M. Schwaiger and S. Aime, *Cancer Res.*, 2016, **76**, 6463.
42. A. Lahooti, S. Sarkar, S. Laurent and S. Shanehsazzadeh, *Contrast Media Mol. Imaging*, 2016, **11**, 428.

Part E

Tracers

CHAPTER 20

Radiotracers for PET and MR-PET Imaging

J. ERMERT*[a] AND B. NEUMAIER[a,b]

[a]Forschungszentrum Jülich, Institute of Neuroscience and Medicine – 5, Wilhelm-Johnen-Straße 1, 52428 Jülich, Germany; [b]Institute of Radiochemistry and Experimental Molecular Imaging, University Clinic Cologne, Kerpener Str. 62, 50937 Cologne, Germany
*E-mail: j.ermert@fz-juelich.de

20.1 Basic Principle of PET-radiotracer Production

Positron emission tomography (PET) is a non-invasive imaging technique using biologically active molecules labelled with short-lived, positron emitting radionuclides.

These neutron deficient radionuclides (unstable in comparison to the stable isotopes of the element) are commonly produced by the bombardment of stable nuclides with charged particles such as protons, deuterons, or α-particles, using a particle accelerator. For this purpose, a circular accelerator, known as a cyclotron is usually employed. The cyclotron is composed of an ion source, which enables the production of charged particles such as positive ions (H^+ and $^2H^+$) or negative ions (especially H^- and $^2H^-$; hydrogen or deuterium atoms with two electrons). To accelerate the ions, a rapidly alternating electrical field is generated by a pair of semi-circular hollow copper electrodes, historically called 'dees', inside the machine. The particles are forced to traverse a spiral trajectory by a static magnetic field

New Developments in NMR No. 19
Hybrid MR-PET Imaging: Systems, Methods and Applications
Edited by N. Jon Shah
© The Royal Society of Chemistry 2019
Published by the Royal Society of Chemistry, www.rsc.org

acting perpendicular to the plane of the dees with a field strength of 1.5–2.0 T. A high vacuum is used to avoid breakdown of radiofrequency and to prevent particle losses due to molecular collisions with the residual gas.[1,2]

After ionisation in the ion source, the ions are injected into the centre of the gap between the dees. Since an alternating electrical field, changing between positive and negative charge (at 20–30 MHz in the radiofrequency range), is applied to the dees, the ions gain energy at each crossing of the gap between the dees. With additional energy gained at each turn, the radius of the accelerated particle increases. This is because the magnetic field is perpendicular to the plane of the dees and particle motion forces the ion particles on a spiral path.

The accelerated ions are extracted from the cyclotron after reaching a critical radius, *i.e.* energy, and are directed onto a target to induce the desired nuclear reaction. The positively charged accelerated ions are commonly extracted by means of an electrostatic deflector that creates an outwardly directed electric field between two electrodes, changing the direction of the beam. The negatively charged accelerated ions (H^- and $^2H^-$) are extracted by passing the negatively charged ion beam through an ultrathin stripper foil (graphite), removing the electrons. The positively charged particles will then rotate in the opposite direction and can easily be extracted from the cyclotron. Both methods enable the extraction of the beam out of the cyclotron in order to bombard an appropriate target material. The simple extraction of negatively charged particles from the cyclotron enables the extraction of beams with different energies and obviates the need for a complex beam extraction system. Moreover, it allows the bombardment of two (even different) targets simultaneously.[2,3]

The bombardment of stable elements by accelerated, positively charged ion particles (projectile) leads to a nuclear reaction. The projectile has a high kinetic energy allowing it to overcome Coulomb repulsion and form a highly excited compound nucleus. This compound nucleus is very unstable, and the excess energy is discharged by the emission of radiation (photons) and/or particles. This finally leads to the formation of an unstable neutron-deficient radionuclide. The positively charged particles for bombarding must have kinetic energies sufficient to overcome both the Coulomb barrier and the negative Q value ('Q value' is the difference in the energy levels of the nuclei of the reactants and products) of the target nucleus.[1,4]

There are several aspects specific to the production radionuclides in a cyclotron. These relate to nuclear data (optimal energy range) and target (engineering of heat dissipation) for high current irradiations. Today the technology of the production of the most common PET-nuclides is well established, facilitating the routine application of radionuclides and radiopharmaceuticals in clinical practice worldwide.

The most commonly used positron emitters, ^{11}C (carbon-11, $T_{½}$ = 20.3 min), ^{13}N (nitrogen-13, $T_{½}$ = 10.0 min), ^{15}O (oxygen-15 $T_{½}$ = 2.0 min) and ^{18}F (fluorine-18, $T_{½}$ = 110 min), are produced *via* low-energy nuclear reactions in medium-sized cyclotrons (≤18 MeV proton). In addition to these cyclotron-produced radionuclides, the positron emitter ^{68}Ga (gallium-68, $T_{½}$ =

67.6 min) is widely used in nuclear medicine and it is easily obtained using a generator system. The generator contains the long-lived parent nuclide ^{68}Ge (germanium-68, $T_{½}$ = 271 days), which can be produced *via* high-energy nuclear reactions. ^{68}Ge decomposes to short-lived ^{68}Ga. The latter can be eluted from the generator in certain intervals for further applications.

The development of PET has stimulated extensive research into the radio chemistry of short-lived positron emitters. The separation of the desired radionuclide from the irradiated target and, above all, tagging of the short-lived positron-emitting radionuclide to an organic compound, necessitates novel concepts. The most critical constraints of radiochemistry with short-lived radionuclides are the minute masses to be handled and the limited time available for the preparation of labelled molecules, termed as radiotracers or radioligands.[5] Presently, a serious limitation in radiochemistry is a lack of methods with which to facilitate easy and robust access to radiotracers. Thus, emerging methods for the preparation of novel radiotracers are highly sought after. These methods also involve the preparation and quality control of the final product within a very short timescale. Moreover, the production of radiotracers must be carried out under good manufacturing conditions (GMP), which is a challenge in PET chemistry. A further challenge in radiochemistry is the transfer of the synthesis procedure into a remotely controlled synthesis module.

20.2 Common PET-radiotracers and Their Applicability Medical Applications

The syntheses of important clinically-relevant PET-radiopharmaceuticals are described in the following sections. The production processes of these radiopharmaceuticals have been optimised in numerous studies, and usually the radiosyntheses are carried out on commercially available synthesis devices.

The elements oxygen, nitrogen and carbon are essential parts of endogenous molecules. Therefore, their corresponding positron emitters, ^{11}C, ^{13}N, and ^{15}O, with short half-lives, are ideal candidates for radiolabelling biomolecules.[6] However, the short half-life limits the syntheses possibilities with all of these radionuclides and generally restricts their application. In contrast, the longer half-life of fluorine-18 is advantageous and allows extended multi-step radiosynthesis, monitoring of slow biological processes over PET-scanning periods exceeding 2 h and, furthermore, enables the shipment of the radiotracer to distant imaging facilites, according to the satellite concept.[7] Fluorine-18 is the most widely used PET-nuclide in nuclear medicine, although the element fluorine is not found in endogenous compounds. However, lead structures of pharmaceuticals often contain fluorine in their structure due to improvements in pharmacokinetics and the metabolic stability of these compounds.[8] The generator-based ^{68}Ga is an appealing alternative to fluorine-18, since no on-site cyclotron is required. Therefore, ^{68}Ga has gained importance as a PET radionuclide in recent years.[9–11]

20.2.1 [^{15}O]Water

[^{15}O]Water is an important radiotracer for studying regional cerebral oxygen extraction and metabolism[12,13] and is readily used for functional activation studies with PET.[14] The main production route of [^{15}O]water proceeds *via* the reduction of [^{15}O]O_2-gas with elemental hydrogen at a platinum[15] or palladium catalyst, at high temperatures. The main starting material, [^{15}O]O_2 gas, is obtained by the ^{14}N(d,n)^{15}O nuclear reaction on nitrogen gas in the presence of an oxygen content of less than 5% in the target, leading to a rather low molar activity.[16] Alternatively, [^{15}O]water can be produced by the bombardment of natural water using the ^{16}O(p,pn)^{15}O nuclear reaction.[17,18] This nuclear reaction, however, necessitates cyclotrons with proton energies of >26 MeV (Figure 20.1).

20.2.2 [^{13}N]Ammonia

The two main clinical applications of [^{13}N]ammonia are PET studies of myocardial perfusion[19,20] and in the examination of glutamate dehydrogenase activity.[21] Furthermore, [^{13}N]ammonia is a starting material of further ^{13}N-labelled compounds.[22,23] Several methods have been established for the synthesis of pure [^{13}N]-ammonia:

- an in-target production by irradiation of water *via* ^{16}O(p,α)^{13}N nuclear reaction in the presence of ethanol,[24,25] saturated with oxygen gas[26] or acetic acid[27] or
- an outside target production by reduction of in-target produced [^{13}N]NO_x by Devarda's alloy[26,28,29] or titanium trichloride.[30]

The preferred production method regarding yield, molar activity and automation is the in-target irradiation of water in the presence of ethanol (Figure 20.2).[26]

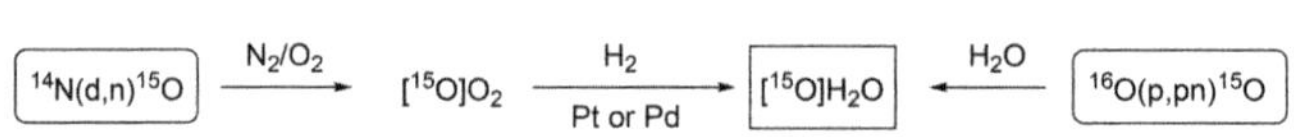

Figure 20.1 The main radio synthesis routes of [^{15}O]water.

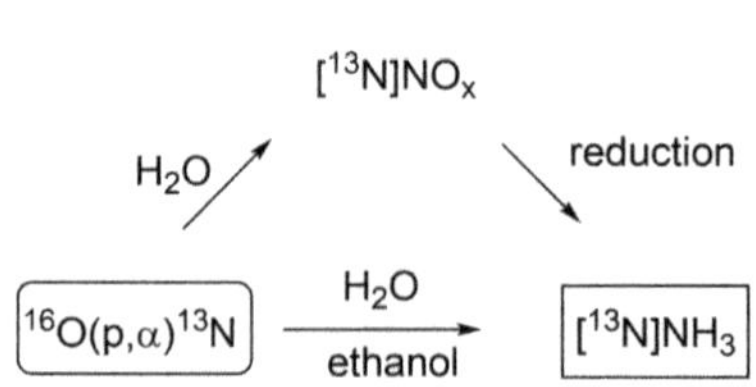

Figure 20.2 The main radio synthesis routes of [^{13}N]ammonia.

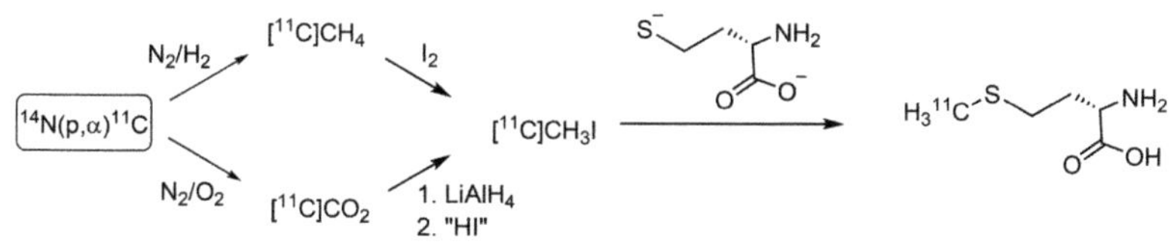

Figure 20.3 The radiosynthesis of L-[S-methyl-^{11}C]methionine.

20.2.3 L-[*S*-methyl-^{11}C]methionine

A clinically well-established ^{11}C-labelled compound is the amino acid tracer L-[*S*-methyl-^{11}C]methionine, which has been extensively used for human brain tumour PET imaging.[31] The synthesis of L-[*S*-methyl-^{11}C]methionine is straightforward compared to other ^{11}C-labelled amino acids,[32,33] because it is based on a well-established ^{11}C-methylation reaction.[34] However, this initially necessitates the synthesis of [^{11}C]CH_3I, which is also used for the preparation of many other ^{11}C-labelled radiopharmaceuticals.[35] Two principal methods established are:

- a 'wet' method starting from [^{11}C]CO_2, which is reduced to [^{11}C]methanol and then converted to [^{11}C]CH_3I using hydrogen iodide[36] or online using solid-phase bonded triphenylphosphine diiodide[37] or
- a 'dry' method starting from [^{11}C]methane, which is converted to [^{11}C]CH_3I with elemental iodine in a gas phase reaction at high temperatures.[38]

The production of [^{11}C]CO_2 or [^{11}C]CH_4 is performed by irradiation of nitrogen gas *via* the ^{14}N(p,α)^{11}C nuclear reaction, in the presence of 0.5–1% oxygen leading to [^{11}C]CO_2 or 5% hydrogen leading to [^{11}C]CH_4.

The volatile [^{11}C]CH_3I is then transferred into a reaction vessel with a sodium salt of L-3-amino-3-carboxypropane-1-thiolate, immediately obtained before radiosynthesis by basic hydrolysis of L-homocysteine thiolactone. Subsequent hydrolysis provides the product, which is purified by HPLC[39] or solid phase extraction (Figure 20.3).[40]

Several other ^{11}C-labelled radiotracers are obtained by this procedure and are well established on commercially available synthesis modules.[41]

20.2.4 2-[^{18}F]Fluoro-2-deoxy-D-glucose

2-[^{18}F]Fluoro-2-deoxy-D-glucose ([^{18}F]FDG) is the most widely used PET-radiopharmaceutical worldwide. This tracer is applied in neuroscience, cardiology, and oncology.[31] The importance of [^{18}F]FDG can be explained by its biochemical behaviour. [^{18}F]FDG is a glucose derivative that reflects glucose uptake and consumption in cells; glucose metabolism is altered in several diseases.

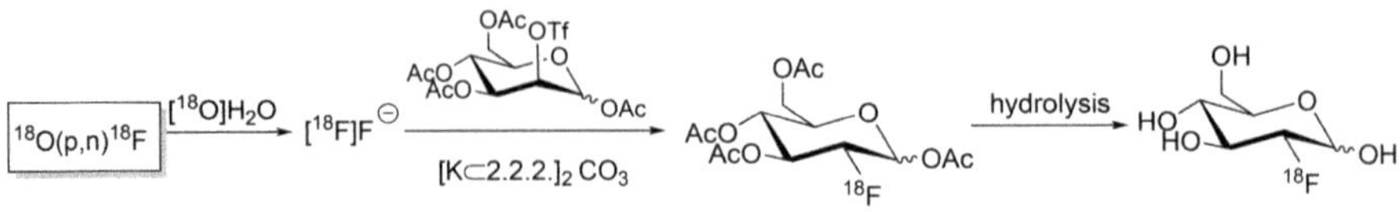

Figure 20.4 Radiosynthesis of [^{18}F]FDG.

The synthesis of [^{18}F]FDG is performed by a nucleophilic ^{18}F-substitution reaction starting from [^{18}F]fluoride. This is obtained by the bombardment of highly enriched (<95%) [^{18}O]water with protons *via* the ^{18}O(p,n)^{18}F nuclear reaction.[42] To enhance the nucleophilicity of the [^{18}F]fluoride anion, activators are added to the [^{18}F]fluoride solution, such as a combination of the cryptand Kryptofix®2.2.2 with K_2CO_3 ((K ⊂ 2.2.2)$_2$CO$_3$) or the phase transfer catalyst tetrabutylammonium(bi)carbonate Subsequently, a drying step, commonly performed by azeotropic distillation, is carried out.[43] The synthesis of [^{18}F]FDG is carried out in the polar aprotic solvent acetonitrile. An acetyl protected mannose derivative is employed as an ^{18}F-labelling precursor which contains triflate as leaving group.[44] The ^{18}F-displacement reaction *via* an S_N2 substitution mechanism (termed as Walden inversion) leads to an inversion of the configuration. Free [^{18}F]FDG is obtained by hydrolysis with hydrochloric acid[44] or with sodium hydroxide.[45] The purification proceeds *via* solid phase extraction. The whole sequence is extensively optimised and established on automated synthesis devices including kit preparation (Figure 20.4).[46]

20.2.5 *O*-[^{18}F]Fluoroethyl-L-tyrosine

^{18}F-labelled amino acids play an important role in tumour detection *via* PET. The longer half-life of fluorine-18 allows a better availability of these classes of tracers compared to their ^{11}C-labelled derivatives. Therefore, *O*-[^{18}F]fluoroethyl-L-tyrosine ([^{18}F]FET) was developed and today it plays a leading role in brain tumour imaging in clinical studies.[47,48] Although [^{18}F]FET is not incorporated into proteins,[49] the high uptake of [^{18}F]FET in brain tumors is closely related to the upregulated activities of the L-type amino acid transporter, resulting in a rather selective tracer accumulation and favourable tumour-to-background ratios.

The synthesis of [^{18}F]FET is realised by two different procedures. The first method makes use of bis(tosyloxy)ethane, which is effectively ^{18}F-labelled by an aliphatic nucleophilic reaction with [^{18}F]fluoride, using the abovementioned conditions. After separation and purification, [^{18}F]fluoroethyl tosylate is coupled to the sodium salt of tyrosine.[50] The whole process includes two HPLC separations, which are time-consuming procedures and difficult to automate. Optimisations were performed using solid-phase extraction methods.[51,52] Alternatively, the synthesis of [^{18}F]FET can also be realised with a one-pot procedure, using a protected tyrosine

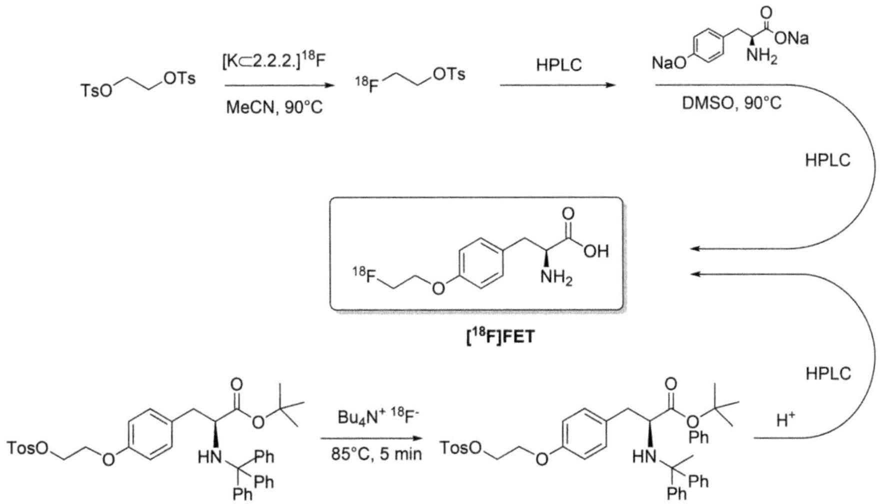

Figure 20.5 The different radio synthesis methods for [^{18}F]FET.

derivative bearing an *O*-ethyl moiety with tosyl as the leaving group.[53] The bulky protecting groups—triphenylmethyl at the amino group and *tert*-butyl as ester function—circumvent a racemisation in α-position of the amino acid. Acidic hydrolysis of the protecting groups and a final HPLC-separation provides pure [^{18}F]FET. The process was further optimised in detail and established in a common synthesis module (Figure 20.5).[54]

20.2.6 6-[^{18}F]Fluoro-3,4-dihydroxy-L-phenylalanine

6-[^{18}F]Fluoro-3,4-dihydroxy-L-phenylalanine (6-[^{18}F]FDOPA) is a fluorinated analogue of the naturally-occurring amino acid 3,4-dihydroxy-L-phenylalanine, the precusor of dopamine. It was developed for the evaluation of the dopaminergic system in the brain with PET, particularly in Parkinson's disease.[55,56] Furthermore, 6-[^{18}F]FDOPA also shows high uptake in neuroendocrine tumours and cerebral gliomas.[57]

The synthesis of 6-[^{18}F]FDOPA was clinically established using an electrophilic ^{18}F-fluorination method. However, serious disadvantages of this procedure are the relatively low radiochemical yield (<25%), the low molar activity, and the limited starting activity of [^{18}F]F_2.[58] A nucleophilic ^{18}F-approach has the advantage that the radiosynthesis can start with high radioactivity amounts (>400 GBq) and it provides a product of high molar activities. Several nucleophilic syntheses of 6-[^{18}F]FDOPA have been developed since 1990 to make use of [^{18}F]fluoride as a starting material.[43] A promising approach uses a very similar stannyl precursor like that used for the electrophilic method. The synthesis is effectively mediated by a copper catalyst and after deprotection using hydrogen bromide, 6-[^{18}F]FDOPA is received in 54% RCY (Figure 20.6).[59]

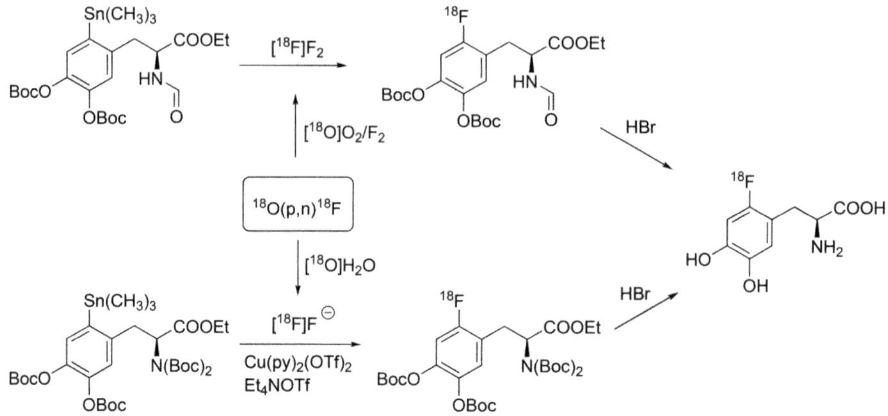

Figure 20.6 The radiosyntheses of 6-[^{18}F]FDOPA *via* electrophilic and nucleophilic ^{18}F-fluorination.

20.2.7 ^{68}Ga-labelled Prostate Specific Membrane Antigen Inhibitors

The prostate specific membrane antigen (PSMA) is overexpressed in nearly all prostate cancers at all stages of disease. The imaging of PSMA using PET has, so far, been presented as the most sensitive and specific method with regard to prostate cancer detection.[60] ^{68}Ga-labelled PSMA-based radiotracers ([^{68}Ga]PSMA) are currently most frequently used in clinical applications.[61,62]

The reason for this is the easy accessibility of gallium-68 *via* the ^{68}Ge/^{68}Ga radionuclide generator system, which facilitates the clinical application comparable to the established ^{99}Mo/^{99m}Tc generator system for SPECT.[10,11,63] In general, a radionuclide generator is a device that enables the radiochemical separation of a short-lived daughter radionuclide from the decaying parent radionuclide. The daughter is obtained in a pure radionuclide and radiochemical form. The longer-lived parent permanently generates the short-lived daughter, which can be obtained repeatedly after corresponding time intervals.[64]

Several matrices have been studied for the adsorbent material of the parent nuclide.[65] On commercially available ^{68}Ge/^{68}Ga generators the parent is immobilised on either a TiO_2- or SnO_2-based material or using an organic resin.[66,67]

Several derivatives of [^{68}Ga]PSMA radiotracers have been developed.[60] All of these consist of an urea-based peptidomimetic structure.[68] The most prominent derivatives, which are presently under clinical examination, are [^{68}Ga]PSMA-11 [69] and [^{68}Ga]PSMA-617.[70]

The synthesis of ^{68}Ga derivatives is based on complexation of ^{68}Ga with a chelating agent, leading to a stable *in vivo* complex. A general problem is the radionuclidic and chemical purity of the ^{68}Ga-eluate and several processes for purification were published.[9,67] A typical labelling example, *e.g.* of

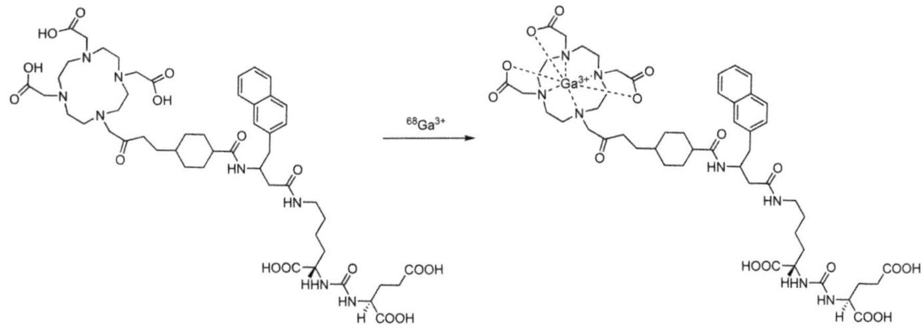

Figure 20.7 Radiosynthesis of [^{68}Ga]Ga-PSMA-617.

[^{68}Ga]PSMA-617 is performed with the [^{68}Ga]Ga^{3+} eluate, heated in HEPES buffer, resulting in a radiochemical yield of more than 90% (Figure 20.7).[70]

20.3 Bimodal MR-PET Probes

MR images are generated from proton NMR signals, mainly derived from water protons, so the signal intensity depends largely on the concentration and relaxation times (T_1 and T_2) of water molecules. Contrast agents for MR imaging are based on paramagnetic metals, such as gadolinium and iron. They effectively work by shortening the T_1 (spin–lattice) relaxation time of water protons by promoting rapid exchange of inner-sphere water molecules with bulk water and enhancing the signal intensity in T_1-weighted MR images.[71] However, to obtain a sufficient contrast, large amounts of paramagnetic ions in the millimolar range are required. All intravenous contrast agents approved by the US Food and Drug Administration (FDA) for use in MRI in a broad spectrum of diseases are gadolinium-based contrast agents: gadopentetate dimeglumine (Magnevist), gadodiamine (Omniscan), gadobenate dimeglumine (Multi-Hance), gadoversetamide (OptiMARK), and gadoteridol (ProHance).[72,73] This is due to the favourable combination of a large magnetic moment and a long electron spin relaxation time of the gadolinium(III) ion, as well as the relatively high *in vivo* stability of gadolinium complexes.[74]

The discovery of nephrogenic systemic fibrosis (NSF), a disease associated with the use of DTPA-bisamide Gd chelates,[75] has led to an active search for non-Gd alternatives.[76] Therefore, present research on contrast agents for MRI increasingly focuses on paramagnetic Mn(II) complexes as alternatives to Gd(III) chelates.[77]

The rather new approach of bimodal hybrid MR-PET imaging offers a valuable and promising tool for the non-invasive diagnosis of various diseases. This is because it benefits from the advantages of both individual imaging modalities. On one hand the superb soft tissue contrast and exceptional spatial, sub-millimetre resolution of MRI with no radiation (compared to PET and computed tomography, CT), and on the other the excellent sensitivity and ability of quantification of PET. With hybrid PET-MR scanners already available, the corresponding development of suitable bimodal MR-PET imaging probes has become of particular interest in recent years. These molecular probes are considered to provide an improved morphological correlation to molecular function, biomarker-responsive imaging and time-saving 'PET-guided' high-resolution MRI. However, despite these advantages, the design of such probes is challenging, owing mainly to an enormous sensitivity difference between PET and MRI.[78,79]

The availability of MR-PET scanners enables the combination of both techniques (PET-radiotracer and paramagnetic compound) in one 'contrast agent'. The use of 'contrast agents' in MR-PET imaging can be classified into two methods (according to Sandiford and de Rosales[72]):

- the unimodal approach, which means that the specific 'contrast agents' for one of both techniques are given non-simultaneously or simultaneously (cocktail) to determine multi-parametric information, and
- the bimodal approach, which means, that a single imaging probe, bearing both "contrast properties", is generated with the goal of providing an enhanced contrast agent for a specific purpose.

20.3.1 Unimodal Approach

The unimodal approach is performed with clinically approved 'contrast agents' and provides different types of information within a single imaging session, which together present an advantage over the bimodal one. The use of an MR-PET hybrid scanner can be advantageous, depending on the scientific question. The growing installation of hybrid PET systems in clinics has led to a multitude of clinical examinations in cardiovascular diseases, paediatrics, oncology, and neurology, simultaneously studying the advantages of this combination of morphological and functional imaging methods.[80,81]

An example of a non-simultaneously performed MR-PET study is brain tumour imaging using perfusion weighted imaging (PWI), with gadolinium-based contrast agents, in combination with PET using the radiolabelled amino acids [^{18}F]FET or 6-[^{18}F]FDOPA. These methods are based on different biochemical-physiological mechanisms, but provide diagnostic information beyond the conventional, mere morphological MRI examination, depending on the respective clinical question.[82]

The usefulness of brain hybrid MR-PET scanners was shown for amide proton transfer (APT) imaging with gadolinium-enhanced MRI and [^{18}F]FDG.[83] The results demonstrate that APT imaging is in concordance with Gd-enhanced MRI and, furthermore, gives complementary information about endogenous cellular proteins and peptides in addition to [^{18}F]FDG metabolism, which might strengthen and extend the clinical applications of MR-PET in tumour diagnosis, treatment planning, and therapy response assessment.[83]

20.3.2 Bimodal Approach

The bimodal MR-PET approach is currently at the preclinical evaluation stage, and so far, has not been clinically investigated. The aim of the development is the enhancement of the properties of a given contrast agent with the second modality component, quantitating physological function.[78] This has led to the development of bimodal molecules consisting of two imaging reporters in the same molecule (a radioactive and a paramagnetic ion). The advantages of adding a radionuclide to an MR contrast agent are the better quantification and enhanced whole-body detectability, which is beneficial for guiding high-resolution MR imaging.[72,84,85] The main disadvantage of MRI is the sensitivity limitation, as it requires millimolar concentrations, whereas PET radiotracers are measured in the nanomolar range. PET is greater than 10^6 times more sensitive than MRI. Therefore, cocktail mixtures of radioactive and non-radioactive probes are used. Both components of the 'cocktail' (MRI and PET) should be chemically equivalent to ensure identical *in vivo* behaviour.

The different strategies for the development of such a bimodal MR-PET probe can be divided into three methods: (1) use of nanoparticles or liposomes, containing paramagnetic ions. The positron-emitting radionuclide is either coupled on the surface or incorporated in the core of the particles; (2) use of 'standard' small molecules MR contrast agents by coupling a radionuclide to the molecule or replacing the paramagnetic ion by a positron-emitting radiometal; and (3) the use of matching pairs of paramagnetic and radioactive nuclides of the same element, *i.e.* isotopes, which are bound by chelators to a biological active molecule.

(1) Several efforts have been made to use nanoparticles as a basis for bimodal probes.[86] A central role is played by the superparamagnetic iron oxide an (SPIO) nanoparticles, containing a paramagnetic iron core, which were originally developed as MRI contrast agents for liver and lymph nodes.[87]
The first literature examples of radionuclide coupling by modification of the surface of an SPIO have been the labelling of the macrocyclic chelating agent 1,4,7,10-tetraazacyclododecane-*N*,*N*′,*N*″,*N*‴-tetraacetic acid (DOTA) with the positron-emitter copper-64 on polyaspartic acid—coated SPIO. This was conjugated with cyclic arginine-glycineaspartic (RGD) peptides for integrin $\alpha_v\beta_3$ recognition.[88] The other used labelling of serum albumin modified Mn-doped magnetism-engineered iron oxide nanoparticle with iodine-124.[89] A further example is the modification of the surface to link the desferrioxamine chelate (DFO) to ferumoxytol, a carbohydrate-coated SPIO, to generate the MR-PET contrast agent DFO-ferumoxytol labelled with zirconium-89 ($t_{½}$ = 3.27 days).[90]
The long-lived positron emitter germanium-69 (t½ = 39.05 h) was incorporated within the core of the SPIO, and the ^{69}Ge-SPION was coated with a layer of poly(ethylene glycol) (PEG) to increase the *in vivo* stability.[91]

However, it also succeeded to incorporate the relatively short-lived gallium-68 into the core of the SPIO by co-precipitation.[92]

An alternative method is the chemisorption of radionuclides on the surface of the SPIO due to their high binding affinity, which was demonstrated by labelling of SPIO with positron emitters of arsenic (As-71, As-72 and As-74).[93]

Simple metal ion salts of gallium-68 or zirconium-89 with chloride or oxalate anions can react thermally with SPIO-based nanoparticles under basic conditions to induce metal ion association into the central iron oxide core without using a chelate.[94]

Besides the use of inorganic nanoparticles, pegylated liposome formulations, housing an MR T1 contrast agent (Gd) and the positron-emitting ^{89}Zr for simultaneous PET and MR tumour imaging capabilities, were conjugated with octreotide in order to selectively target neuroendocrine tumours *via* human somatostatin receptor subtype-2.[95]

(2) In contrast to MR-PET probes based on nanoparticles, only a few examples are available for small molecules. This is due to the challenging design of such tracers, which is hampered by the enormous sensitivity difference between PET and MRI.

A first small molecule dual probe for use in MR-PET was prepared starting from EP-2104R, a well-established fibrin-targeted gadolinium-based MR contrast agent, by partial exchange of the paramagnetic gadolinium by copper-64.[96] However, the drawback of this approach is that the resulting pairs of MRI and PET probes are not chemically equivalent, which can result in different pharmacokinetics.

Another bimodal MR-PET agent was designed for quantitative pH imaging at concentrations commonly used for *in vivo* MRI (0.1–1 mM). For this, the pH-sensitive MR agent GdDOTA-4AMP was modified by adding a side chain containing a fluorine atom which was partly replaced by fluorine-18,[97] which enables the quantitative and non-invasive reporting of pH values *in vivo* (Figure 20.8).

Another possibility to combine MR and PET functionality is to bring two different chelator systems with different affinity behaviour for metal ions together in one small molecule. The chelator DOTA serves as a

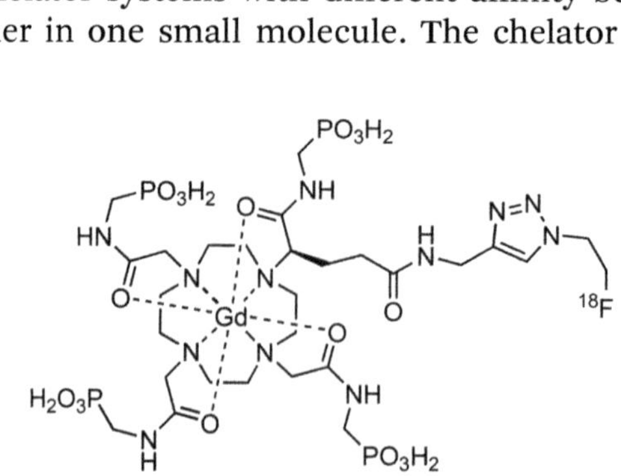

Figure 20.8 The structure of ^{18}F-labelled GdDOTA-4AMP.

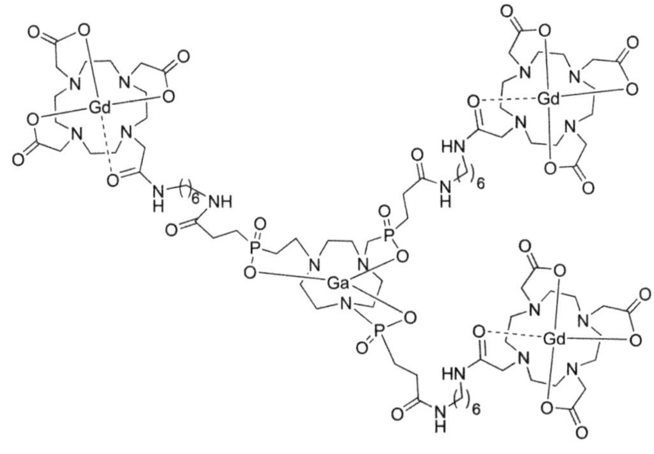

Figure 20.9 The structure of ^{68}Ga-and Gd-labelled TRAP(HM-Da-DOTA)$_3$.

scaffold for the majority of lanthanide-based MRI probes, like gadolinium, but also for the positron emitter gallium-68. This chelator system was combined with of triazacyclononanetriphosphinate (TRAP), characterised by a higher selectivity for Ga(III) ions. The MR-PET bimodal agent was composed by adding three DOTA chelators units to one TRAP unit. The complex is loaded with an excess of Gd(III) to achieve saturation of all DOTA cages. In order to remove all TRAP bound Gd(III), the intermediate is washed with an excess of sodium diethylenetriamine-pentaacetate (DTPA). Adding of generator-produced ^{68}gallium(III) leads to the ^{68}Ga-labelling of the bimodal probe (Figure 20.9).[98]

(3) The sensitivity and resolution mismatch, *i.e.* the huge mass difference, between the modalities can be counterbalanced by applying mixtures of paramagnetic MR contrast agents in millimolar amounts, spiked with the corresponding isotopically labelled PET-compound. Both compounds will exhibit the same *in vivo* biodistribution pattern, since both are chemically identical. For the development of such combinations with matching pairs of paramagnetic and radioactive isotopes, the most commonly used paramagnetic ion, gadolinium(III), cannot be used, because there is no useful positron emitter of gadolinium.[99]

The manganese(II) ion appears to be an excellent choice for replacing gadolinium(III) ion in magnetic resonance applications:[100] in addition to its high spin and slow electronic relaxation, Mn^{2+} is an endogenous element showing toxic effects only at rather high concentrations. Currently, $[Mn(DPDP)]^{4-}$ (Teslascan) is the only Mn^{2+}-based contrast agent used in clinical practice (liver or cardiac imaging) and is associated with a low relaxivity (R_1) due to the absence of water molecules coordinated to the paramagnetic centre ($q = 0$). *In vivo*, the observed relaxation effect results from the slow release of Mn^{2+} ions. Thus, the ligand is only aimed at controlling the release rate.

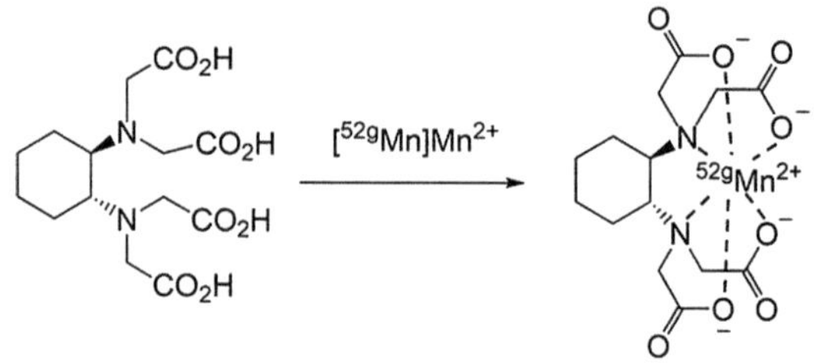

Figure 20.10 The radiosynthesis of [^{52g}Mn(CDTA)]$^{2-}$.

A useful positron-emitter of manganese is ^{52g}Mn ($t_{½}$: 5.6 days). *Trans*-1,2-diaminocyclohexane-*N*,*N*,*N'*,*N'*-tetraacetic acid (CDTA) offer as a suitable chelator for Mn(II) for *in vivo* MR-PET applications. It has high thermodynamic stability and kinetic inertness, on the one hand, and a fair T_1 contrast enhancement due to the presence of at least one water molecule coordinated to the Mn(II) cation, on the other hand. This chelator was labelled with a mixture of ^{55}Mn and its radioisotope ^{52g}Mn(II) in >99% RCY, resulting in the first Mn-based bimodal MR-PET probe.[101,102] The [^{52g}Mn]CDTA chelate was shown to be very stable against air oxidation and sufficiently inert to decomplexation in blood serum.[103,104]

This technique ensures that both the PET and MR reporters are present in the required concentration for each imaging modality and that they are chemically identical, guaranteeing the same *in vivo* pharmacokinetic behaviour (Figure 20.10).

References

1. S. Guhlke, A. M. Verbruggen and S. Vallabhajosula, in *Clinical Nuclear Medicine*, ed. H. J. Biersack and L. M. Freeman, Springer, Berlin; New York, 2007.
2. T. J. Ruth, in *Handbook of Radiopharmaceuticals, Radiochemistry and Applications*, ed. M. J. Welch and C. S. Redvanly, John Wiley & Sons, New York, 2003, p. 71.
3. N. Satyamurthy, M. E. Phelps and J. R. Barrio, *Clin. Positron Imaging*, 1999, **2**, 233.
4. D. J. Schlyer, in *Handbook of Radiopharmaceuticals*, ed. M. J. Welch and C. S. Redvanly, John Wiley & Sons, New York, 2003, p. 1.
5. J. S. Fowler and A. P. Wolf, *Acc. Chem. Res.*, 1997, **30**, 181.
6. B. Långström, F. Karimi and Y. Watanabe, *J. Labelled Compd. Radiopharm.*, 2013, **56**, 251.
7. H. H. Coenen, in *Ernst Schering Research Foundation Workshop 62*, ed. P. A. Schubiger, L. Lehmann and M. Friebe, Springer, Berlin Heidelberg, 2007, p. 15.

8. S. Purser, P. R. Moore, S. Swallow and V. Gouverneur, *Chem. Soc. Rev.*, 2008, **37**, 320.
9. L. Martiniova, L. De Palatis, E. Etchebehere and G. Ravizzini, *Curr. Radiopharm.*, 2016, **9**, 87.
10. S. R. Banerjee and M. G. Pomper, *Appl. Radiat. Isot.*, 2013, **76**, 2.
11. F. Roesch and P. J. Riss, *Curr. Top. Med. Chem.*, 2010, **10**, 1633.
12. J. C. Baron and T. Jones, *Neuroimage*, 2012, **61**, 492.
13. K. Zhang, H. Herzog, J. Mauler, C. Filss, T. W. Okell, E. Rota Kops, L. Tellmann, T. Fischer, B. Brocke, W. Sturm, H. H. Coenen and N. J. Shah, *J. Cereb. Blood Flow Metab.*, 2014, **34**, 1373.
14. M. Sajjad, J. S. Liow and J. Moreno-Cantu, *Appl. Radiat. Isot.*, 2000, **52**, 205.
15. M. S. Berridge, A. H. Terris and E. H. Cassidy, *Int. J. Appl. Radiat. Isot.*, 1990, **41**, 1173.
16. S. M. Qaim, *Nucl. Med. Biol.*, 2017, **44**, 31–49.
17. G. K. Mulholland, M. R. Kilbourn and J. J. Moskwa, *Int. J. Appl. Radiat. Isot.*, 1990, **41**, 1193.
18. J. Van Naemen, M. Monclus, P. Damhaut, A. Luxen and S. Goldman, *Nucl. Med. Biol.*, 1996, **23**, 413.
19. H. R. Schelbert, M. E. Phelps, S. C. Huang, N. S. Macdonald, H. Hansen and D. E. Kuhl, *Circulation*, 1981, **63**, 1259.
20. M. Suda, M. Onoguchi, T. Tomiyama, K. Ishihara, N. Takahashi, M. Sakurai, K. Matsumoto and S. I. Kumita, *J. Nucl. Cardiol.*, 2016, **23**, 457.
21. A. J. L. Cooper, *Neurochem. Int.*, 2011, **59**, 456.
22. J. Ermert, *J. Labelled Compd. Radiopharm.*, 2018, **61**, 179.
23. V. Gómez-Vallejo, V. Gaja, K. B. Gona and J. Llop, *J. Labelled Compd. Radiopharm.*, 2014, **57**, 244.
24. M. Statuto, E. Galli, F. Bertagna, E. Migliorati, I. Zanella, D. Di Lorenzo, A. De Agostini, C. Rodella, P. Apostoli, L. Caimi, R. Giubbini and G. Biasiotto, *Nucl. Med. Commun.*, 2016, **37**, 412.
25. G. Bormans, W. Langendries, L. Mortelmans and A. Verbruggen, *Appl. Radiat. Isot.*, 1995, **46**, 83.
26. K. Suzuki and Y. Yoshida, *Appl. Radiat. Isot.*, 1999, **50**, 497.
27. B. Wieland, G. Bida, H. Padgett, G. Hendry, E. Zippi, G. Kabalka, J. L. Morelle, R. Verbruggen and M. Ghyoot, *Int. J. Appl. Radiat. Isot.*, 1991, **42**, 1095.
28. W. Vaalburg, J. A. A. Kamphuis, M. D. Beerling-van der Molen, S. Reiffers, A. Rijskamp and M. G. Woldring, *Int. J. Appl. Radiat. Isot.*, 1975, **26**, 316.
29. M. S. Berridge and B. J. Landmeier, *Appl. Radiat. Isot.*, 1993, **44**, 1433.
30. H. Krizek, R. Lembares, I. Dinwoodie, I. Gloria, K. Lathrop and P. V. Harper, *J. Nucl. Med.*, 1973, **14**, 629.
31. S. Vallabhajosula, L. Solnes and B. Vallabhajosula, *Semin. Nucl. Med.*, 2011, **41**, 246.
32. J. Ermert and H. H. Coenen, *J. Labelled Compd. Radiopharm.*, 2013, **56**, 225.

33. A. Pekošak, U. Filp, A. J. Poot and A. D. Windhorst, *Mol. Imaging Biol.*, 2018, **20**, 510–532.
34. D. Comar, J. C. Cartron, M. Maziere and C. Marazano, *Eur. J. Nucl. Med.*, 1976, **1**, 11.
35. P. W. Miller, K. Kato and B. Långström, Carbon-11, Nitrogen-13, and Oxygen-15 Chemistry: An introduction to Chemistry with short-lived radioisotopes, *The Chemistry of Molecular Imaging*, 2014, p. 79.
36. C. Crouzel, B. Långström, V. W. Pike and H. H. Coenen, *Appl. Radiat. Isot.*, 1987, **38**, 601.
37. M. Holschbach and M. Schüller, *Appl. Radiat. Isot.*, 1993, **44**, 779.
38. P. Larsen, J. Ulin, K. Dahlstrøm and M. Jensen, *Appl. Radiat. Isot.*, 1997, **48**, 153.
39. K. I. Mizuno, S. Yamazaki, R. Iwata, C. Pascali and T. Ido, *Appl. Radiat. Isot.*, 1993, **44**, 788.
40. M. Mitterhauser, W. Wadsak, A. Krcal, J. Schmaljohann, H. Eidherr, A. Schmid, H. Viernstein, R. Dudczak and K. Kletter, *Appl. Radiat. Isot.*, 2005, **62**, 441.
41. X. Shao, R. Hoareau, A. C. Runkle, L. J. M. Tluczek, B. G. Hockley, B. D. Henderson and P. J. H. Scott, *J. Labelled Compd. Radiopharm.*, 2011, **54**, 819.
42. S. M. Qaim, J. C. Clark, C. Crouzel, M. Guillaume, H. J. Helmeke, B. Nebeling, V. W. Pike and G. Stöcklin, in *Radiopharmaceuticals for Positron Emission Tomography – Methodological Aspects*, ed. G. Stöcklin and V. W. PikeW, Kluwer Academic Publishers, Dordrecht, Boston, London, 1993, p. 1.
43. H. H. Coenen and J. Ermert, *Clin. Transl. Imaging*, 2018, **6**, 169–193.
44. K. Hamacher, H. H. Coenen and G. Stöcklin, *J. Nucl. Med.*, 1986, **27**, 235.
45. F. Füchtner, J. Steinbach, P. Mäding and B. Johannsen, *Appl. Radiat. Isot.*, 1996, **47**, 61.
46. R. Krasikova, *Curr. Org. Chem.*, 2013, **17**, 2097.
47. K. J. Langen, N. Galldiks, E. Hattingen and N. J. Shah, *Nat. Rev. Neurol.*, 2017, **13**, 279.
48. W. A. Weber, H. J. Wester, A. L. Grosu, M. Herz, B. Dzewas, H. J. Feldmann, M. Molls, G. Stocklin and M. Schwaiger, *Eur. J. Nucl. Med.*, 2000, **27**, 542.
49. C. Juhász, S. Dwivedi, D. O. Kamson, S. K. Michelhaugh and S. Mittal, *Mol. Imaging*, 2014, **13**, 7290.2014.00015.
50. H. J. Wester, M. Herz, W. Weber, P. Heiss, R. Senekowitsch-Schmidtke, M. Schwaiger and G. Stöcklin, *J. Nucl. Med.*, 1999, **40**, 205.
51. M. Zuhayra, A. Alfteimi, C. V. Forstner, U. Lützen, B. Meller and E. Henze, *Bioorg. Med. Chem.*, 2009, **17**, 7441.
52. D. Mueller, I. Klette, F. Kalb and R. P. Baum, *Nucl. Med. Biol.*, 2011, **38**, 653.
53. K. Hamacher and H. H. Coenen, *Appl. Radiat. Isot.*, 2002, **57**, 853.

54. T. Bourdier, C. J. R. Fookes, T. Q. Pham, I. Greguric and A. Katsifis, *J. Labelled Compd. Radiopharm.*, 2008, **51**, 369.
55. E. S. Garnett, G. Firnau and C. Nahmias, *Nature*, 1983, **305**, 137.
56. M. Politis, *Nat. Rev. Neurol.*, 2014, **10**, 708.
57. H. Minn, S. Kauhanen, M. Seppanen and P. Nuutila, *J. Nucl. Med.*, 2009, **50**, 1915.
58. M. Pretze, C. Wängler and B. Wängler, *BioMed Res. Int.*, 2014, **2014**, 674063.
59. F. Zarrad, B. Zlatopolskiy, P. Krapf, J. Zischler and B. Neumaier, *Molecules*, 2017, **22**, 2231.
60. E. Gourni and G. Henriksen, *Molecules*, 2017, **22**, 523.
61. M. Perera, N. Papa, D. Christidis, D. Wetherell, M. S. Hofman, D. G. Murphy, D. Bolton and N. Lawrentschuk, *Eur. Urol.*, 2016, **70**, 926.
62. C. Udovicich, M. Perera, M. S. Hofman, S. Siva, A. Del Rio, D. G. Murphy and N. Lawrentschuk, *Prostate Int.*, 2017, **5**, 125–129.
63. R. Alberto and H. Braband, SPECT/PET imaging with Technetium, Gallium, Copper, and other metallic radionuclides, *Compr. Inorg. Chem. II Second Ed.*, 2013, 785.
64. F. Rösch and F. F. R. Knapp, in *Handbook of Nuclear Chemistry*, ed. A. Vertes, S. Nagy, Z. Klencsár, R. G. Lovas and F. Rösch, Springer, Dordrecht, Heidelberg, 2011.
65. H. R. Maecke and J. P. Andre, in *Pet Chemistry: The Driving Force in Molecular Imaging*, ed. P. A. Schubiger, L. Lehmann and M. Friebe, 2007, p. 215.
66. M. D. Bartholomä, A. S. Louie, J. F. Valliant and J. Zubieta, *Chem. Rev.*, 2010, **110**, 2903.
67. F. Rösch, *Appl. Radiat. Isot.*, 2013, **76**, 24.
68. C. A. Foss, R. C. Mease, H. Fan, Y. Wang, H. T. Ravert, R. F. Dannals, R. T. Olszewski, W. D. Heston, A. P. Kozikowski and M. G. Pomper, *Clin. Cancer Res.*, 2005, **11**, 4022.
69. M. Eder, M. Schäfer, U. Bauder-Wüst, W. E. Hull, C. Wängler, W. Mier, U. Haberkorn and M. Eisenhut, *Bioconjugate Chem.*, 2012, **23**, 688.
70. M. Benešová, U. Bauder-Wüst, M. Schäfer, K. D. Klika, W. Mier, U. Haberkorn, K. Kopka and M. Eder, *J. Med. Chem.*, 2016, **59**, 1761.
71. S. Iwaki, K. Hokamura, M. Ogawa, Y. Takehara, Y. Muramatsu, T. Yamane, K. Hirabayashi, Y. Morimoto, K. Hagisawa, K. Nakahara, T. Mineno, T. Terai, T. Komatsu, T. Ueno, K. Tamura, Y. Adachi, Y. Hirata, M. Arita, H. Arai, K. Umemura, T. Nagano and K. Hanaoka, *Org. Biomol. Chem.*, 2014, **12**, 8611.
72. L. Sandiford and R. T. M. de Rosales, *PET Clin.*, 2016, **11**, 119.
73. S. P. Kalva, M. A. Blake and D. V. Sahani, *Appl. Radiol.*, 2006, **35**, 18.
74. C. T. Yang and K. H. Chuang, *MedChemComm*, 2012, **3**, 552.
75. E. Kanal and M. F. Tweedle, *Radiology*, 2015, **275**, 630.
76. E. M. Gale and P. Caravan, *ACS Chem. Neurosci.*, 2018, **9**, 395.
77. B. Drahoš, I. Lukeš and E. Tóth, *Eur. J. Inorg. Chem.*, 2012, **2012**, 1975.
78. R. T. M. de Rosales, *J. Labelled Compd. Radiopharm.*, 2014, **57**, 298.

79. A. Louie, *Chem. Rev.*, 2010, **110**, 3146.
80. D. L. Bailey, G. Antoch, P. Bartenstein, H. Barthel, A. J. Beer, S. Bisdas, D. A. Bluemke, R. Boellaard, C. D. Claussen, C. Franzius, M. Hacker, H. Hricak, C. la Fougère, B. Gückel, S. G. Nekolla, B. J. Pichler, S. Purz, H. H. Quick, O. Sabri, B. Sattler, J. Schäfer, H. Schmidt, J. van den Hoff, S. Voss, W. Weber, H. F. Wehrl and T. Beyer, *Mol. Imaging Biol.*, 2015, **17**, 297.
81. T. J. Fraum, K. J. Fowler and J. McConathy, *Acad. Radiol.*, 2016, **23**, 220.
82. C. P. Filss, F. Cicone, N. J. Shah, N. Galldiks and K. J. Langen, *Clin. Transl. Imaging*, 2017, **5**, 209.
83. H. Sun, J. Xin, J. Zhou, Z. Lu and Q. Guo, *Mol. Imaging Biol.*, 2018, **20**, 473.
84. M. Sollini, R. Berchiolli, M. Kirienko, A. Rossi, A. W. J. M. Glaudemans, R. Slart and P. A. Erba, *Semin. Nucl. Med.*, 2018, **48**, 225.
85. A. Lahooti, S. Sarkar, S. Laurent and S. Shanehsazzadeh, *Contrast Media Mol. Imaging*, 2016, **11**, 428.
86. J. Lamb and J. P. Holland, *J. Nucl. Med.*, 2018, **59**, 382.
87. Y. X. J. Wang, S. Xuan, M. Port and J. M. Idee, *Curr. Pharm. Des.*, 2013, **19**, 6575.
88. H. Y. Lee, Z. Li, K. Chen, A. R. Hsu, C. Xu, J. Xie, S. Sun and X. Chen, *J. Nucl. Med.*, 2008, **49**, 1371.
89. J. S. Choi, J. C. Park, H. Nah, S. Woo, J. Oh, K. M. Kim, G. J. Cheon, Y. Chang, J. Yoo and J. Cheon, *Angew. Chem. Int. Ed.*, 2008, **47**, 6259.
90. D. L. J. Thorek, D. Ulmert, N. F. M. Diop, M. E. Lupu, M. G. Doran, R. M. Huang, D. S. Abou, S. M. Larson and J. Grimm, *Nat. Commun.*, 2014, **5**, 3097.
91. R. Chakravarty, H. F. Valdovinos, F. Chen, C. M. Lewis, P. A. Ellison, H. Luo, M. E. Meyerand, R. J. Nickles and W. Cai, *Adv. Mater.*, 2014, **26**, 5119.
92. J. Pellico, J. Ruiz-Cabello, M. Saiz-Alia, G. del Rosario, S. Caja, M. Montoya, L. F. de Manuel, M. P. Morales, L. Gutierrez, B. Galiana, J. A. Enriquez and F. Herranz, *Contrast Media Mol. Imaging*, 2016, **11**, 203.
93. F. Chen, P. A. Ellison, C. M. Lewis, H. Hong, Y. Zhang, S. Shi, R. Hernandez, M. E. Meyerand, T. E. Barnhart and W. Cai, *Angew. Chem. Int. Ed.*, 2013, **52**, 13319.
94. E. Boros, A. M. Bowen, L. Josephson, N. Vasdev and J. P. Holland, *Chem. Sci.*, 2015, **6**, 225.
95. D. S. Abou, D. L. J. Thorek, N. N. Ramos, M. W. H. Pinkse, H. T. Wolterbeek, S. D. Carlin, B. J. Beattie and J. S. Lewis, *J. Pharm. Res.*, 2013, **30**, 878.
96. R. Uppal, C. Catana, I. Ay, T. Benner, A. G. Sorensen and P. Caravan, *Radiology*, 2011, **258**, 812.
97. L. Frullano, C. Catana, T. Benner, A. D. Sherry and P. Caravan, *Angew. Chem. Int. Ed.*, 2010, **49**, 2382.

98. J. Notni, P. Hermann, I. Dregely and H. J. Wester, *Chem. Eur. J.*, 2013, **19**, 12602.
99. F. O. Denzler, N. A. Lebedev, A. F. Novgorodov, F. Rösch and S. M. Qaim, *Appl. Radiat. Isot.*, 1997, **48**, 319.
100. D. Pan, S. D. Caruthers, A. Senpan, A. H. Schmieder, S. A. Wickline and G. M. Lanza, *Wiley Interdiscip. Rev. Nanomed. Nanobiotechnol.*, 2011, **3**, 162.
101. H. H. Coenen, M. Buchholz, I. Spahn, C. Vanasschen, J. Ermert and B. Neumaier, *EJNMMI Phys.*, 2014, **1**, 1.
102. C. Vanasschen, M. Brandt, J. Ermert, B. Neumaier and H. H. Coenen, *EJNMMI Phys.*, 2015, **2**, A85.
103. C. Vanasschen, M. Brandt, J. Ermert and H. H. Coenen, *Dalton Trans.*, 2016, **45**, 1315.
104. C. Vanasschen, E. Molnár, G. Tircsó, F. K. Kálmán, E. Tóth, M. Brandt, H. H. Coenen and B. Neumaier, *Inorg. Chem.*, 2017, **56**, 7746.

Subject Index